SDGSAT-1卫星热红外影像图集

Atlas of SDGSAT-1 Satellite Thermal Infrared Image

郭华东　主编

Guo Huadong Chief Editor

可持续发展大数据国际研究中心

International Research Center of Big Data for Sustainable Development Goals

内 容 简 介

本图集以 SDGSAT-1 卫星热红外成像仪获取的 30m 分辨率的三波段与单波段昼夜热红外图像为主要内容，分别收录了涉及到水体、山脉、农业用地、沙漠、火灾、海冰、工业热源、城市、船只和数据产品的共计 118 个地点的热红外影像，并配文字介绍，展示自然环境中的热分布和其中人类活动的影响，反应了地表热环境的丰富动态。本图集既是一本具有科学价值和实用价值的热红外遥感影像集，也是一本具有艺术魅力和审美趣味的自然景观与人居环境图集，通过 SDGSAT-1 卫星的眼睛，为读者提供一个独特的视角观察我们的地球。

本书可供遥感卫星技术和可持续发展科学研究相关的学者参考，也可以作为地球科学及社会科学爱好者的收藏、兴趣读物。

审图号：GS京（2024）1707号

图书在版编目（CIP）数据

SDGSAT-1 卫星热红外影像图集 / 郭华东主编 . 北京 : 科学出版社 , 2024. 9.
—ISBN 978-7-03-079290-7
Ⅰ . TP75-64
中国国家版本馆CIP数据核字第2024C6S710号

责任编辑：董　墨 / 责任校对：郝甜甜
责任印制：徐晓晨　书籍设计：北京美光设计制版有限公司

科 学 出 版 社 出版
北京东黄城根北街16号
邮政编码：100717
http://www.sciencep.com

北京中科印刷有限公司印刷
科学出版社发行　各地新华书店经销
*
2024年9月第　一　版　开本：889 × 1194　1/16
2024年9月第一次印刷　印张：19　1/2
字数：450　000

定价：398.00元

（如有印装质量问题，我社负责调换）

图集编辑委员会

Editorial Board

序

世界各国领导人于 2015 年通过的《联合国 2030 年可持续发展议程》（简称“2030 年议程”）以其 17 项雄心勃勃的可持续发展目标为指导，引领全球实现向社会经济发展与自然环境的和谐共生的转变。该议程涵盖了广泛的领域，计划利用 15 年时间，通过相互依存、紧密联结的指标体系实现可持续发展目标。在 2030 年议程的前半段历程中，尽管国际社会做出了引人注目的贡献，但多数可持续发展目标的实施远未步入正轨。不断加剧的气候变化引发了极端热浪、洪水和干旱等灾害，不可持续的消费和生产破坏了环境，粮食危机和全球供应短缺加剧了部分发展中国家的饥饿程度，海洋酸化和海平面上升，生物多样性持续丧失……，所有这些问题都需要“通过团结、可持续理念和科学技术带来解决方案”。

中国作为推动可持续发展的坚定力量，自 2030 年议程实施以来，始终坚持创新、协调、绿色、开放、共享的新发展理念。中国国家主席习近平在第 75 届联合国大会宣布成立的可持续发展大数据国际研究中心（SDG 中心），以先进技术（特别是地球大数据和人工智能）促进 2030 年议程实施，为可持续发展目标实现注入了新动力。

联合国《2023 年可持续发展目标报告：特别版》显示，由于健康危机、气候变化和社会动荡等因素的叠加影响，超过 50% 的可持续发展目标进展不足，甚至约 30% 的目标停滞不前或倒退，这些危机威胁到人类社会为实现可持续发展未来所做的努力。在联合国可持续发展技术促进机制指引下，科学、技术和创新，尤其是以空间技术、大数据和人工智能为代表的前沿技术，可以有效促进全球可持续发展目标的实现。

空间对地观测技术具有宏观、动态和客观的监测能力，不仅是全球数据获取的高效工具，亦是研究地球系统及其与人类活动相互作用的重要手段，因此可在支持可持续发展目标的监测、评估和实现方面发挥重要作用。为践行空间技术服务可持续发展的理念，SDG 中心设计并研制了全球首颗专门服务 2030 年议程的科学卫星——SDGSAT-1。自 2021 年 11 月 5 日发射以来，SDGSAT-1 已在轨运行近三年，累计服务近百个国家、地区、联合国机构及其他国际组织，促进了全球可持续发展的研究和决策。SDGSAT-1 在空间技术领域取得了突破性进展，在支撑全球可持续发展领域发挥了引领作用，为人类共同事业作出了卓越贡献。SDGSAT-1 的成功实践证明了空间对地观测技术在填补可持续发展目标监测和评估数据缺口方面的潜力。

SDGSAT-1 是可持续发展卫星星座的首发星，为全球可持续发展目标监测、评估与研究提供了大量宝贵数据。为了直观展示 SDGSAT-1 精确描绘地球表面的热环境分布与变化的能力，《SDGSAT-1 卫星热红外影像图集》展示了全球范围内超过 100 个自然景观、人类活动和城市的热红外图像。这些影像数据反映了全球不同地区在昼夜和不同季节的温度分布与变化。该图集对于促进全球可持续发展目标的顺利实现具有重要意义和参考价值。希望该图集的出版能够为决策者、致力于热红外数据应用的研究人员以及地理爱好者提供有价值的信息，为实现全球可持续发展提供有益借鉴。

Csaba Kőrösi

第 77 届联合国大会主席

2024 年 8 月

Foreword

The United Nations' Agenda for Sustainable Development (2030 Agenda), adopted by world leaders in 2015, serves as a guide with its 17 ambitious Sustainable Development Goals (SDGs) to lead to the transition of the world towards a harmonious coexistence between socio-economic development and the natural environment. The 2030 Agenda not only covers a wide range of areas but also spans 15 years, relying on the interdependence and close coupling of various walks of life to achieve the world's sustainability. Despite the impressive engagement from different stakeholders around SDGs during the first half of the journey, the implementation of the 2030 Agenda is far from its right track towards the achievement of SDGs. Accelerating climate change has caused deadly heatwaves, floods, and droughts; unsustainable consumption and production have scarred the environment; food crises and global supply shortages have escalated the hunger level in some developing countries; ocean acidification and rising sea levels are worsening, biodiversity loss continues...... all these hinders need "solutions through solidarity, sustainability and science".

China, as a staunch force in promoting sustainable development, has consistently upheld the philosophy of innovative, coordinated, green, open and shared development since the beginning of the 2030 Agenda. For instance, the establishment of the International Research Center of Big Data for Sustainable Development Goals (CBAS) announced by Chinese President Xi Jinping at the 75th session of the UN General Assembly to facilitate the implementation of the 2030 Agenda with cutting-edge technologies, especially Big Earth Data and Artificial Intelligence, is injecting new vitalities into global achievement of SDGs.

The Sustainable Development Goals Report 2023: Special Edition shows more than 50% of SDGs targets are weak and insufficient, and even 30% have stalled or gone into reverse, due to the synchronous and often interlinked effect of health, climate and social crises. These challenges threaten to derail all efforts made by human society to achieve a sustainable future for everyone. On the momentum that's being generated by the UN Technology

Facilitation Mechanism, the science, technology and innovation, especially the cutting-edge technologies represented by space technology, big data and artificial intelligence, can be fully leveraged to facilitate the implementation of global SDGs. Earth observation from space, with its macroscopic, dynamic, and objective monitoring capabilities, can provide us not only an efficient tool of global data acquisition but also a scientific means for the study of Earth systems and their interactions with human activities, thus can play an important role in supporting the monitoring and evaluation of the results of implementation of the of SDGs. To practice the concept of space technology in support of SDGs, SDGSAT-1, the world's first scientific satellite dedicated to serving the 2030 Agenda, was designed and developed by CBAS. After almost three years of operation on orbit since its launching on 5 November 2021, SDGSAT-1 has served almost 100 countries and regions, UN Entities, and international organizations to facilitate the SDGs-related study and decision-making. As a remarkable contribution to humanity's joint endeavor, SDGSAT-1 makes the breakthrough and takes the leading role in the field of space technology to support the global sustainable development. The successful practice of SDGSAT-1 demonstrates the power of Earth Observation in data gap filling for SDGs monitoring and evaluation.

As the pioneer of the future Sustainable Development Satellite Constellation, SDGSAT-1 has provided invaluable data to the global SDGs monitoring, evaluation and research. To vividly demonstrate SDGSAT-1's ability in capturing thermal-related human activities and natural environments, the *Atlas of SDGSAT-1 Satellite Thermal Infrared Image* showcases the thermal infrared images of over 100 natural scenes, human activities and cities. It reflects the day-night temperature distribution and variations of locations from different parts of the Earth across seasons. This atlas has significant importance and reference value for promoting the smooth implementation of global sustainable development. It is hoped that the publication of this atlas will provide valuable information for decision-makers, researchers dedicated to thermal infrared data applications, and geography enthusiasts, contributing to the global transition towards sustainable development.

Csaba Kőrösi

77th President of the UN General Assembly

August 2024

前言

可持续发展是人类社会的永恒主题。自 1987 年可持续发展理念首次提出，到 2000 年联合国千年发展目标，再到 2015 年《联合国 2030 年可持续发展议程》，我们见证了人类在追求可持续发展道路上的坚定决心与不懈探索。然而，当今世界百年变局加速演进，全球经济复苏缓慢，极端气候和自然灾害频发，2030 年可持续发展目标的实现面临巨大的挑战。

以大数据、人工智能和空间技术等为代表的前沿科技正在重塑我们的生活，其中空间观测技术以其独特的优势，正在成为理解地球的新钥匙和知识挖掘的新手段，为服务全球可持续发展目标实现提供了全新视角。2018 年伊始，中国科学院设立了“地球大数据科学工程”A 类战略性先导科技专项（地球大数据专项），专项设置了“可持续发展科学卫星 1 号（SDGSAT-1）”项目，研制一颗专门服务 2030 年议程的科学卫星，为服务全球可持续发展目标监测与评估提供支撑。

SDGSAT-1 是全球首颗可持续发展科学卫星，于 2021 年 11 月 5 日成功发射。该卫星搭载了高性能微光、热红外和多谱段成像仪。通过三个载荷昼夜协同观测，实现对“人类活动痕迹的精细刻画”，揭示与人类活动和自然环境的相关可持续发展指标间的关联和耦合，以及人类活动引起的环境变化和演变规律，为表征人与自然交互作用的可持续发展目标研究提供支撑。这也标志着继我国气象卫星、海洋卫星、资源卫星、环境卫星和高分卫星等系列卫星之后，又一新的系列卫星——可持续发展系列卫星问世。

自然地物和人类活动均表现出不同的热特征，地表热能的分布和动态变化与人类社会经济活动、工业生产状况、地貌特征等息息相关，可以用于广泛的可持续发展研究。SDGSAT-1 的技术特点可为热相关可持续发展研究提供有力支撑：热红外成像仪可通过 30m 分辨率识别地表 0.2℃的温度变化，微光成像仪具备全球首创的同时获取 10m 全色与 40m 彩色微光影像的能力，多谱段成像仪可通过 1 个红边与 2 个深蓝波段的设计，分别监测植被生长状态和水质情况。同时，三个载荷的成像幅宽均为 300km，保证了全球数据的获取效能。

2022 年 9 月，可持续发展大数据国际研究中心发起了“SDGSAT-1 开放科学计划”，同时中国政府宣布其数据向全球开放共享。截至 2024 年 7 月，已有 35 余万景影像数据面向全球开放共享，已为来自近 100 个国家和地区的用户和科研人员提供了数据支撑，助力各国可持续发展目标监测和决策支持。

通过全球首部遥感三波段热红外影像图集，读者将有机会一窥 SDGSAT-1 热红外视角下的江河湖海、山川丘陵、戈壁荒漠等自然景观，以及工业排放、港口机场与城市变迁等与人类活动密切相关的场景。这些热红外图像为全球气候变化、城市热岛效应及自然资源管理等研究提供了宝贵资料和数据支持，展现出空间观测技术服务 2030 年议程实施的巨大潜力。

图集主要内容由十个部分组成，收录了全球 118 处自然景观，人类活动与城市的热红外影像。“水体”部分展示了 16 处水体热红外影像；“山脉”部分选择了 6 处山脉热红外影像；“农业用地”部分为 8 处农业用地的热红外影像；“沙漠”部分展示了 17 处沙漠的热红外影像；“火灾”部分为 5 处火灾的代表性热红外影像；“海冰”部分展示了 6 处海冰的热红外影像；“工业热源”部分为 14 处工业热源的热红外影像；“城市”部分展示了 31 处城市的热红外影像；“船只”部分为 8 处船只集中海域的热红外影像；“数据产品”部分展示了 7 处热红外数据产品影像。本图集全面展示了典型的人工和自然地貌的热分布特征，为更直观地了解全球不同地区城市和自然地貌提供了机会，这将有助于读者更好地理解人类社会经济发展和自然环境在热特征方面的表现及其与相关可持续发展目标的关联。

本图集是集体劳动的结晶。值全书付梓之际，笔者衷心感谢地球大数据专项领导小组组长白春礼、侯建国院长和副组长张亚平、张涛副院长，SDGSAT-1 卫星工程总指挥阴和俊和相里斌副院长、常务副总指挥于英杰局长、卫星工程总设计师樊士伟研究员，地球大数据专项各参研单位及地球大数据专项项目一研究集体，SDGSAT-1 卫星工程研制队伍与运行团队，并特别感谢窦长勇、丁海峰、唐韵玮、李晓明、陈凡胜、贾立、钱永刚、马彩虹、蒋倪君等科研专家与制图人员的工作，感谢为本书做出贡献的所有人员。

受有关条件限制，书中影像质量存在差异，另由于时间仓促，书中难免存在疏漏与不妥之处，敬请同行专家和读者不吝指正。

中国科学院院士
SDGSAT–1 卫星首席科学家
2024 年 *8* 月于北京

Preface

Sustainable development is the eternal pursuit of human society. Since the concept of sustainable development was first proposed in 1987, to the United Nations (UN) Millennium Development Goals in 2000, and then to *Transforming our world: the 2030 Agenda for Sustainable Development* in 2015, we have witnessed humankind's firm determination and tireless exploration on the path of pursuing sustainable development. However, the world is now facing great changes that have not been seen in a century with setbacks in the recovery of the global economy, escalating extreme climate changes, and frequent natural disasters, hindering global progress towards the Sustainable Development Goals (SDGs).

Cutting-edge digital technologies, represented by Big Data, Artificial Intelligence and Space Technology, are reshaping our lives. Among them, Earth observation technology, with its unique advantages, has gradually become a new key to understanding the Earth system and a new means of knowledge mining, providing a new perspective to serve the implementation of the 2030 Agenda. At the beginning of 2018, the Chinese Academy of Sciences (CAS) set up the Strategic Priority Research Program "Big Earth Data Science Engineering Program (CASEarth)", in which the project "Sustainable Development Science Satellite 1 (SDGSAT-1)" is tailored to develop a specific scientific satellite to serve the needs for monitoring and assessing the SDGs.

SDGSAT-1, the world's first scientific satellite dedicated to serving the 2030 Agenda, was successfully launched on 5 November 2021. SDGSAT-1 is designed to carry three payloads, including an advanced Glimmer Imager (GLI), Thermal Infrared Spectrometer (TIS) and Multi-spectral Imager (MSI). Through the coordinated day and night operations of the three payloads, the aim is to achieve a fine portrayal of the "traces of human activities" and provide data support for the study of SDGs characterizing the interaction between human beings and the natural environment. The SDGSAT-1 also marks the launch of a new series of satellites for sustainable development, following China's successive launch the series satellites of meteorological satellites, resources satellites, environmental Satellites, marine satellites and high-resolution satellites.

The distribution and dynamic changes of surface thermal energy exhibit distinct characteristics influenced by both natural phenomena and human activities. These thermal patterns are closely intertwined with socio-economic activities, industrial productions, and geographical features. They can be extensively studied for applications in sustainable development. The technological features of SDGSAT-1 provide robust support for research into thermal-related sustainable development initiatives: the TIS can identify the temperature change of 0.2°C on the Earth's surface with 30m spatial resolution, the GLI is the world's first spaceborne system that can acquire 10m panchromatic and 40m muti-band nighttime light at the same time, and the MSI monitors the growth status of the vegetation and the water quality through the design of the one red-edge band and the two deep-blue bands, respectively. The data from all three payloads have a swath of 300km, which ensures the effectiveness of global data acquisition. In September 2022, the International Research Center of Big Data for Sustainable Development Goals (CBAS), the operator of SDGSAT-1, launched the "SDGSAT-1 Open Science Program", and till June 2024, more than 350000 images acquired by SDGSAT-1 satellite have been shared globally free-of-charge to support research on sustainable development in various countries, and have provided data for researchers from almost 100 countries and regions to facilitate SDGs monitoring and decision making.

Through the world's first atlas of three-band thermal infrared remote sensing images, readers will have the opportunity to glimpse natural landscapes such as rivers, lakes, seas, mountains, hills, and deserts from the SDGSAT-1 thermal infrared perspective from space. They will also observe scenes closely related to human activities, including industrial emissions, ports, airports, and urban transformations. These thermal infrared images provide valuable information and data support for research on global climate change, urban heat island effect and natural resource management, demonstrating the great potential of space observation technology to serve the implementation of the 2030 Agenda.

The atlas is composed of ten parts, featuring thermal infrared images of 118 natural scenes, human activities and cities from around the world. The "Water Bodies" part showcases thermal infrared images of 16 water bodies; the "Mountain Ranges" part includes thermal infrared images of 6 mountain ranges; the "Agricultural Land" part features thermal infrared images of 8 agricultural areas; the "Desert" part presents thermal infrared images of 17 deserts; the "Fire Incidents" part highlights representative thermal infrared images of 5 fire incidents; the "Sea Ice" part displays thermal infrared images of 6 sea ice areas; the "Industrial Heat Sources" part includes thermal infrared images of 14 industrial heat sources; the "Cities" part showcases thermal infrared images of 31 cities; the "Boats and Ships" part presents thermal infrared images of 8 sea areas concentrated with ships; and the "Data Products" part features 7 thermal infrared data product images. This atlas comprehensively displays the thermal distribution characteristics of typical artificial and natural landscapes on the Earth surface, providing an opportunity for a more intuitive understanding of urban and natural landscapes in different regions around the world, which will contribute to a better understanding of the performance of human socio-economic development and the natural environment in terms of thermal characteristics, as well as their relevance to the relevant SDGs.

The atlas is a production of the team works. As the collection goes to press, the author would like to express his heartfelt thanks to CAS president Bai Chunli and Hou Jianguo, the head of the leaders group of the CASEarth Program; CAS vice president Zhang Yaping and Zhang Tao, the deputy head of leaders group of the CASEarth Program; CAS vice president Yin Hejun and Xiangli Bin, the Chief Director of SDGSAT-1 satellite project; Yu Yingjie, the Deputy Chief Director of SDGSAT-1 satellite project; Professor Fan Shiwei, the Chief Designer of the SDGSAT-1 satellite project; The research, engineering and operation teams of the SDGSAT-1 satellite project; Especially, Dr. Dou Changyong, Dr. Ding Haifeng, Dr. Tang Yunwei, Dr. Li Xiaoming, Dr. Chen Fansheng, Dr. Jia Li, Dr. Qian Yonggang, Dr. Ma Caihong, Mr. Jiang Nijun. The chief editor thanks all contributors to this atlas.

Member of the 2nd 10 Member Group of the United Nations TFM for SDGs
SDGSAT-1 Chief Scientist
August 2024

SDGSAT-1 卫星

2030 年议程的实施迫切需要数据和方法的支撑。空间对地观测作为高效的数据获取手段和研究方法，能够为 2030 年议程做出重要贡献。为此，研制、运行系列可持续发展科学卫星成为可持续发展大数据国际研究中心的一项重要使命。

可持续发展科学卫星 1 号（SDGSAT-1）是全球首颗专门服务 2030 年议程的科学卫星，也是中国科学院研制并发射的首颗地球科学卫星。该卫星由中国科学院“地球大数据科学工程”A 类战略性先导科技专项（地球大数据专项）研制，是可持续发展大数据国际研究中心的首发星。

针对全球可持续发展目标（SDGs）监测、评估和科学研究的需求，SDGSAT-1 卫星通过多载荷全天时协同观测，旨在实现“人类活动痕迹的精细刻画”，服务全球 SDGs 的实现，为表征人与自然交互作用的指标研究提供支撑。

SDGSAT-1 卫星通过探测人类活动与地球表层环境交互影响的地物参量，实现综合探测数据向 SDGs 应用信息的转化，研究跟人类活动和自然环境相关指标间的关联和耦合。充分利用 SDGSAT-1 卫星对地表进行宏观、动态、大范围、多载荷昼夜协同探测的优势，可以实现人居格局（SDG2、SDG6）、城市化水平（SDG11）、能源消耗（SDG13）、近海生态（SDG14、SDG15）等以人类活动为主引起的环境变化和演变规律研究，探索夜间灯光或月光等微光条件下地表环境要素探测的新方法与新途径，服务 SDGs 相关领域的研究。

SDGSAT-1 卫星为太阳同步轨道设计，搭载了高分辨率宽幅热红外、微光及多谱段成像仪三种载荷，轨道高度为 505km，倾角为 97.5°，空间分辨率分别为 30m、10m/40m 和 10m，幅宽均为 300km，重访周期约 11 天。设计有“热红外 + 多谱段”“热红外 + 微光”以及单载荷观测等普查观测模式，可实现全天时、多载荷协同探测；同时，SDGSAT-1 拥有月球定标、黑体变温定标、LED 灯定标、一字飞行定标等星上和场地定标模式，保证了精确定量探测的需求。其中，热红外成像仪具有高分辨率宽幅观测能力，尤其在降低载荷功耗的前提下，空间分辨率比国际同类卫星提升 3.3 倍，是中国首次在热红外谱段采用全光路低温光学系统设计，可在大动态范围下实现优于 80mK 的噪声等效温差。卫星可精细探测城市热能分布，为测算全球能源消耗提供基础数据，服务清洁能源、气候行动等可持续发展目标。卫星热红外数据已成功在地表温度反演，土地沙漠化评估，船只与飞机监测和工业热排放监控等场景下得到应用。

微光和多谱段成像仪同样具备 300km 的大幅宽，可实现地球昼夜全天时切换成像。其中微光成像仪是国际首个同时具备全色和彩色高分辨率微光探测载荷，能够以 10m（全色）和 40m（彩色）空间分辨率探测夜间地表灯光的颜色、强度和空间分布，进而提供判识全球经济发展水平及区域发展差异的信息，服务可持续城市与社区、工业创新和基础设施等可持续发展目标。特别值得一提的是，SDGSAT-1 卫星过境时间为当地 21:00 至 22:00 之间，这一时段恰好是夜间人类活动的高峰期。多谱段成像仪拥有 7 个波段，具有高信噪比的特点，空间分辨率为 10m，其中设计有 2 个深蓝波段和 1 个红边波段，特别有利于水体质量和植被生长状态的监测，可服务清洁饮水和卫生设施、陆地生物等可持续发展目标。

SDGSAT-1 卫星具有以下特点：

（1）卫星平台：高载荷平台比，高灵敏度一体化设计与多模态高精度操控，高速数据技术及自主智能任务规划和标定模式。

（2）热红外成像仪：大幅宽（300km）、高分辨率（30m），幅宽 / 分辨率综合指标国际领先；高动态范围（220 ~ 340K），高探测灵敏度（0.2K@300K）。

（3）微光 / 多谱段成像仪：采用多模共光路成像设计，地球昼夜全天时切换成像；微光载荷探测动态范围 60dB；多谱段载荷 7 个波段信噪比大于 130。

SDGSAT-1 卫星主要技术指标

类别	指标项	具体指标
轨道	轨道类型	太阳同步轨道
	轨道高度	505km
	轨道倾角	97.5°
热红外成像仪	幅宽	300km
	探测谱段	8~10.5μm 10.3~11.3μm 11.5~12.5μm
	空间分辨率	30m
微光 / 多谱段成像仪	幅宽	300km
	微光探测谱段	P: 444~910nm B: 424~526nm G: 506~612nm R: 600~894nm
	微光空间分辨率	全色 10m，彩色 40m
	多谱段探测谱段	B1: 374~427nm B2: 410~467nm B3: 457~529nm B4: 510~597nm B5: 618~696nm B6: 744~813nm B7: 798~911nm
	多谱段空间分辨率	10m

SDGSAT-1 Satellite

The 2030 Agenda urgently demands refined data and methodologies. Earth observation, an effective data collection and research method, holds significant potential for contributing to the 2030 Agenda. In view of this, the International Research Center of Big Data for Sustainable Development Goals (CBAS) has prioritized the development and operation of a series of scientific satellites dedicated to serving the global sustainable development.

The Sustainable Development Science Satellite-1 (SDGSAT-1) is a landmark achievement, being the first scientific satellite globally that is devoted to supporting the 2030 Agenda, and the premier Earth science satellite developed and launched by the Chinese Academy of Sciences (CAS). The development of SDGSAT-1 was facilitated by the "Big Earth Data Science Engineering" program of the CAS. SDGSAT-1 is the inaugural satellite in a series planned by CBAS.

SDGSAT-1 was designed for monitoring and evaluating the global sustainable development goals (SDGs), as well as facilitating related scientific research. Through multi-payload, round-the-clock collaborative observation, the satellite captures fine details of "human activity traces", aiding the realization of global SDGs and supporting research into the interplay between humanity and nature.

SDGSAT-1 effectively transforms comprehensive observation data into SDGs application information by detecting ground feature parameters concerning the interaction between human activities and the surface environment of earth. It also examines the correlation and coupling with human activities and natural environment-related indicators. The satellite enables macroscopic, dynamic, wide-range, multi-payload, and 24-hour observation of the earth's surface. This aids in studying environmental changes and evolution patterns brought about by human activities, such as human settlement patterns (SDG2, SDG6), urbanization levels (SDG11), energy consumption (SDG13), and near-shore ecology (SDG14, SDG15). It also explores innovative methods for detecting ground surface environmental elements under dim-light conditions, such as moonlight or nocturnal light, thereby assisting research related to the SDGs.

Operating in a sun-synchronous orbit, SDGSAT-1 is equipped with three types of payloads: high-resolution Thermal Infrared Spectrometer (TIS), Glimmer Imager (GLI), and Multispectral Imager (MSI). The satellite orbits at an altitude of 505km, with an inclination angle of 97.5°, spatial resolutions of 30m, 10/40m and 10m respectively, and a swath width of 300km and revisit cycle of 11 days. It employs various observation modes, including a combination of TIS and MSI, TIS and GLI, as well as single-payload mode. These modes facilitate multi-payload coordinated observation throughout the day and night. Additionally, SDGSAT-1 utilizes on-orbit and vicarious calibration modes such as lunar calibration, blackbody calibration, LED calibration, and yaw maneuver calibration to ensure precise and quantitative applications.

The TIS has the ability of performing high-resolution (30m) and wide-swath (300km) observation, and its comprehensive performance has reached the international leading level. Remarkably, the spatial resolution is 3.3 times higher than that of similar international satellites, while also reducing payload power consumption. This is the first instance of China adopting an all-optical low-temperature system design in the thermal infrared spectrum, capable of distinguishing a temperature difference better than 80mK in a large dynamic range. It can accurately measure urban thermal energy distribution, offering essential data for global energy consumption calculations and supporting SDGs like clean energy and climate action. The SDGSAT-1 thermal infrared data has been successfully applied in fields, including ground temperature retrieval, land desertification evaluation, ships, aeroplanes and industrial emissions monitoring.

The GLI and MSI also have a large swath width of 300km, enabling round-the-clock imaging of the Earth through day and night. In particular, the GLI is the world's first color high-resolution night-light detection payload, capable of detecting nighttime light at a spatial resolution of 10m. In comparison, the resolution of glimmer images from other similar satellites is about 100~1000m at present. It can provide night light intensity information to identify global economic development levels and regional development disparities, serving SDGs such as sustainable cities and communities and industrial innovation and infrastructure, etc. It's worth noting that SDGSAT-1's primary imaging time for GLI is between 21:00 and 22:00 local time, coinciding with the peak of local human activities at night, making it particularly significant in terms of representativeness compared with other satellites.

The MSI is distinguished by its high signal-to-noise ratio and a spatial resolution of 10m. It features seven bands, including two optimized for the deep blue spectrum to effectively monitor water color index, transparency, and suspended solids across various turbid water bodies. Additionally, it has a red-edge band that ensures precise tracking of vegetation growth on the ground.

The SDGSAT-1 satellite encompasses the following performance features:

Satellite Platform: It has a high payload-to-platform ratio, a high-sensitivity integrated design, multi-mode high-precision control, high-speed data downloading technology, and autonomous intelligent task planning and calibration modes.

TIS: It has a wide swath width (300km) and high resolution (30m), leading the world in swath width/resolution. The imager has a wide dynamic range (220~340K) and high detection sensitivity (0.2K@300K).

GLI/MSI: The imager adopts a multi-mode common-path imaging design and can alternatively switch imaging between the day and night. The dynamic range of glimmer payload detection is 60dB. The signal-to-noise ratio of the 7 bands of the multispectral payload is greater than 130.

In summary, SDGSAT-1 embodies a significant step forward in sustainable development research. By providing a more nuanced understanding of the interaction between human activities and the natural environment, it offers invaluable insights and data to guide global efforts towards achieving the SDGs.

Technical Parameters of SDGSAT-1

Orbit / Sensors	Parameter	Specifications
Orbit	Type	Sun-synchronous
	Altitude	505km
	Inclination	97.5°
Thermal Infrared Spectrometer	Swath Width	300km
	Bands of TIS	B1: 8~10.5μm B2: 10.3~11.3μm B3: 11.5~12.5μm
	Spatial Resolution	30m
Glimmer / Multispectral Imager	Swath Width	300km
	Bands of GLI	P: 444~910nm B: 424~526nm G: 506~612nm R: 600~894nm
	Spatial Resolution of GLI	Panchromatic: 10m, RGB: 40m
	Bands of MSI	B1: 374~427nm B2: 410~467nm B3: 457~529nm B4: 510~597nm B5: 618~696nm B6: 744~813nm B7: 798~911nm
	Spatial Resolution of MSI	10m

CONTENTS

目录

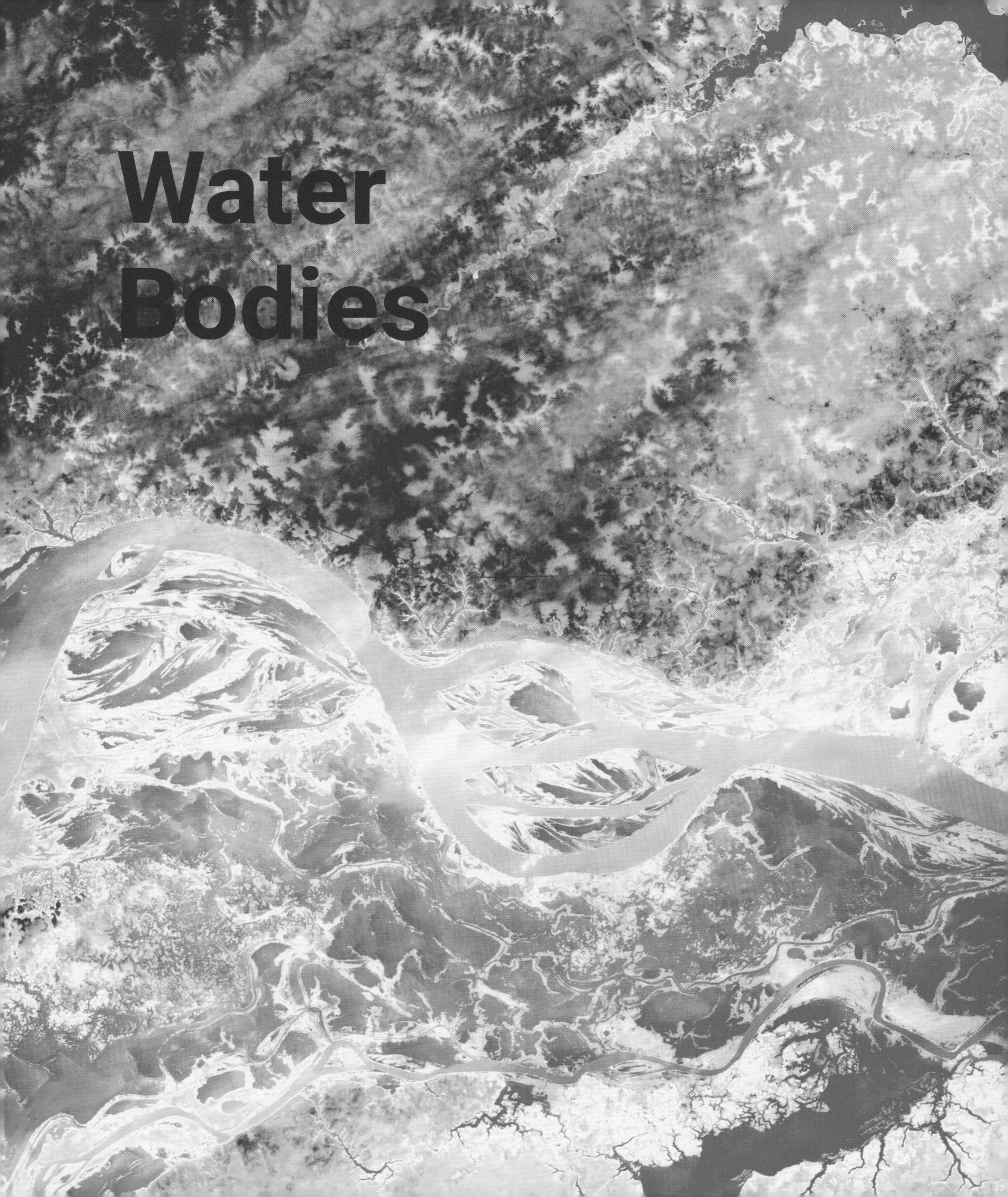
Water
Bodies

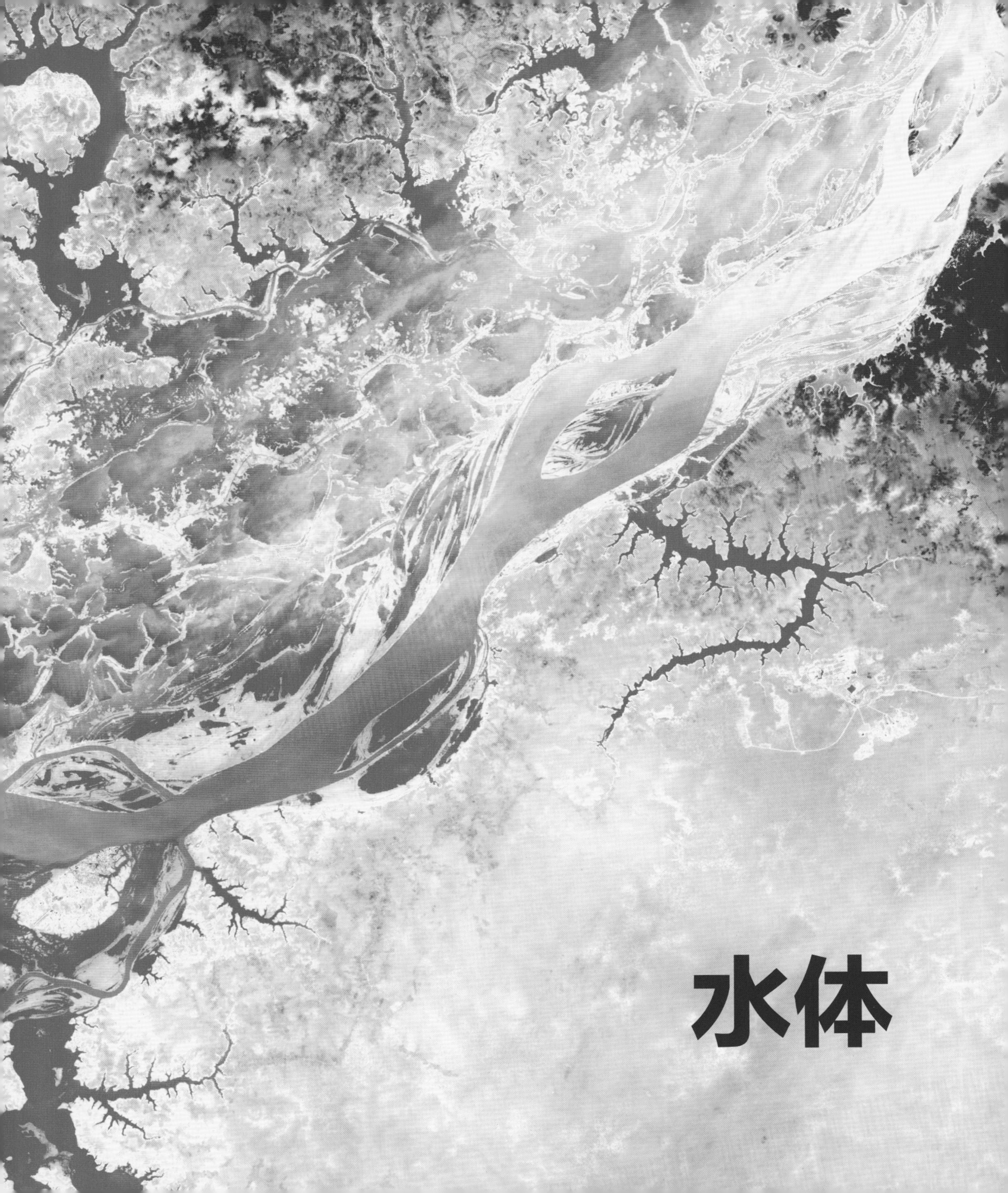

水体

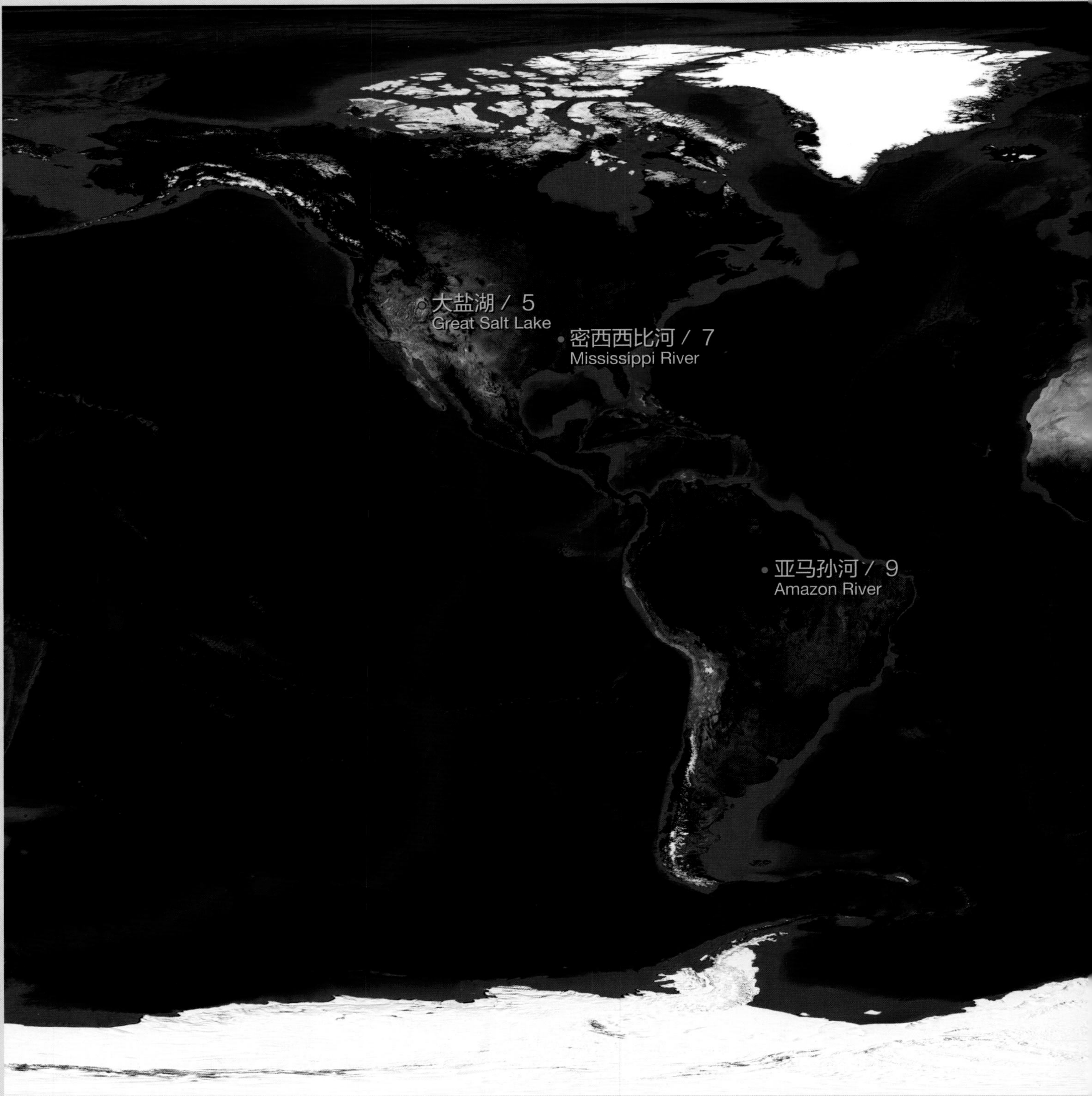

大盐湖 / 5
Great Salt Lake

密西西比河 / 7
Mississippi River

亚马孙河 / 9
Amazon River

卡累利阿地峡 / 11
Karelian Isthmus

第聂伯河 / 13
Dnieper River

日内瓦湖 / 14
ake Geneva

咸海 / 16
Aral Sea

青海湖 / 22
Qinghai Lake

长江 / 25
Yangtze River

尼罗河三角洲－苏伊士运河 / 19
Nile Delta-Suez Canal

鄱阳湖 / 27
Poyang Lake

恒河三角洲 / 29
Ganges Delta

乍得湖 / 20
Lake Chad

伊洛瓦底江三角洲 / 30
Irrawaddy Delta

湄公河三角洲 / 32
Mekong Delta

艾尔湖 / 35
Lake Eyre

大盐湖
Great Salt Lake

大盐湖，是北美洲最大的内陆盐湖，西半球最大咸水湖。位于美国犹他州西北部，东面是落基山，西面是沙漠，大盐湖为更新世大冰期大盆地内大淡水湖的残迹湖。大盐湖干燥的自然环境与著名的死海相似，湖水的化学特征与海水相同。在历史上湖的面积变化极大。夜间热红外图像中，湖面温度较周边环境高。

The Great Salt Lake is the largest inland salt lake in North America and the largest saltwater lake in the Western Hemisphere. Located in the northwestern part of the state of Utah, the United States, with the Rocky Mountains to the east and the desert to the west, the Great Salt Lake is a remnant lake of the freshwater lake system within the Great Basin of the Pleistocene Ice Age. The dry natural environment of the Great Salt Lake is similar to that of the famous Dead Sea, and the chemical characteristics of the lake's water are identical to those of seawater. The size of the lake has varied greatly over its history. Water surface temperatures are higher than their surrounding environments in nighttime thermal infrared images.

单波段热红外图像（波段 2）　成像时间：2024-04-03 **夜间**

Single band thermal infrared image (band 2)

Imaging time: 2024-04-03 Nighttime

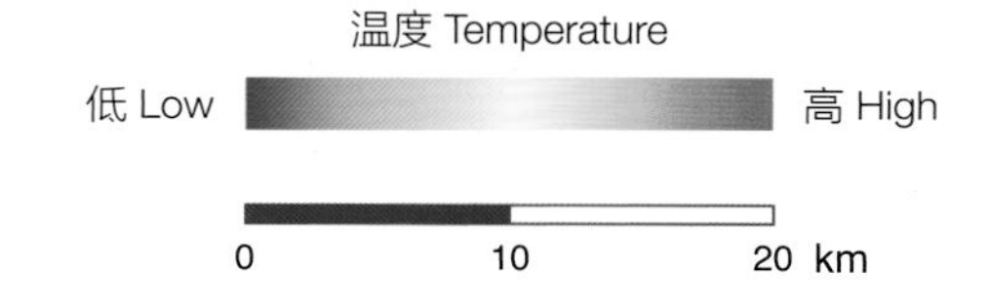

单波段热红外图像（波段 2） 成像时间：2024-04-13 夜间
Single band thermal infrared image (band 2) Imaging time: 2024-04-13 Nighttime

密西西比河
Mississippi River

密西西比河自北向南跨越北美大陆，河流干流与支流滋养着两岸的农业生产区，同时提供着航运价值。河流春夏季水位较高，冬季较低。热红外图像中，河流温度与陆地明显不同。

The Mississippi River spans the North American continent from north to south, and the river's main stem and tributaries nourish agriculturally productive areas on both sides of the river, as well as providing navigation value. River levels are high in the spring and summer and low in the winter. River temperatures are distinctly different from land in thermal infrared imagery.

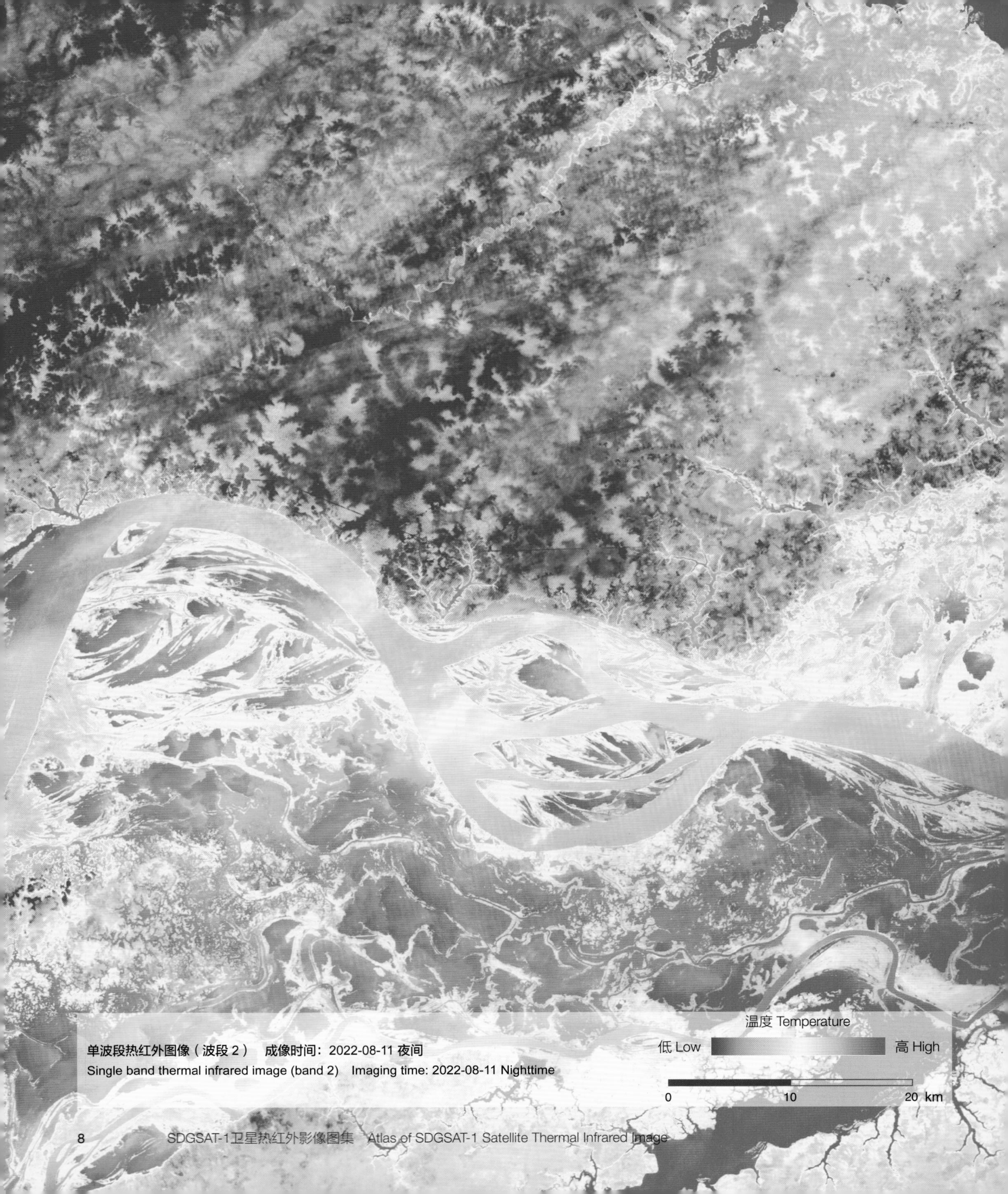

单波段热红外图像（波段 2）　成像时间：2022-08-11 夜间

Single band thermal infrared image (band 2)　Imaging time: 2022-08-11 Nighttime

亚马孙河
Amazon River

亚马孙河位于南美洲北部，是世界上流量、流域最大、支流最多的河流。亚马孙河河长有着多种不同的说法，其中之一即为常见的6400km。该河流共有1.5万条支流，分布在南美洲大片土地上，流域面积几乎大如整个澳大利亚。夜间热红外图像中，河水温度较周边雨林较高。

The Amazon River, located in the northern part of South America, is the world's largest river in terms of flow rate, basin size and a number of tributaries. There are various estimates for the length of the Amazon River, with one common figure being 6400km. The river has a total of 15000 tributaries spread across a vast area of South America, with a basin size almost as large as the entire Australia. The river is warmer than the surrounding rainforest in the nighttime thermal infrared image.

单波段热红外图像（波段 2） 成像时间：2022-04-22 日间
Single band thermal infrared image (band 2) Imaging time: 2022-04-22 Daytime

卡累利阿地峡
Karelian Isthmus

卡累利阿地峡位于俄罗斯圣彼得堡北方，东临拉多加湖水西接芬兰湾。地峡位于温带大陆性湿润气候覆盖范围内，年降水可达 800mm。湖泊密布河网纵横，构成了复杂的北方针叶林生态环境。在四月日间热红外图像中，水体热红外信号较低，陆地信号高，南方圣彼得堡热信号强度较周边环境尤其明显，西侧芬兰湾中可见部分浮冰，精细地展示该地区热环境分布与对比情况。

The Karelian Isthmus is located north of Saint Petersburg, Russia, with Ladoga Lake to the east and the Gulf of Finland to the west. The isthmus lies within the coverage area of a humid continental climate, with annual precipitation reaching up to 800mm. The area is characterized by numerous lakes and a network of rivers, forming a complex ecosystem of northern coniferous forests. In the thermal infrared image taken during daylight hours in April, water bodies exhibit lower thermal infrared signals compared with land, while the built-up areas of Saint Petersburg shows notably stronger thermal signals than its surrounding environment, especially towards the west where parts of the Gulf of Finland still have some floating ice. This finely illustrates the distribution and contrast of the thermal environment in the region.

真彩色多光谱图像（波段组合：5-4-3） 成像时间：2022-03-23 日间

True color multispectral image (band combination: 5-4-3) Imaging time: 2022-03-23 Daytime

第聂伯河
Dnieper River

第聂伯河是欧洲东部的第二大河，欧洲第四大河，源出俄罗斯瓦尔代丘陵南麓。第聂伯河向南流经白俄罗斯、乌克兰，注入黑海。第聂伯河长 2200km，流域面积 50.4 万 km^2，主要支流有杰斯纳河、索日河、普里皮亚季河等。上游有运河同涅曼河相通。日间热红外图像中，河流相较于陆地，尤其与建成区相比，温度较低。

Dnieper River, the second largest river in Eastern Europe and the fourth largest in Europe, originates from the southern foothills of the Valdai Hills in Russia. It flows southward through Belarus and Ukraine before emptying into the Black Sea. With a length of 2200km and a basin area of 504000km^2, the Dnieper River is characterized by several major tributaries, including the Desna River, the Sozh River and the Pripyat River etc. In the upstream, it connects the Neman River, with canals. The daytime thermal infrared images show lower temperatures in the river compared with the land, especially the built-up area.

单波段热红外图像（波段 2） 成像时间：2022-03-23 日间
Single band thermal infrared image (band 2) Imaging time: 2022-03-23 Daytime

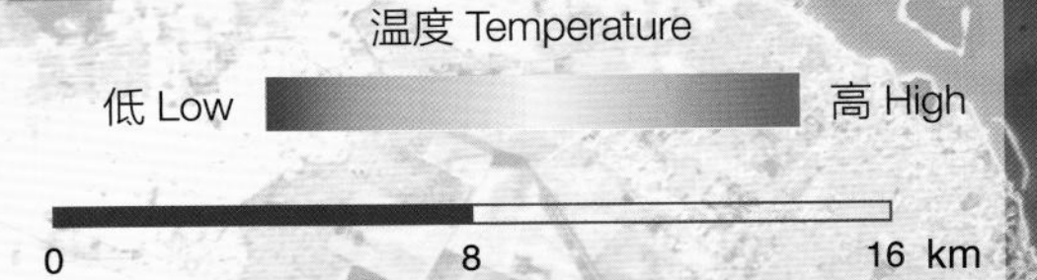

日内瓦湖
Lake Geneva

日内瓦湖，是阿尔卑斯湖群中最大的一个，位于瑞士与法国的边境。日内瓦湖是罗纳冰川形成的。湖周边气候温和。湖身为弓形，湖的凹处朝南。夜间热红外图像中水面温度高于陆地。

Lake Geneva, the largest of the Alpine lakes, is located on the Swiss-French border. Lake Geneva was formed by the Rhone Glacier. The body of the lake is bow-shaped, with the concave part of the lake facing south. Water surface temperatures are higher than land in nighttime thermal infrared images.

单波段热红外图像（波段 2） 成像时间：2023-10-06 夜间

Single band thermal infrared image (band 2) Imaging time: 2023-10-06 Nighttime

咸海
Aral Sea

咸海是位于中亚的咸水湖，北接哈萨克斯坦，南接乌兹别克斯坦。咸海浅水区曾经是世界第四大内陆水域。咸海地区的特点是沙漠大陆性气候，昼夜温差大，冬季寒冷，夏季炎热，降雨稀少。如今，咸海已大面积干涸，水域面积明显缩小，湖底盐碱裸露，周围地区沙化严重。2005 年，哈萨克斯坦建造了科卡拉尔大坝，提高了咸海北部的水位。

The Aral Sea is a saltwater lake in Central Asia, bordeing Kazakhstan to the north and Uzbekistan to the south. The shallow Aral Sea was once the world's fourth largest body of inland water. The Aral Sea area is characterized by a desert-continental climate that features wide-ranging diurnal air temperatures, cold winters, hot summers, and sparse rainfall. Today, the Aral Sea has dried up considerably, with a significantly reduced water surface area, exposed saline bottom, and a heavily sandy surrounding area. In 2005, Kazakhstan built the Kok-Aral Dam, which raised the water level in the northern part of the Aral Sea.

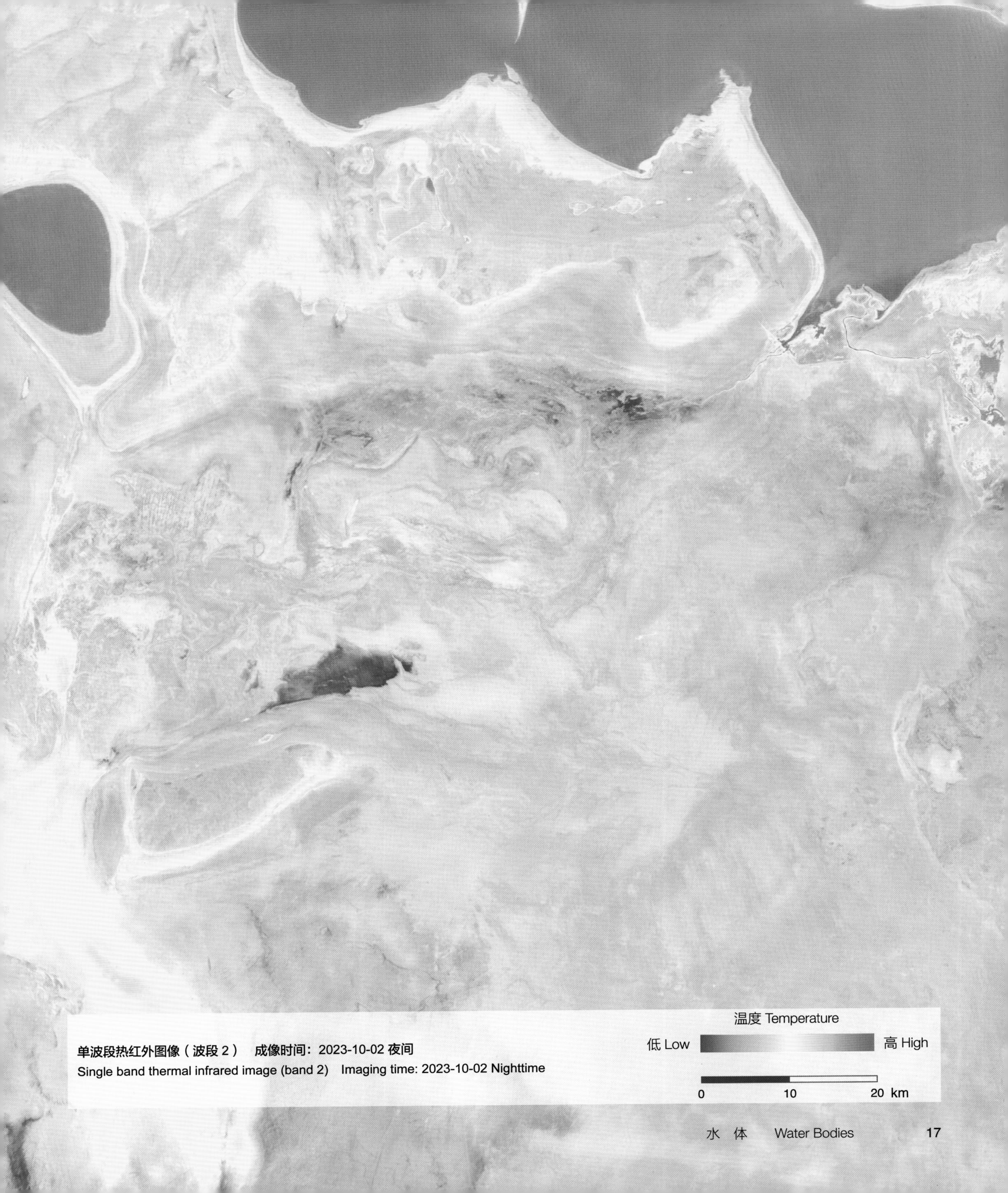

单波段热红外图像（波段 2）　成像时间：2023-10-02 夜间
Single band thermal infrared image (band 2)　Imaging time: 2023-10-02 Nighttime

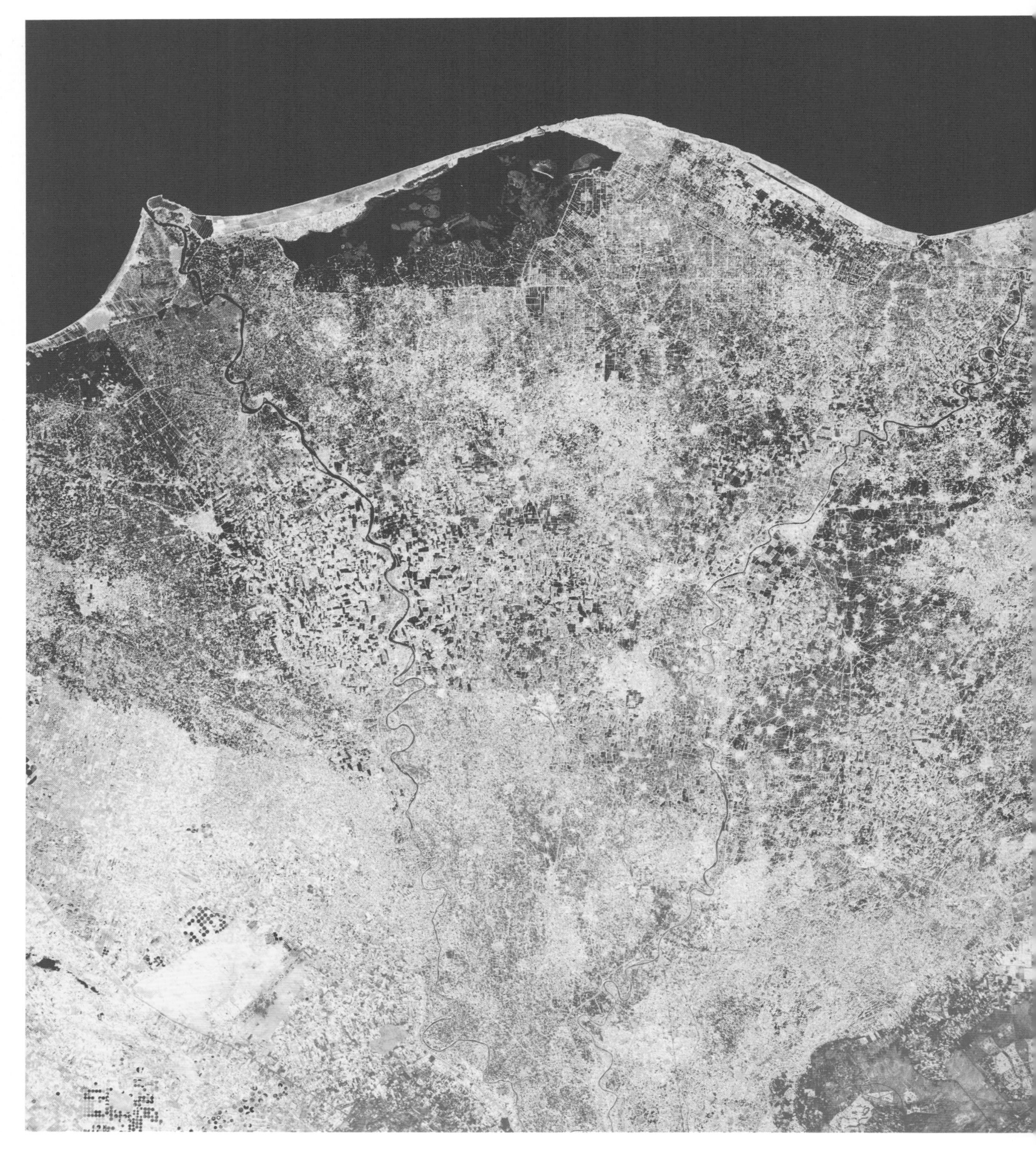

尼罗河三角洲－苏伊士运河
Nile Delta-Suez Canal

尼罗河三角洲位于埃及北部，是尼罗河在注入地中海前形成的三角洲地区。这个地区以其肥沃的土壤、广阔的农田和丰富的历史文化遗产而闻名。尼罗河三角洲是埃及的农业中心，因其灌溉便利和丰富的水资源而适合种植大米、小麦和棉花等作物。此外，三角洲地区还拥有丰富的渔业资源，为当地居民提供了重要的食物来源。苏伊士运河紧靠尼罗河三角洲，是国际重要航道。日间热红外图像中，水体温度明显低于周边沙漠干旱地带。

The Nile Delta is located in northern Egypt, formed by the Nile River before it flows into the Mediterranean Sea. This region is known for its fertile soil, vast farmland and rich historical and cultural heritage. The Nile Delta is the agricultural heartland of Egypt, suitable for cultivating crops such as rice, wheat and cotton due to its convenient irrigation and abundant water resources. Additionally, the delta region boasts rich fisheries resources, providing a significant source of food for local residents. The Suez Canal, an important international waterway, is adjacent to the Nile Delta. In the daytime thermal infrared image, the temperature of the water body is significantly lower than that of the surrounding desert arid zone.

单波段热红外图像（波段 2）　成像时间：2023-09-11 日间
Single band thermal infrared image (band 2)
Imaging time: 2023-09-11 Daytime

温度 Temperature
低 Low　高 High

0　10　20 km

乍得湖

Lake Chad

乍德湖是非洲第四大湖，位于非洲中西部萨赫勒地区，乍得、喀麦隆、尼日尔和尼日利亚四国交界处。乍得湖主要由三条河流的水汇合而成，它们是科马杜古约贝河、洛贡河和沙里河。乍德湖位于内陆盆地，该盆地以前被一个更大的古代海洋所占据，也被称为“巨型乍得”。乍得湖的水文贡献和生物多样性是当地重要的自然瑰宝。日间热红外图像中，湖面温度较地面低。

Lake Chad is the fourth largest lake in Africa and is located in the Sahel region of west-central Africa, on the border between Chad, Cameroon, Niger and Nigeria. Lake Chad is formed by the confluence of the waters of three major rivers, the Komadugu Yobe , the Logone and the Chari. It is situated in an interior basin formerly occupied by a much larger ancient sea that is sometimes called Mega-Chad. The hydrologic contributions and biological diversity of Lake Chad are important regional assets. In the daytime thermal infrared image, the lake temperature is lower than its surrounding landmass.

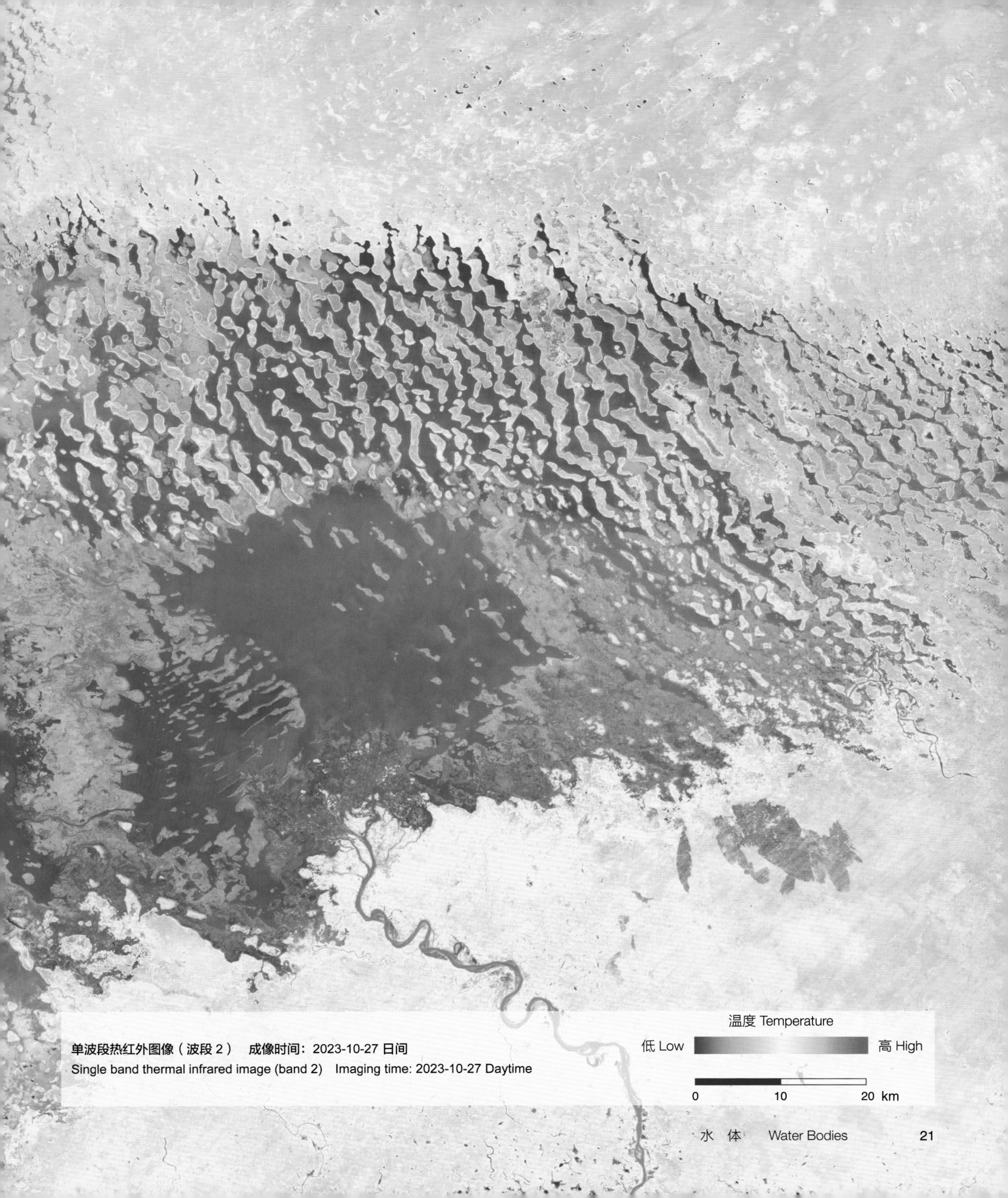

单波段热红外图像（波段 2）　成像时间：2023-10-27 日间

Single band thermal infrared image (band 2)　Imaging time: 2023-10-27 Daytime

青海湖
Qinghai Lake

青海湖地处青藏高原东北部，是中国面积最大的高原内陆咸水湖。青海湖四周被四座高山所环抱，是世界上海拔最高的湖泊之一。气候类型是典型高原大陆性气候。青海湖是具有国际性重要意义的内陆盐碱湿地，巨大的水体和周围的水系孕育了广阔而独特的高原内陆湿地生态系统，为许多野生动物提供了理想的栖息地。冬季青海湖湖面封冻，热红外图像中可清晰分辨冰面与温度较高的湖水。

Located in the northeastern part of the Qinghai-Tibet Plateau, Qinghai Lake is the largest plateau inland saltwater lake in China. Surrounded by four high mountains, Qinghai Lake is one of the highest lakes in the world. Its climate type is a typical highland continental climate. Qinghai Lake is an inland saline wetland of international importance. Its huge water body and the surrounding water system have given birth to the vast and unique plateau inland wetland ecosystem, providing ideal habitat for many wild animals. In winter, when the surface of Qinghai Lake is partially frozen, the ice surface and the warmer lake water can be clearly distinguished in the thermal infrared image.

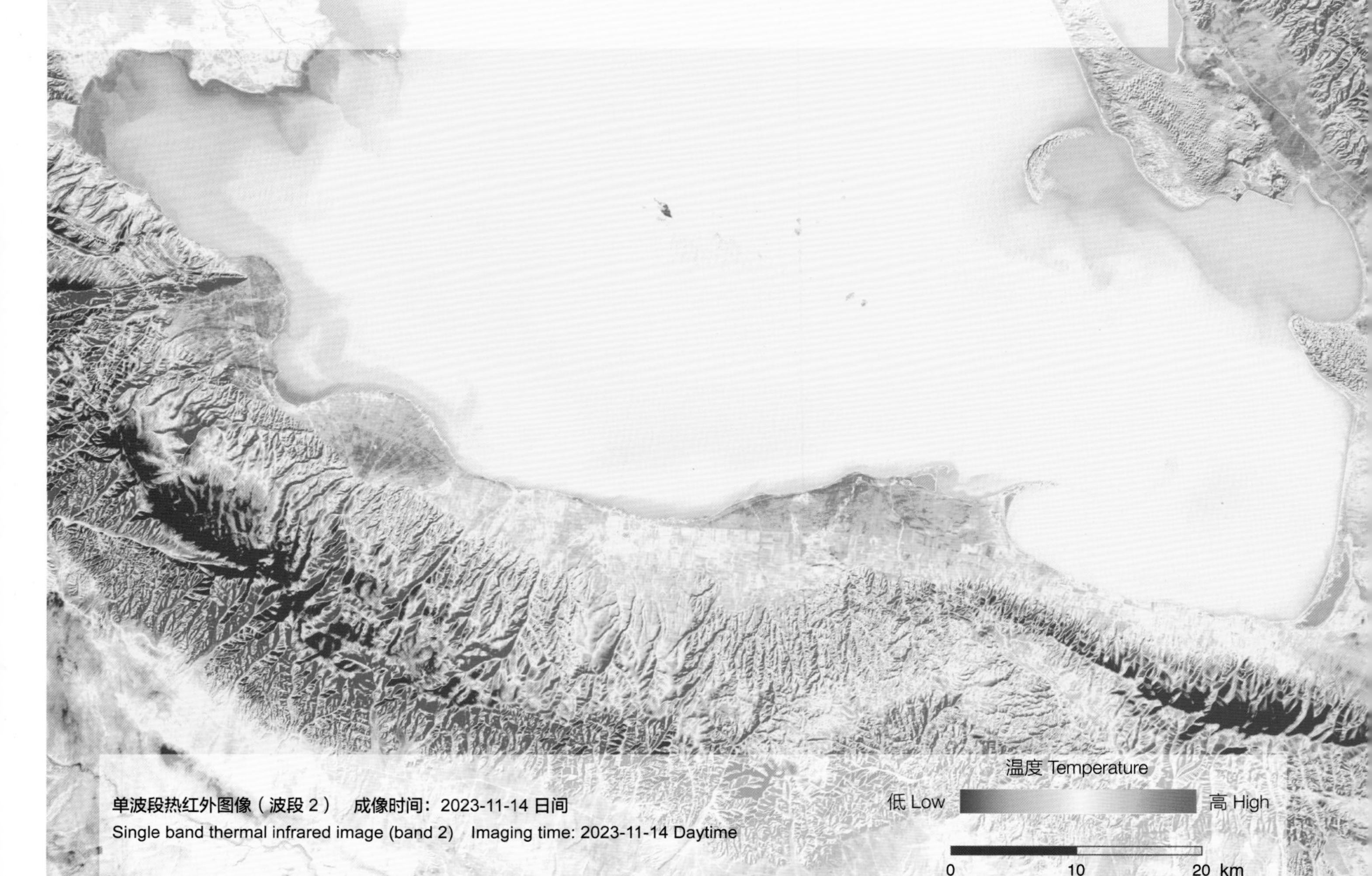

单波段热红外图像（波段 2）　成像时间：2023-11-14 日间

Single band thermal infrared image (band 2)　Imaging time: 2023-11-14 Daytime

单波段热红外图像（波段 2） 成像时间：2023-01-17 日间
Single band thermal infrared image (band 2) Imaging time: 2023-01-17 Daytime

单波段热红外图像（波段 2） 成像时间：2024-01-24 日间

Single band thermal infrared image (band 2) Imaging time: 2024-01-24 Daytime

长江
Yangtze River

长江是亚洲第一长河和世界第三长河，也是世界上完全在一国境内的最长河流，全长6300km，干流发源于青藏高原唐古拉山脉，穿越中国西南、中部、东部，在上海市汇入东海。长江流域覆盖中国五分之一陆地面积，养育中国三分之一的人口，长江经济带也是中国最大的经济带之一。长江流域生态类型多样，水生生物资源丰富。此冬季热红外图像中，水的温度较高。

The Yangtze River, also known as the Chang Jiang, is the longest river in Asia and the third longest river in the world. It holds the distinction of being the longest river contained entirely within one country, stretching over 6300km. Originating from the Tanggula Mountains on the Qinghai-Tibet Plateau, it traverses southwest, central and eastern China before emptying into the East China Sea in Shanghai. The Yangtze River basin covers one-fifth of China's land area and supports one-third of the country's population. Its economic belt is also one of the largest in China. With diverse ecological habitats, the Yangtze River basin boasts abundant aquatic resources. The temperature of the water is higher in this winter thermal infrared image.

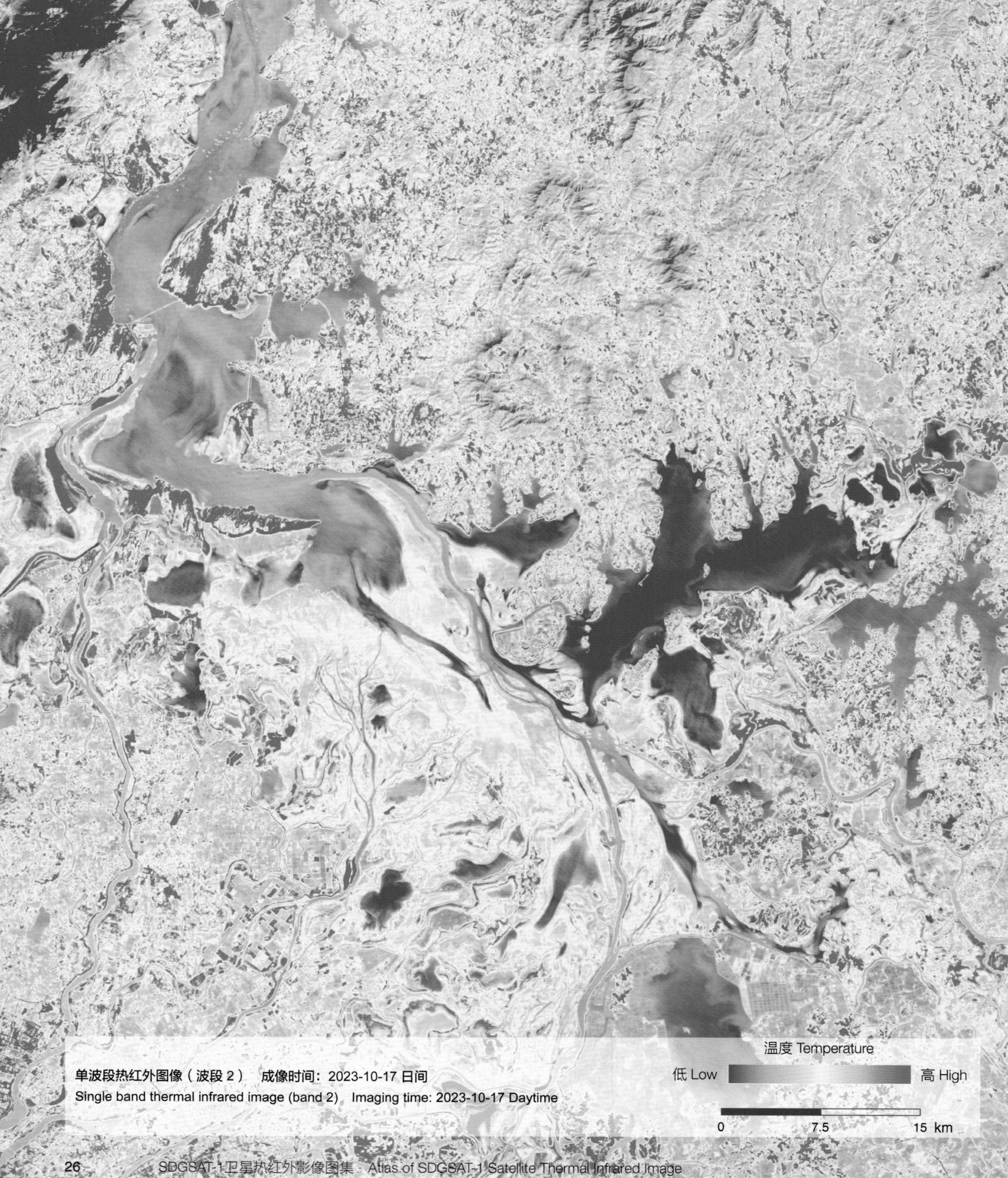

单波段热红外图像（波段 2）　成像时间：2023-10-17 日间

Single band thermal infrared image (band 2)　Imaging time: 2023-10-17 Daytime

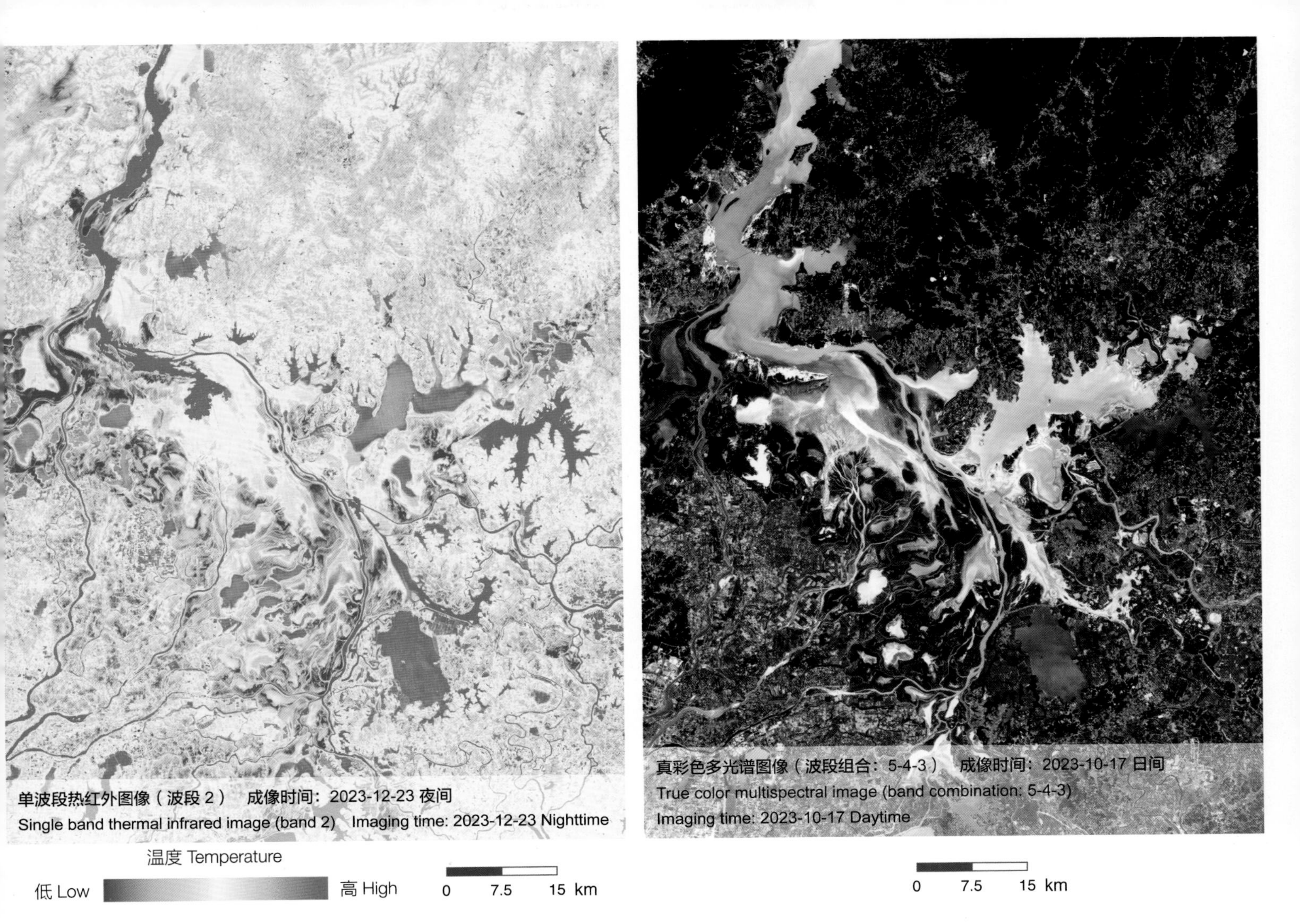

鄱阳湖
Poyang Lake

鄱阳湖，位于江西省北部，是中国第一大淡水湖，也是中国第二大湖。同样是江西省的水文枢纽，五河汇流并通过湖口与长江相连，形成了一个巨大的水收集盆地和转运站。鄱阳湖是一个浅水湖泊，地势平坦，微地形复杂。这个浅水湖泊的水文特性呈现出明显的季节性和年际变化，水位波动幅度极大。

Poyang Lake, located in the northern part of Jiangxi Province, is the largest freshwater lake in China and the second largest lake in China. Also serving as a hydrological hub in Jiangxi Province, the five rivers converge and are connected to the Yangtze River through the mouth of the lake, forming a huge water collection basin and transfer station. Poyang Lake is a shallow lake with flat terrain and complex microtopography. The hydrological characteristics of this shallow lake show obvious seasonal and inter-annual variations, and the water level fluctuates greatly.

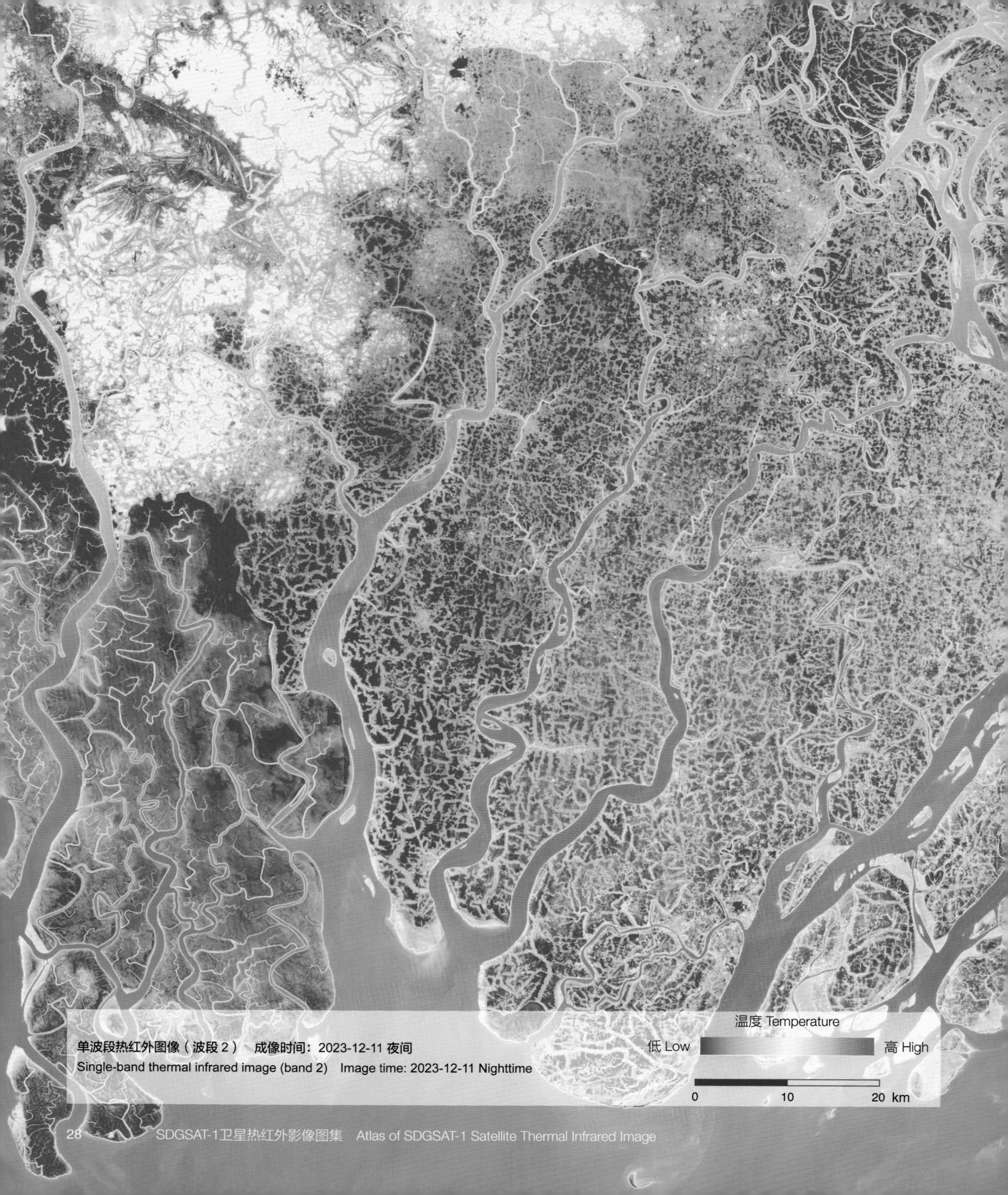

单波段热红外图像（波段 2） 成像时间：2023-12-11 夜间

Single-band thermal infrared image (band 2) Image time: 2023-12-11 Nighttime

恒河三角洲
Ganges Delta

恒河三角洲大部位于孟加拉国境内，是世界上最大的河口三角洲，总面积超过 10 万 km^2。三角洲位于热带，年降水可达 2000 ~ 3000mm，冲积平原土壤肥沃人口密集，散布有大量沼泽湖泊，包括濒危的孟加拉虎在内的众多珍稀野生动植物也分布于三角洲流域。该夜间热红外图像中包括了河流入海区域，相较于陆地，河流与海面在夜间温度较高，产生了明显对比，有效展示了入海口区域的热环境情况。

The majority of the Ganges Delta is located in Bangladesh, and it is the largest river delta in the world, with a total area exceeding 100000km^2. The delta is situated in a tropical region where annual precipitation can reach 2000~3000mm. With fertile plains and numerous marshes and lakes, the delta is home to 280 million people. Many rare and endangered wildlife species, including the Bengal tiger, are also distributed throughout the delta region. This thermal infrared image includes the river-mouth area where the river meets the sea. Compared with land, both the river and the sea exhibit higher temperature at night, creating a noticeable contrast and effectively showcasing the thermal environmental conditions of the estuarine region.

伊洛瓦底江三角洲
Irrawaddy Delta

伊洛瓦底江三角洲位于缅甸南部的热带地区，雨量充沛，全年温度较高。得益于优良的水热条件，三角洲区域广泛种植大米，种植区域总体产量较高，是全缅甸的主要产粮地之一。然而密集的农业生产活动也不可避免地对当地的红树林生态系统造成了影响。日间热红外图像中，入海的河流与海面温度较低，地面温度较高。

Located in the southern part of Myanmar, the Irrawaddy Delta is situated in the tropics with abundant rainfall and high temperatures throughout the year. Thanks to the excellent thermal and hydrological conditions, rice is widely cultivated in the delta, and the overall yield of the cultivated area is high, making it one of the major food-producing areas in the whole of Myanmar. However, intensive agricultural activities have inevitably affected the local mangrove ecosystem. In the daytime thermal infrared image, the temperature of the river into the sea and sea surface is lower, and the ground temperature is higher.

单波段热红外图像（波段 2） 成像时间：2023-09-11 日间
Single band thermal infrared image (band 2) Imaging time: 2023-09-11 Daytime

湄公河三角洲
Mekong Delta

湄公河三角洲位于东南亚，是由湄公河与其支流潘切河、拜公河和苏岱河在越南、柬埔寨和老挝三国交界处形成的一个广阔三角洲地区。该三角洲被认为是世界上最大的三角洲之一。湄公河三角洲地区的地形是由河流泥沙沉积形成的，土壤肥沃，适宜农业生产。除了农业，湄公河三角洲还以其丰富的渔业资源而闻名。夜间热红外图像中，海水与河水温度较高。

The Mekong Delta is located in Southeast Asia, formed by the confluence of the Mekong River and its tributaries, including the Bassac, Tien and Hau Rivers, spanning across Vietnam, Cambodia and Laos. It is considered one of the world's largest deltas. The topography of the Mekong Delta is characterized by sediment deposits from the river, resulting in fertile soils ideal for agriculture. In addition to agriculture, the Mekong Delta is renowned for its abundant fisheries. Higher sea and river temperatures can be observed in thermal infrared images at night.

单波段热红外图像（波段 2） 成像时间：2024-01-22 夜间
Single band thermal infrared image (band 2) Imaging time: 2024-01-22 Nighttime
温度 Temperature
低 Low
高 High
0
10
20 km

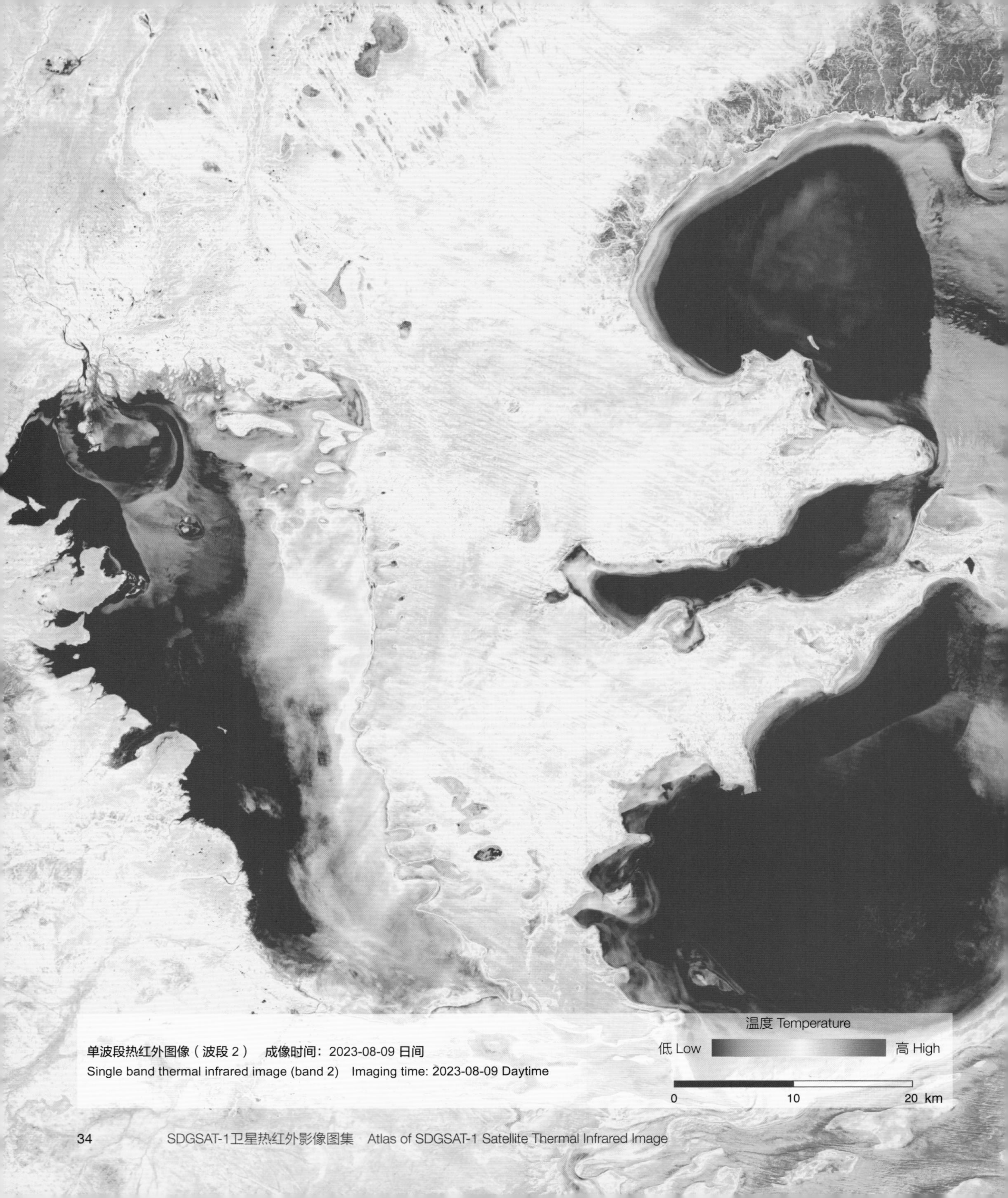

单波段热红外图像（波段 2）　成像时间：2023-08-09 日间

Single band thermal infrared image (band 2)　Imaging time: 2023-08-09 Daytime

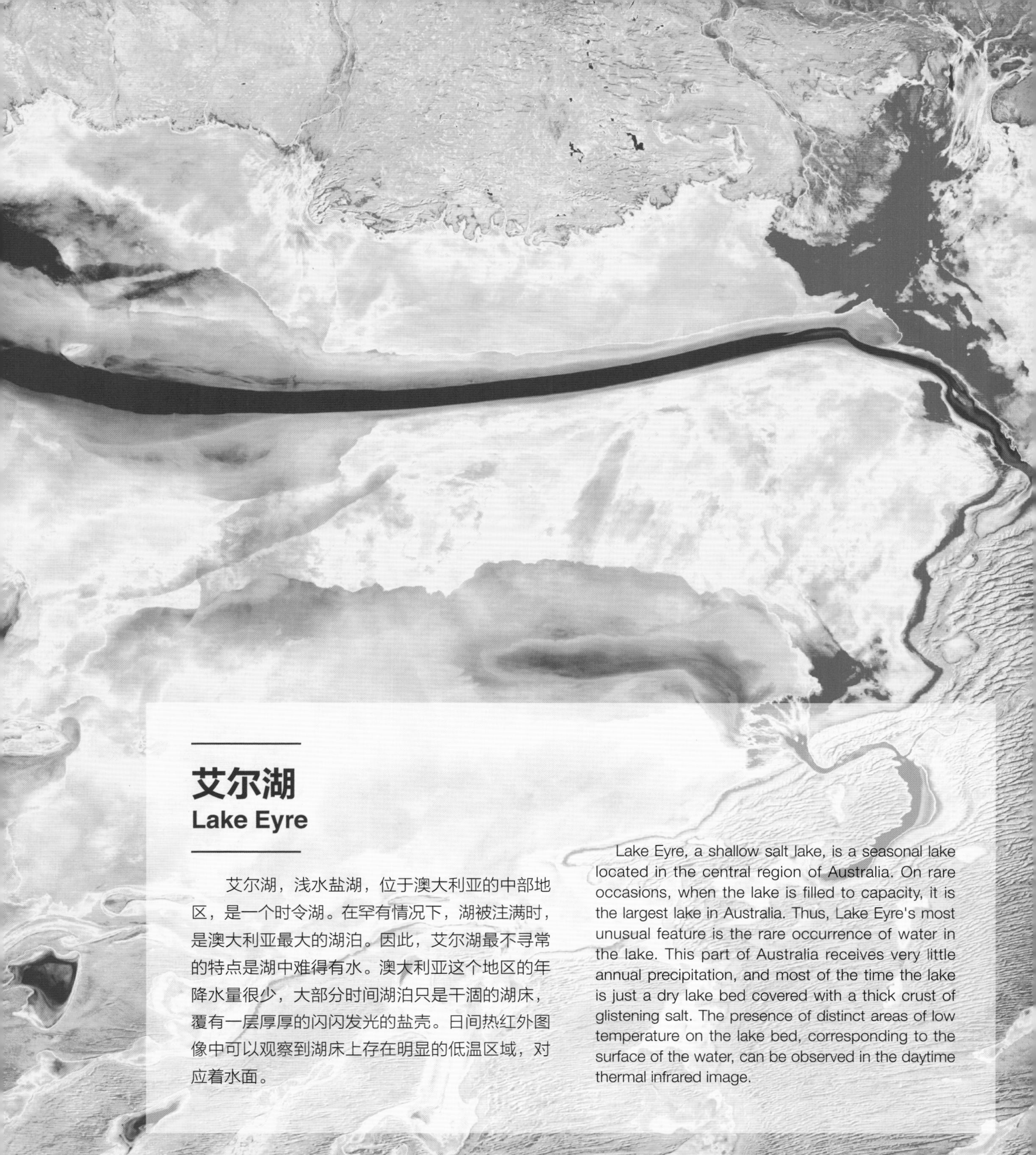

艾尔湖
Lake Eyre

艾尔湖，浅水盐湖，位于澳大利亚的中部地区，是一个时令湖。在罕有情况下，湖被注满时，是澳大利亚最大的湖泊。因此，艾尔湖最不寻常的特点是湖中难得有水。澳大利亚这个地区的年降水量很少，大部分时间湖泊只是干涸的湖床，覆有一层厚厚的闪闪发光的盐壳。日间热红外图像中可以观察到湖床上存在明显的低温区域，对应着水面。

Lake Eyre, a shallow salt lake, is a seasonal lake located in the central region of Australia. On rare occasions, when the lake is filled to capacity, it is the largest lake in Australia. Thus, Lake Eyre's most unusual feature is the rare occurrence of water in the lake. This part of Australia receives very little annual precipitation, and most of the time the lake is just a dry lake bed covered with a thick crust of glistening salt. The presence of distinct areas of low temperature on the lake bed, corresponding to the surface of the water, can be observed in the daytime thermal infrared image.

山脉

Mountain Ranges

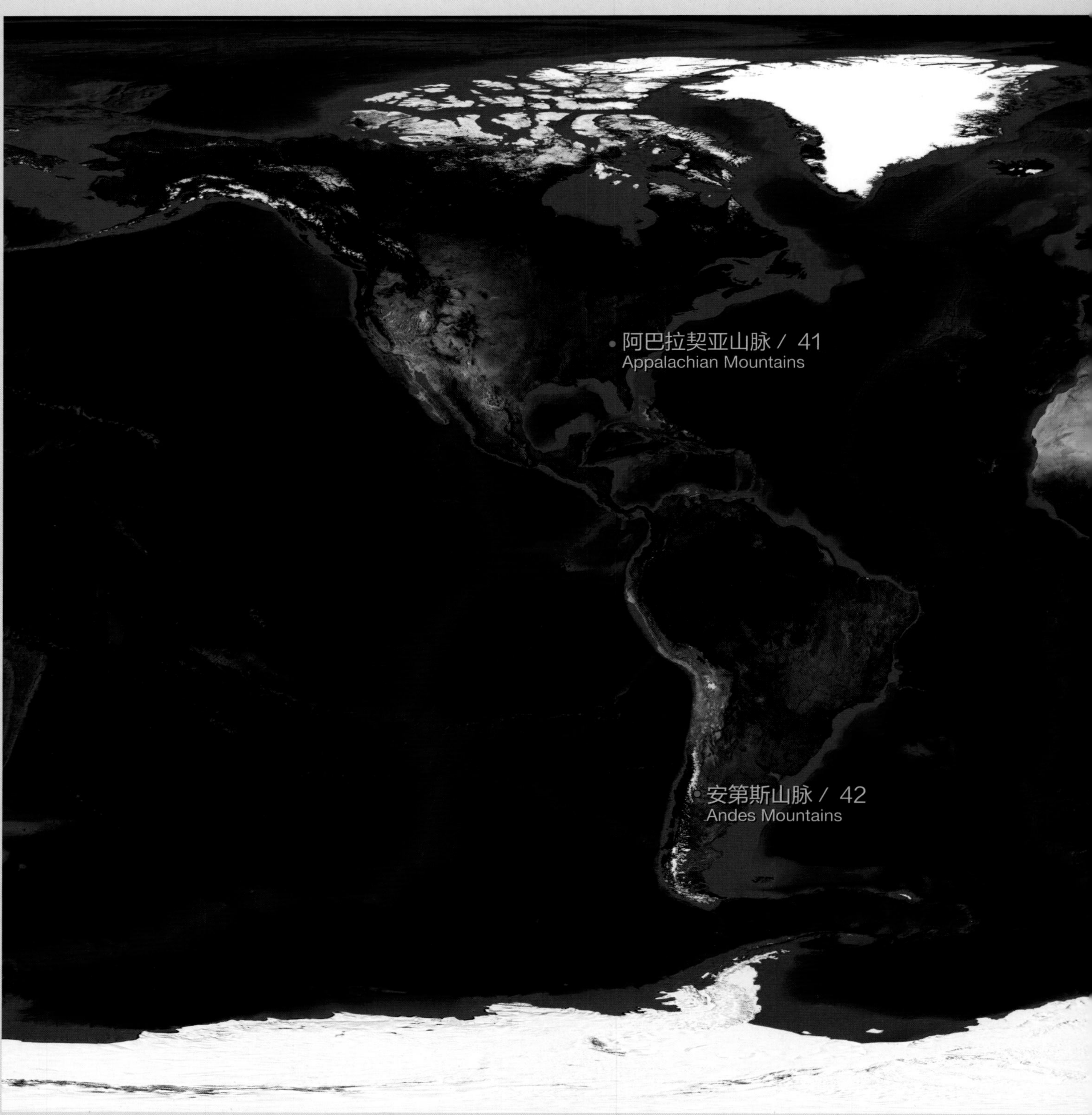

阿巴拉契亚山脉 / 41
Appalachian Mountains

安第斯山脉 / 42
Andes Mountains

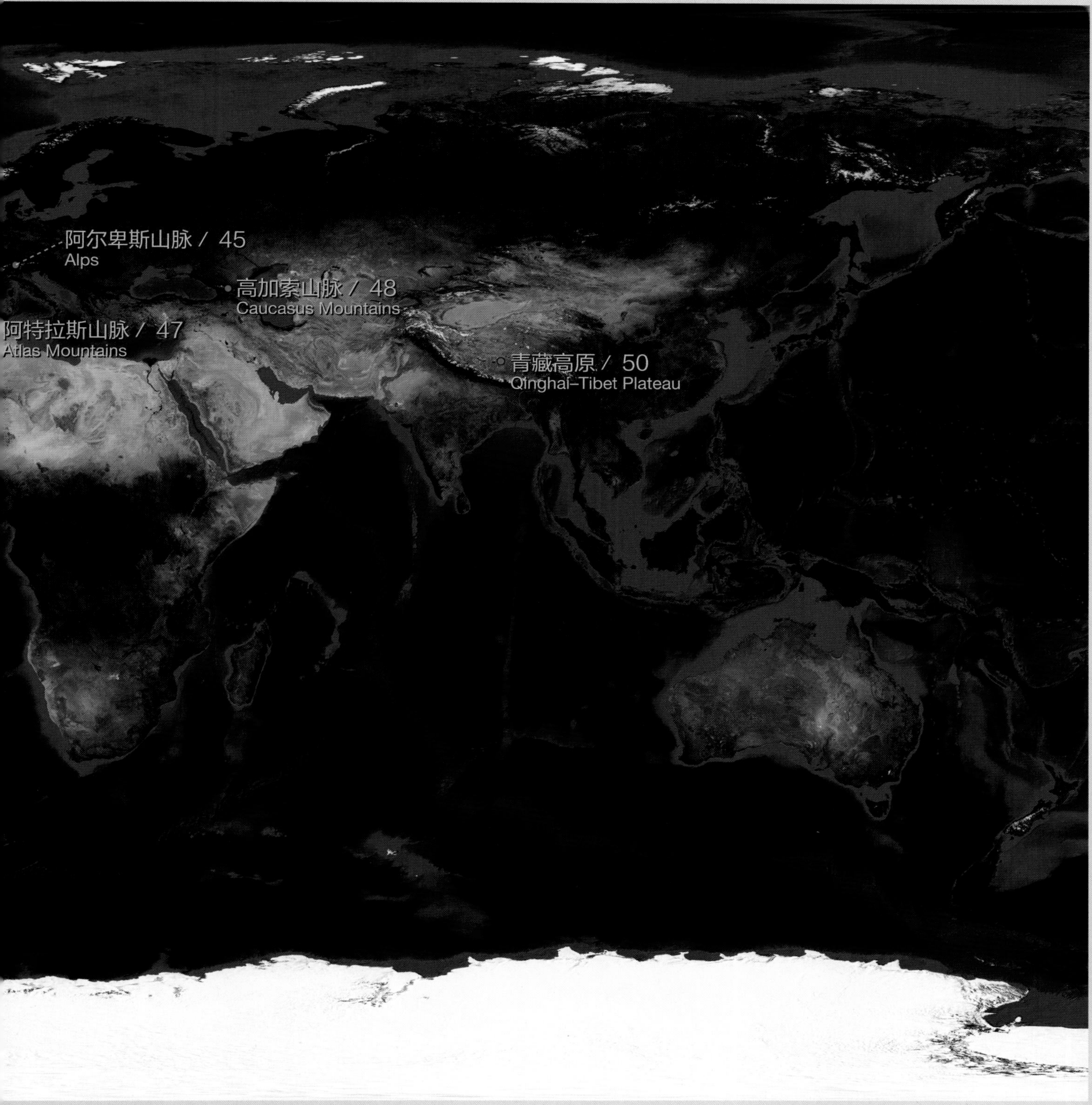

阿尔卑斯山脉 / 45
Alps

高加索山脉 / 48
Caucasus Mountains

阿特拉斯山脉 / 47
Atlas Mountains

青藏高原 / 50
Qinghai–Tibet Plateau

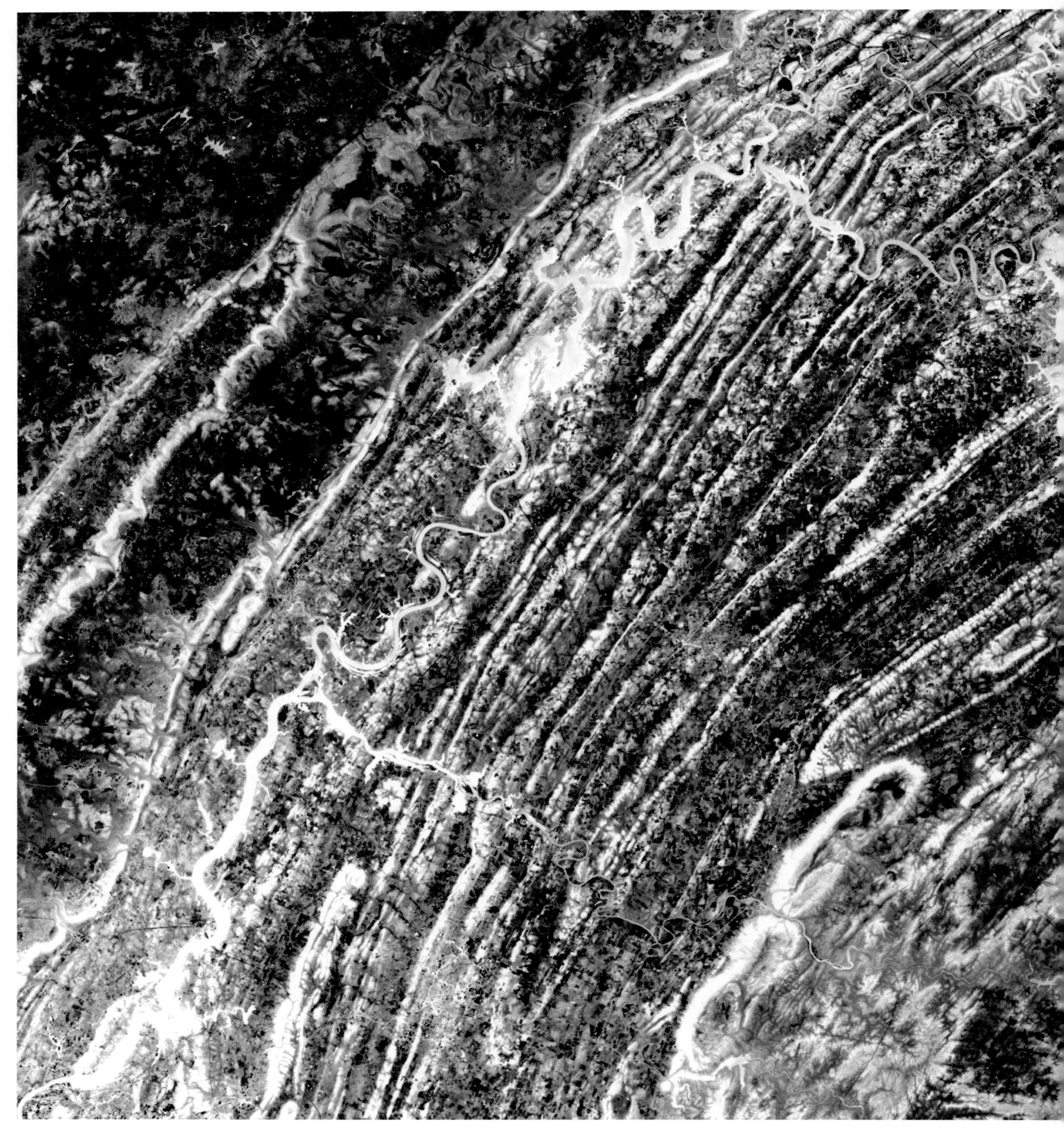

三波段热红外图像（波段组合：3-2-1） 成像时间：2024-04-14 夜间

Three band thermal infrared image (band combination: 3-2-1) Imaging time: 2024-04-14 Nighttime

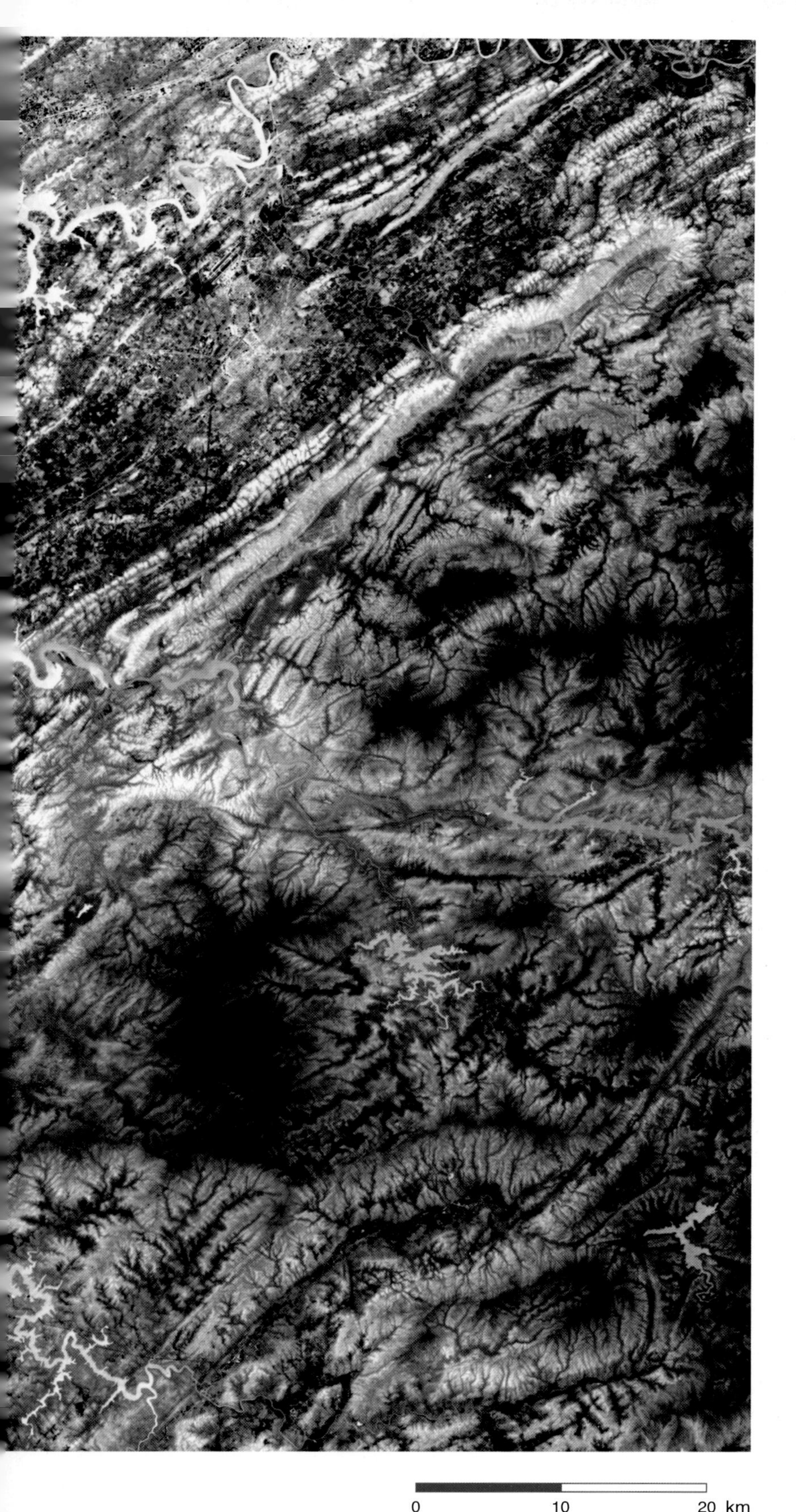

阿巴拉契亚山脉
Appalachian Mountains

阿巴拉契亚山脉位于北美洲东部，靠近大西洋。山脉覆盖地区气候湿润，年降水量较高，支撑了茂密的森林以及纵横交错的水系。阿巴拉契亚山脉一直以其丰富的矿产资源闻名，其煤炭储量居世界前列。阿巴拉契亚山脉山势平缓，景观独特，四季分明，是著名的徒步旅游胜地。三波段热红外图像中山脉结构分明，可见多条河流。

The Appalachian Mountains are located in eastern North America, near the Atlantic Ocean. The region covered by the mountains has a humid climate with high annual precipitation, supporting dense forests as well as crisscrossing waterways. The Appalachian Mountains have long been known for their rich mineral resources, with some of the world's largest coal reserves. The Appalachian Mountains have unique seasonal landscapes, making them a famous hiking destination. The structure of the mountain range is well visible in the three-band thermal infrared image, and many rivers are visible.

安第斯山脉
Andes Mountains

安第斯山脉位于南美洲，山脉以南北向沿太平洋海岸穿越整个大陆。安第斯山脉中含有大量矿石资源，已探明的矿产包括铜矿、锂矿和锡矿。因山脉跨越范围广，不同山脉区域的气候随纬度与海拔有显著变化。山脉北部以及南部较为湿润而中部则比较干燥。三波段热红外图像中，图像中部的山脉高海拔地区与西侧临海和东侧内陆地区有着不同的温度属性。

The Andes are located in South America and run in a north-south direction across the continent along the Pacific coast. The Andes contain significant mineral resources, with prospected ores including copper, lithium and tin. Due to the wide span of the mountain range, the climate in different regions of the mountain range varies significantly with latitude and altitude. The northern and southern parts of the range are wetter while the central part is drier. In the three-band thermal infrared image, the higher elevations of the mountain range in the central part of the image have different temperature properties compared with the sea-facing western and inland areas in the eastern part of the image.

三波段热红外图像（波段组合：3-2-1） 成像时间：2022-11-02 夜间
Three band thermal infrared image (band combination: 3-2-1) Imaging time: 2022-11-02 Nighttime

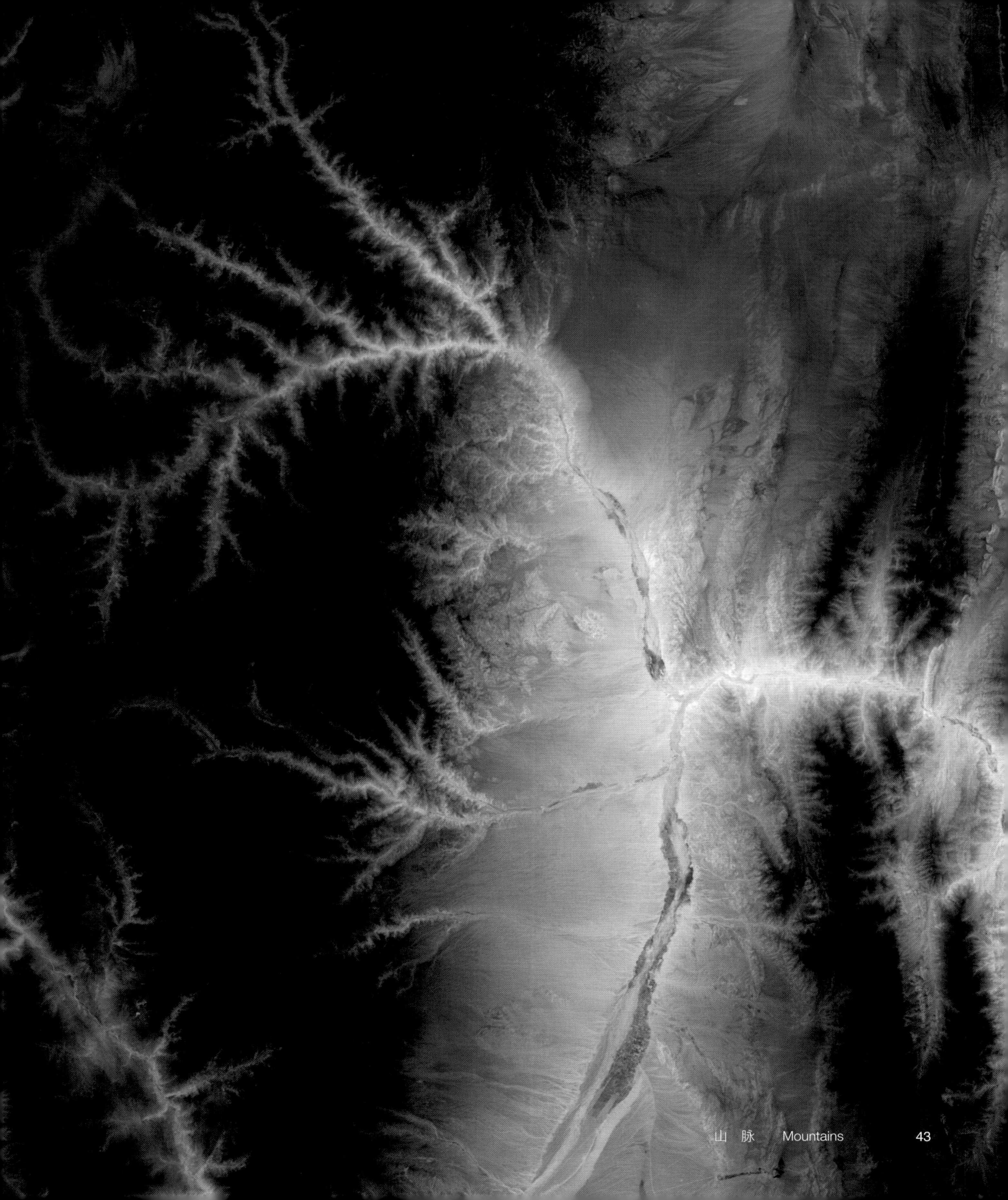

三波段热红外图像（波段组合：3-2-1） 成像时间：2024-04-14 日间

Three band thermal infrared image (band combination: 3-2-1) Imaging time: 2024-04-14 Daytime

阿尔卑斯山脉
Alps

阿尔卑斯山脉是欧洲最具标志性的山脉之一，连接法国、德国、意大利在内的数个欧洲国家。山脉常年有积雪，风景优美，山峰挺拔陡峭。作为多个欧洲文化的交汇地，山脉隘口城镇历史悠久，文旅资源丰富。三波段热红外图像中高海拔地区偏暗，温度明显低于低海拔地区。

The Alps is one of the most iconic mountain ranges in Europe, connecting several European countries including France, Germany and Italy. The mountains are characterized by its distinctive snow cover, beautiful scenery and steep peaks. As a crossroads of several European cultures, the mountain pass towns have a long history and are rich in cultural and tourism resources. The three-band thermal infrared image is darker at high altitude and the temperature is significantly lower than that at low altitude.

三波段热红外图像（波段组合：3-2-1） 成像时间：2024-03-13 日间
Three band thermal infrared image (band combination: 3-2-1) Imaging time: 2024-03-13 Daytime

阿特拉斯山脉
Atlas Mountains

阿特拉斯山脉位于非洲西北部，北临地中海，西接大西洋，横贯摩洛哥、阿尔及利亚和突尼斯三国。与山脉南方严酷干旱的撒哈拉沙漠不同，阿特拉斯山脉中有一定的淡水资源，可以支撑森林覆盖与农业发展。三波段热红外图像中山脉结构清晰，北方为地中海海岸，南方为较干旱地带。

The Atlas Mountains are located in north-western Africa, bordering the Mediterranean Sea to the north and the Atlantic Ocean to the west, and traversing Morocco, Algeria and Tunisia. Unlike the harsh and arid Sahara Desert to the south of the mountain range, the Atlas Mountains have adequate freshwater resources to support forest cover and agriculture. The structure of the mountain range is clear in the three-band thermal infrared composite image, with the Mediterranean coast in the north and the more arid zone in the south.

高加索山脉
Caucasus Mountains

高加索山脉位于亚欧交界，西临黑海，东接里海，数个高峰海拔分布在俄罗斯、格鲁吉亚、阿塞拜疆和亚美尼亚境内。山脉范围内因海拔变化有着复杂多变的生态环境，山脉北坡的低海拔区域常有森林覆盖，而南坡地区草地覆盖较多。三波段热红外图像中高海拔地区偏暗，温度明显低于低海拔地区。

The Caucasus Mountains are located on the border between Asia and Europe, bordered by the Black Sea in the west and the Caspian Sea in the east, with several high peaks spread over the territories of Russia, Georgia, Azerbaijan and Armenia. The mountain range has a complex and varied ecological environment due to the altitude changes, and the low altitude areas on the northern slopes of the mountain range are often covered with forests, while the southern slopes are covered with more grasslands. The three-band thermal infrared image is darker at high altitude and the temperature is significantly lower than that at low altitude.

三波段热红外图像（波段组合：3-2-1） 成像时间：2024-03-31 夜间
Three band thermal infrared image (band combination: 3-2-1) Imaging time: 2024-03-31 Nighttime

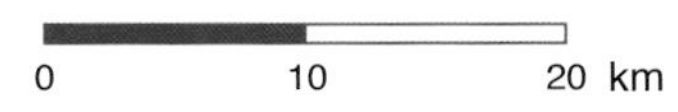

青藏高原
Qinghai-Tibet Plateau

青藏高原位于亚洲，是包括黄河和长江在内的众多重要河流的源头，同时因其广阔的冰川覆盖，青藏高原常被喻为世界第三极。青藏高原大部分全年气温低、太阳辐射高，地质环境复杂，地震频率高，有丰富的矿产和地热资源。三波段热红外图像中可见包括当穹错、达则错和吴如错在内众多高原湖泊。

Located in Asia, the Qinghai-Tibet Plateau is the source of many important rivers, including the Yellow River and Yangtze River, and is often referred to as the world's third pole because of its extensive glacial coverage. The Qinghai-Tibet Plateau is characterized by low temperatures and high solar radiation throughout most of the year, a complex geological environment, a high frequency of earthquakes, and abundant mineral and geothermal resources. Numerous lakes on plateau are visible in the three-band thermal infrared image.

三波段热红外图像（波段组合：3-2-1）　成像时间：2024-04-01 日间
Three band thermal infrared image (band combination: 3-2-1)　Imaging time: 2024-04-01 Daytime

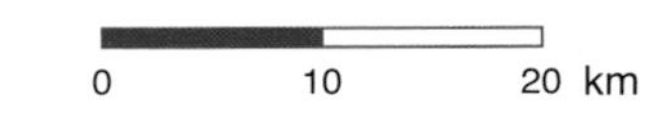

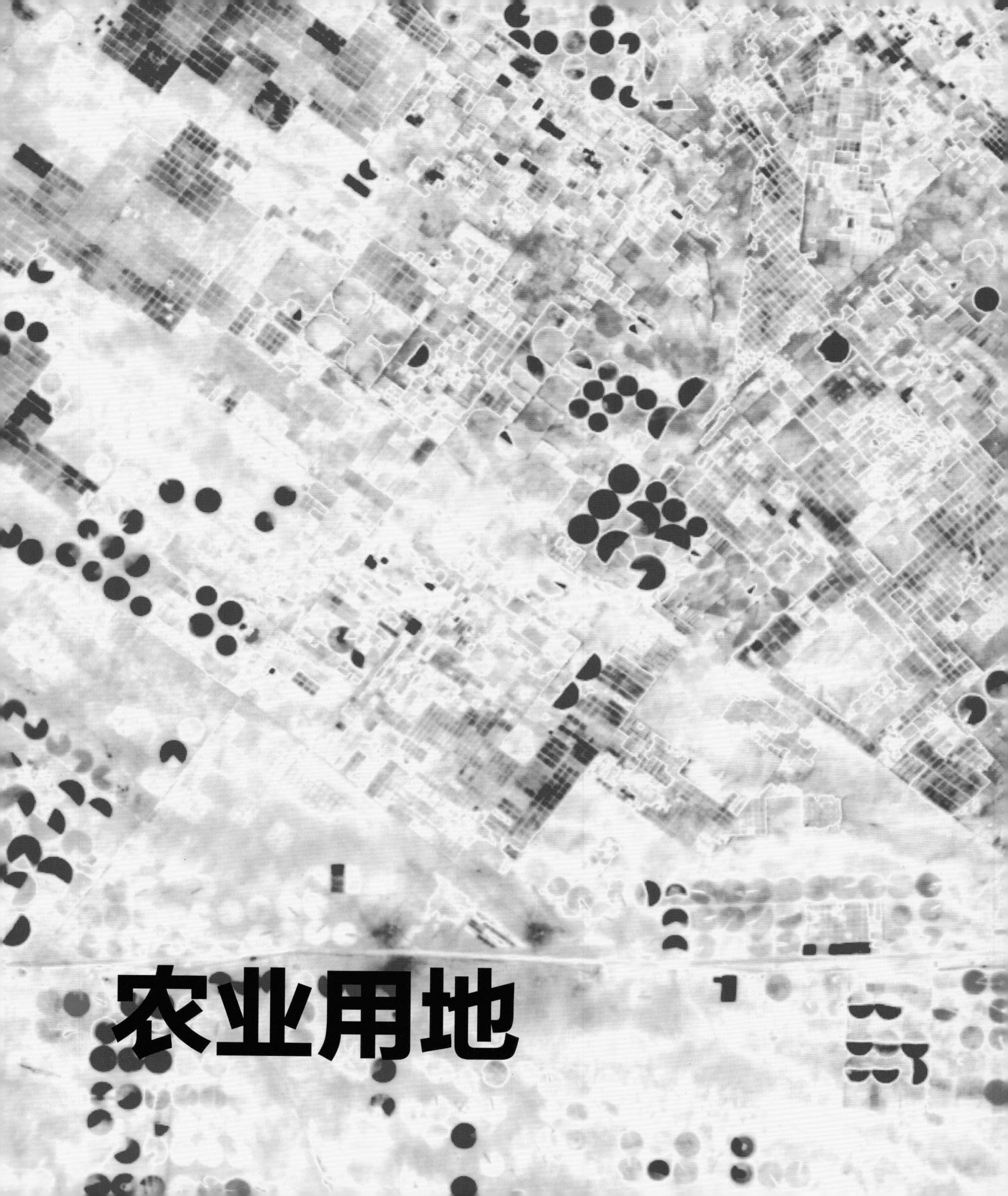

农业用地

Agricultural Land

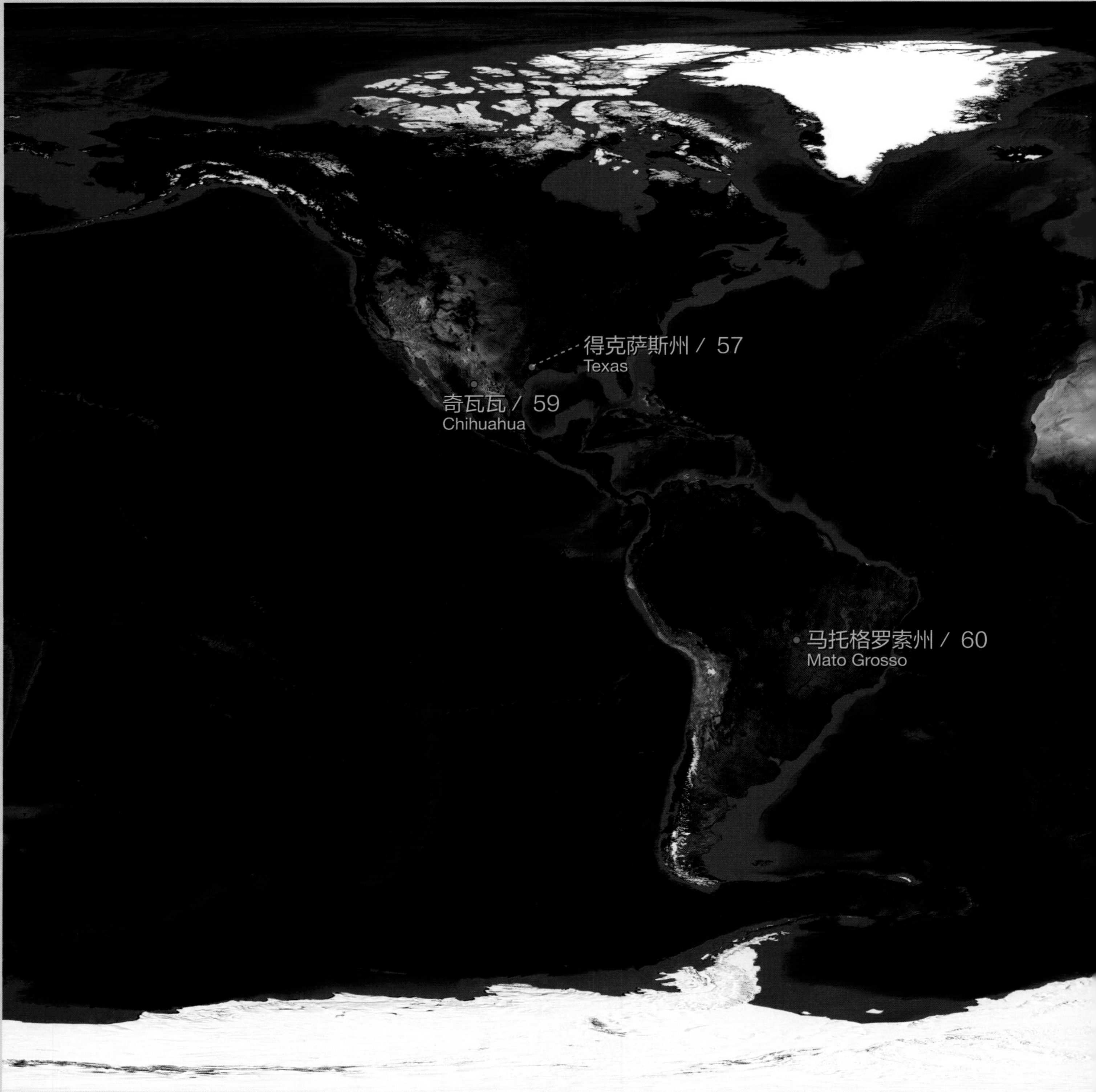

得克萨斯州 / 57
Texas

奇瓦瓦 / 59
Chihuahua

马托格罗索州 / 60
Mato Grosso

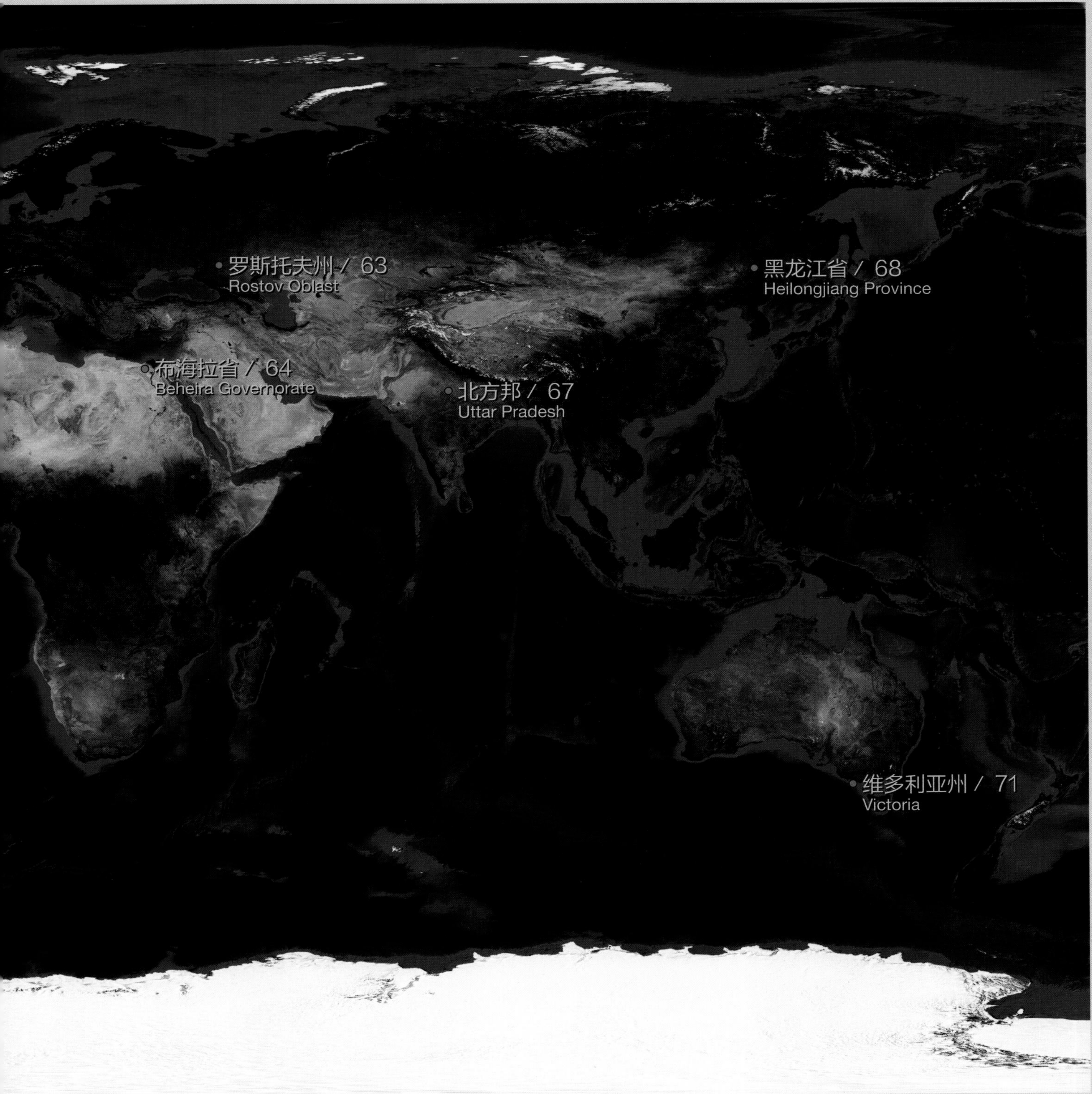

罗斯托夫州 / 63
Rostov Oblast

黑龙江省 / 68
Heilongjiang Province

布海拉省 / 64
Beheira Governorate

北方邦 / 67
Uttar Pradesh

维多利亚州 / 71
Victoria

真彩色多光谱图像（波段组合：5-4-3）　成像时间：2023-12-03 日间
True color multispectral image (band combination: 5-4-3)　Imaging time: 2023-12-03 Daytime

得克萨斯州
Texas

得克萨斯州位于美国南部，在面积上为美国最大州之一。州内基本属温带气候，南部部分地区为亚热带气候，适宜农业生产。得克萨斯州农业发达，居全美前列，主要农产品有棉花、高粱和水稻。其畜牧业为全国之冠，拥有美国最多的农场，其中养牛数量在全国领先。在热红外图像中，得克萨斯州的农业用地展现出整齐的规划，不同状态下的农田的温度也有差异。

Texas is located in the southern part of the United States and is one of the largest states in terms of area. The state's climate is generally temperate, with parts of the southern part of the state having a subtropical climate, which is suitable for agricultural production. Texas has a developed agricultural industry that ranks among the top in the United States, with cotton, sorghum and rice as its main agricultural products. Its livestock industry is nation's top, with the largest number of farms in the United States, including the number of cattle in the nation's leading. In thermal infrared imagery, agricultural land in Texas shows a grid pattern, and the temperature of farmland varies from area to area.

单波段热红外图像（波段 2）　成像时间：2023-12-03 日间

Single band thermal infrared image (band 2)　Imaging time: 2023-12-03 Daytime

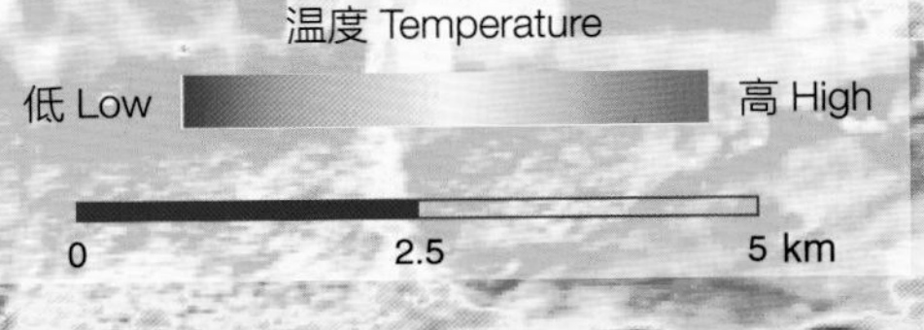

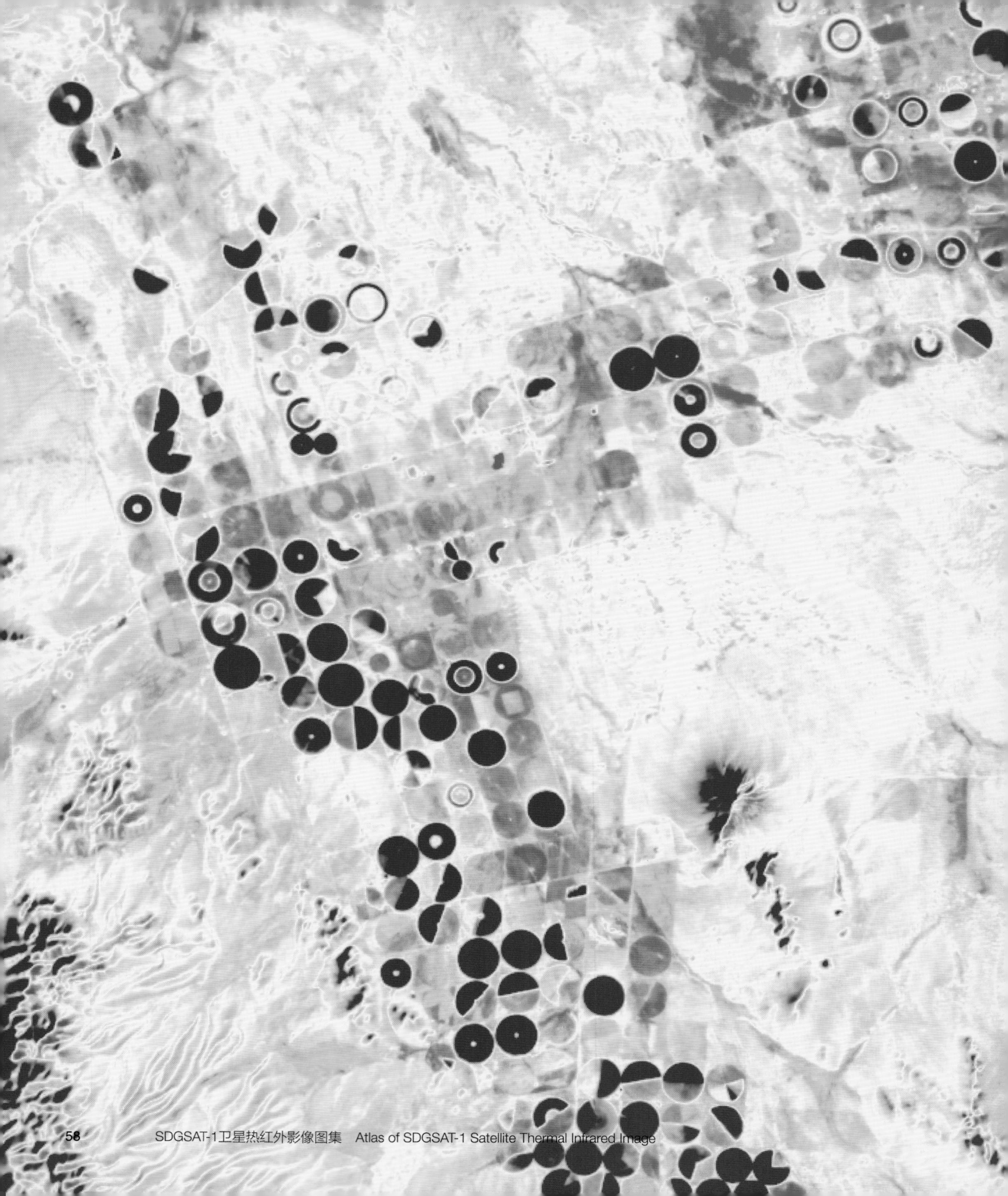

奇瓦瓦
Chihuahua

奇瓦瓦是墨西哥北部边境的一个内陆州，是全国重要的农业生产地。奇瓦瓦境内环境高度多样化，既有森林也有沙漠覆盖。在沙漠覆盖的干旱地带，农业生产主要依靠灌溉系统。日间热红外图像中，圆形的灌溉农业区温度较周围环境低，可见显著的温度差异。

Chihuahua is a landlocked state on Mexico's northern border and is an important agricultural producer for the country. Chihuahua has a highly diverse environment, with both forest and desert cover. In the arid desert-covered areas, agricultural production relies heavily on irrigation systems. Significant temperature differences can be seen in the daytime thermal infrared images where the circular irrigated agricultural areas are cooler than their surroundings.

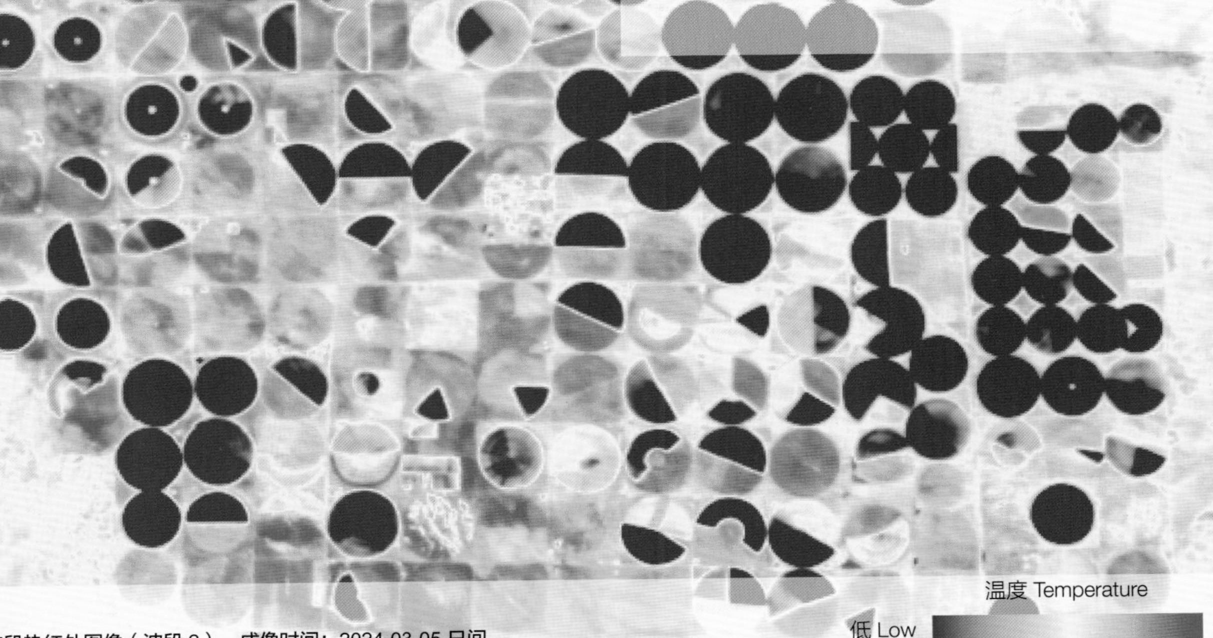

单波段热红外图像（波段 2） 成像时间：2024-03-05 日间
Single band thermal infrared image (band 2) Imaging time: 2024-03-05 Daytime

马托格罗索州

Mato Grosso

马托格罗索州位于巴西西部，是巴西二十六个州之一，是巴西最大的粮食生产州。该州的大部分地区为草地相间的灌丛林地。南部经济以养牛业为主，还产稻谷、玉米、豆类、木薯、甘蔗、马黛茶等。热红外图像中，围绕树林开发的农业用地展现出与原始丛林完全不同的温度特征。

Located in western Brazil, Mato Grosso is one of the country's twenty-six states, the largest food-producing state. Much of the state is scrub woodland interspersed with grasslands. The economy of the south is dominated by cattle ranching, and also produces rice, corn, beans, cassava, sugarcane and mate tea. The agricultural land developed around the forest in the thermal infrared image shows a completely different temperature profile from the original forest.

单波段热红外图像（波段 2）　成像时间：2023-07-18 日间

Single band thermal infrared image (band 2)　Imaging time: 2023-07-18 Daytime

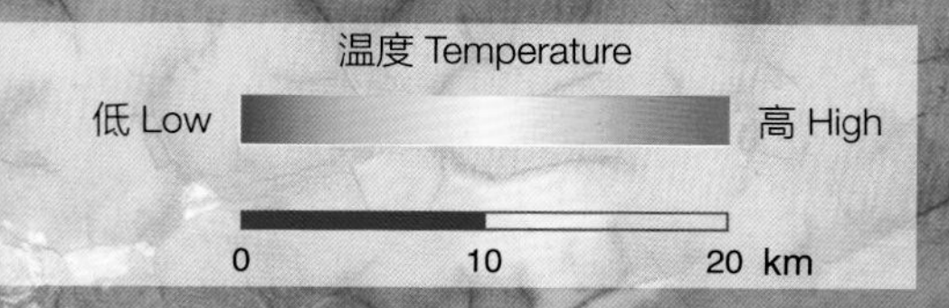

真彩色多光谱图像（波段组合：5-4-3） 成像时间：2022-07-08 日间

True color multispectral image (band combination: 5-4-3) Imaging time: 2022-07-08 Daytime

罗斯托夫州
Rostov Oblast

罗斯托夫州位于俄罗斯南部，高加索山脉以北，是俄罗斯最大的农业生产基地之一，粮食出口是罗斯托夫州的支柱产业。州内气候适宜，水资源供应稳定且有大量黑土地覆盖，非常适合农业生产。将热红外图像与可见光图像进行对比可以发现，有作物生长的地区温度较低，裸露土壤温度较高。热红外数据可以有效支撑农业生产监控。

Located in the south of the country, north of the Caucasus Mountains, the Rostov Oblast is one of the Russia's largest agricultural bases, with grain exports being the pillar of the Oblast's economy. The region's favourable climate, stable water supply and extensive black soil cover make it ideal for agricultural production. The comparison of the thermal infrared image with the visible light image shows lower temperatures in areas with vegetation covers and higher temperatures in bare soil. Based on the thermal infrared data, it is possible to effectively support the monitoring of agricultural production.

波段热红外图像（波段 2）　成像时间：2022-07-08 日间
ngle band thermal infrared image (band 2)　Imaging time: 2022-07-08 Daytime

布海拉省
Beheira Governorate

布海拉省位于埃及北部的尼罗河三角洲区域，毗邻沙漠干旱地带，全年降雨少气温较高，然而依靠着尼罗河支流的供水和中心转轴灌溉系统等浇灌技术的支持，布海拉省农业发达，产出小麦等主粮作物，以及绿豆等经济作物。热红外图像中，浇灌区域和干旱地面区分明显，浇灌土地一般温度较低。

Located in the Nile Delta region of northern Egypt, Beheira Governorate is adjacent to an arid desert area with low rainfall and high temperatures throughout the year. However, with the water supply from the branches of the Nile River and watering techniques such as center pivot irrigation systems, the Governorate has a well-developed agricultural sector, which produces staple crops, such as wheat, as well as cash crops, such as mung beans. In the thermal infrared image, watered areas are clearly distinguished from dry ground, and watered land is generally cooler.

单波段热红外图像（波段 2）　成像时间：2023-09-11 **日间**
Single band thermal infrared image (band 2)
Imaging time: 2023-09-11 Daytime

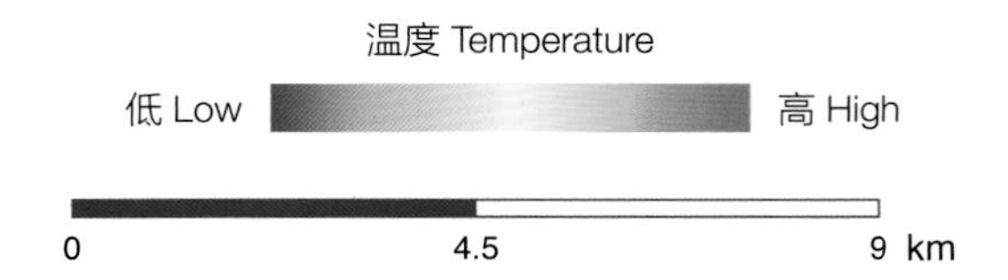

真彩色多光谱图像（波段组合：5-4-3） 成像时间：2023-12-06 日间

True color multispectral image (band combination: 5-4-3) Imaging time: 2023-12-06 Daytime

北方邦
Uttar Pradesh

北方邦，位于印度北部，是印度人口最多的邦，农业为北方邦的支柱性产业。北方邦是印度最大的小麦、豆类、甘蔗、烟草、土豆以及牛奶产地。它的豆类和烟草产量居全国首位。另外，该邦的农业灌溉系统覆盖密度在全国各邦中居于前列。在夜间热红外图像中，城建区域和水体的温度明显高于附近的农业用地。

Uttar Pradesh, located in the north of India, is the most populous state in India with agriculture as the state pillar industry. Uttar Pradesh is India's largest producer of wheat, pulses, sugarcane, tobacco, potatoes and milk. It leads the country in the production of pulses and tobacco. In addition, it has one of the highest densities of agricultural irrigation system coverage among the states in the country. In the nighttime thermal infrared imagery, the temperature of the built-up areas and water bodies were significantly higher than that of the nearby agricultural land.

单波段热红外图像（波段 2）　成像时间：2023-11-12 夜间
Single band thermal infrared image (band 2)　Imaging time: 2023-11-12 Nighttime

温度 Temperature
低 Low　高 High
0　4　8 km

黑龙江省

Heilongjiang Province

黑龙江省地处中国东北部、东北亚中心区域，北部与俄罗斯隔江相望。黑龙江地处世界三大黑土带之一，耕地平坦，黑土地土壤肥沃，是中国最大的农业生产基地之一。该省主要种植玉米、大豆和水稻等农作物。省内河流众多，属大陆性季风气候，适于农作物生长。在日间热红外图像中，不同状态下的农田展现出具有差异的温度特征。

Heilongjiang Province is located in northeastern China and the central region of Northeast Asia, with Russia across the river to the north. Located in one of the world's three major black soil belts, Heilongjiang Province has flat arable land and fertile black soil, making it one of the largest agricultural production bases in the country. Heilongjiang Province mainly grows crops such as corn, soybeans and rice. The province has many rivers and a continental monsoon climate, which is suitable for crop growth. In daytime thermal infrared imagery, farmland in different states exhibits differential temperature characteristics

真彩色多光谱图像（波段组合：5-4-3）　成像时间：2023-08-19 日间

True color multispectral image (band combination: 5-4-3)　Imaging time: 2023-08-19 Daytime

0　4.5　9 km

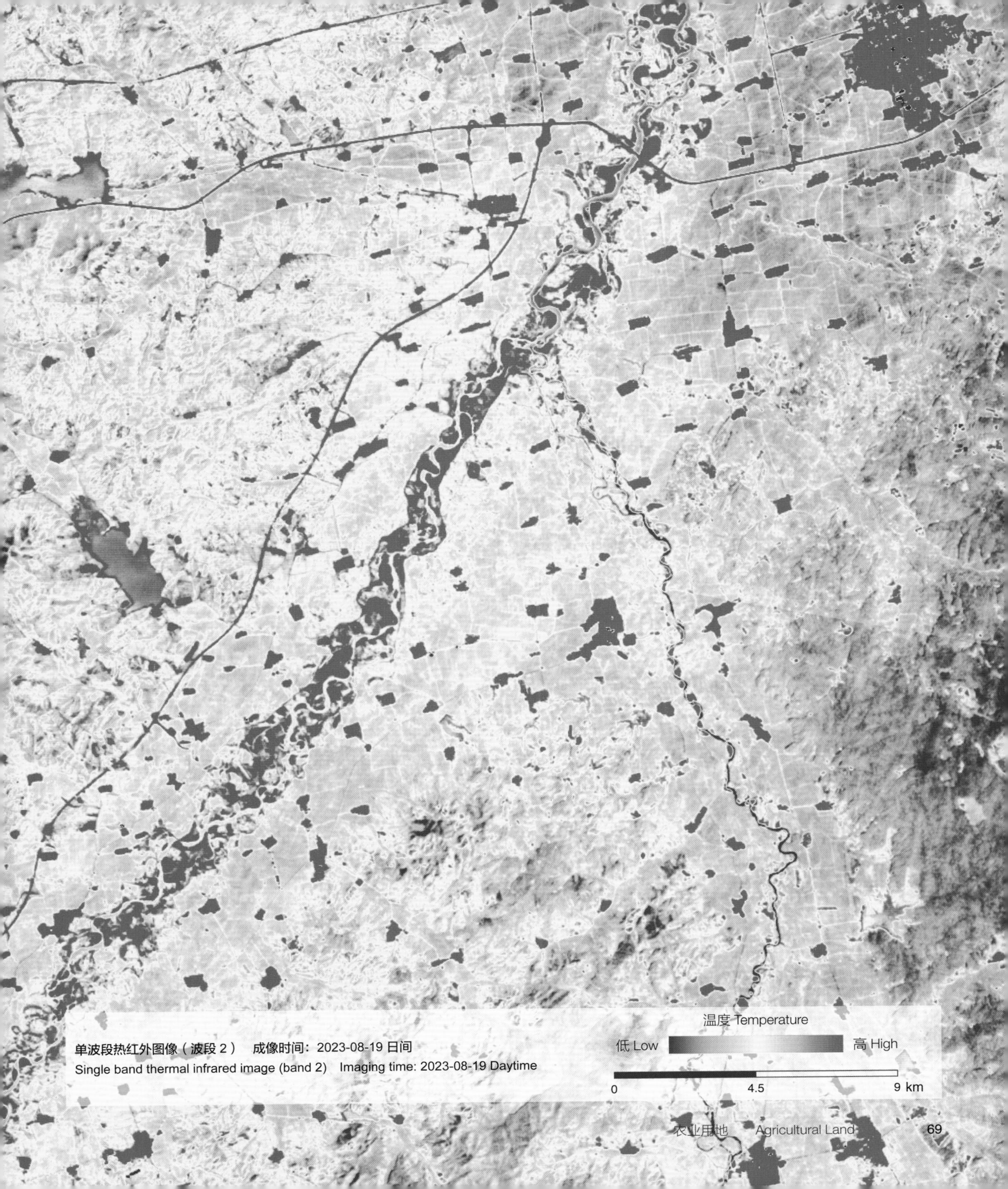

单波段热红外图像（波段 2）　成像时间：2023-08-19 日间

Single band thermal infrared image (band 2)　Imaging time: 2023-08-19 Daytime

真彩色多光谱图像（波段组合：5-4-3）　成像时间：2023-10-26 日间

True color multispectral image (band combination: 5-4-3)

Imaging time: 2023-10-26 Daytime

维多利亚州
Victoria

维多利亚州位于澳大利亚大陆的东南沿海，是农牧业生产最为发达的一个州，出口小麦、肉类、羊毛、乳制品和酒类。全州大多数土地为平原且气候比较湿润，有海风自然调节温度，适合农牧业发展。在维多利亚州南部地带成规模的农业用地尤为集中。热红外图像中农业用地星罗棋布，展现出不同的温度特征。

Located on the south-east coast of mainland Australia, Victoria is one of the most agriculturally productive states, exporting wheat, meat, wool, dairy products and alcohol. Most of the state's land is plain and the climate is relatively humid, with sea air naturally moderating temperatures, making it suitable for agricultural and pastoral development. There is a concentration of large-scale agricultural land in the southern Victoria. In the thermal infrared image, agricultural land is scattered, showing different temperature characteristics.

单波段热红外图像（波段 2） 成像时间：2023-10-26 日间
Single band thermal infrared image (band 2) Imaging time: 2023-10-26 Daytime

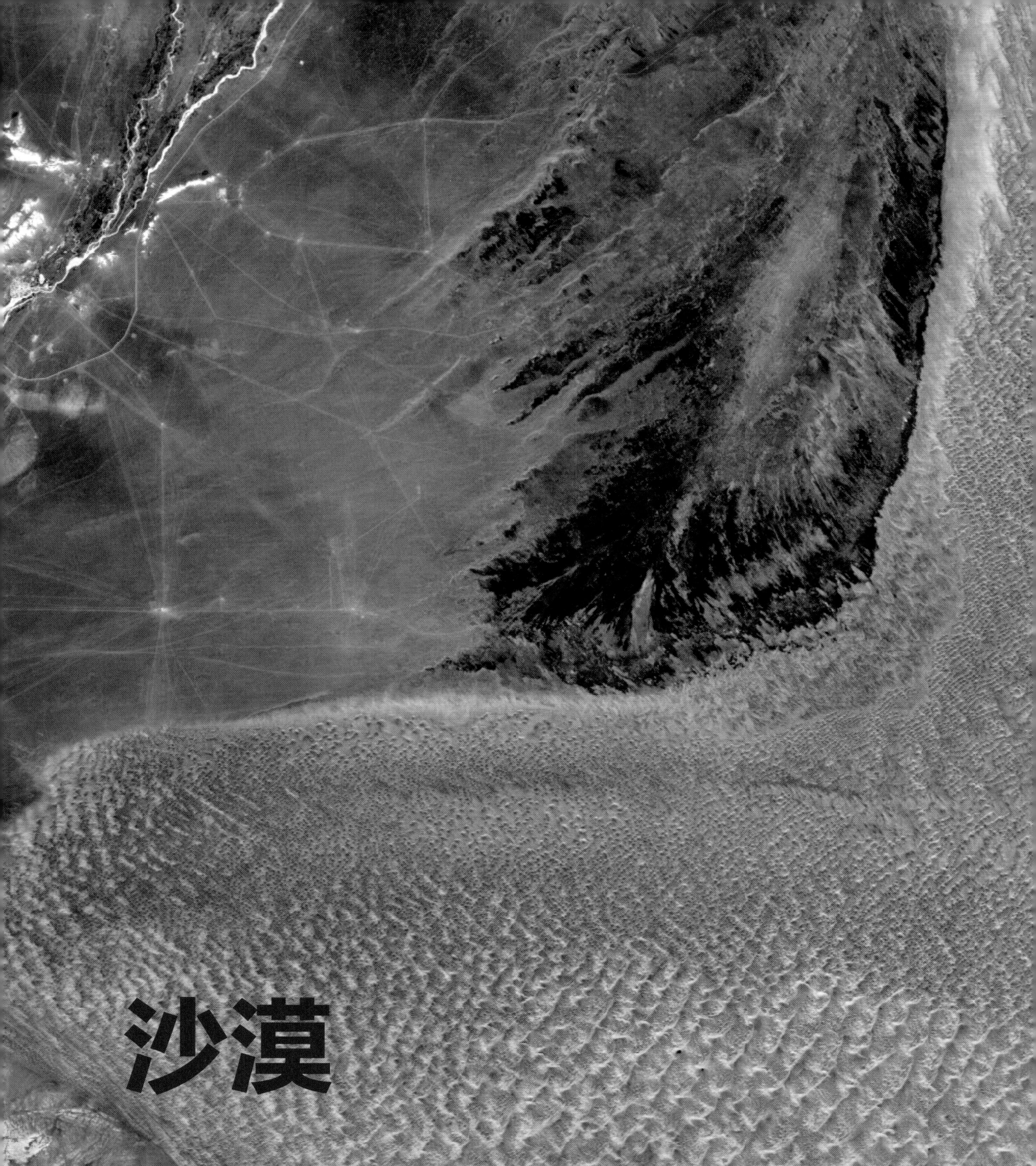

沙漠

Deserts

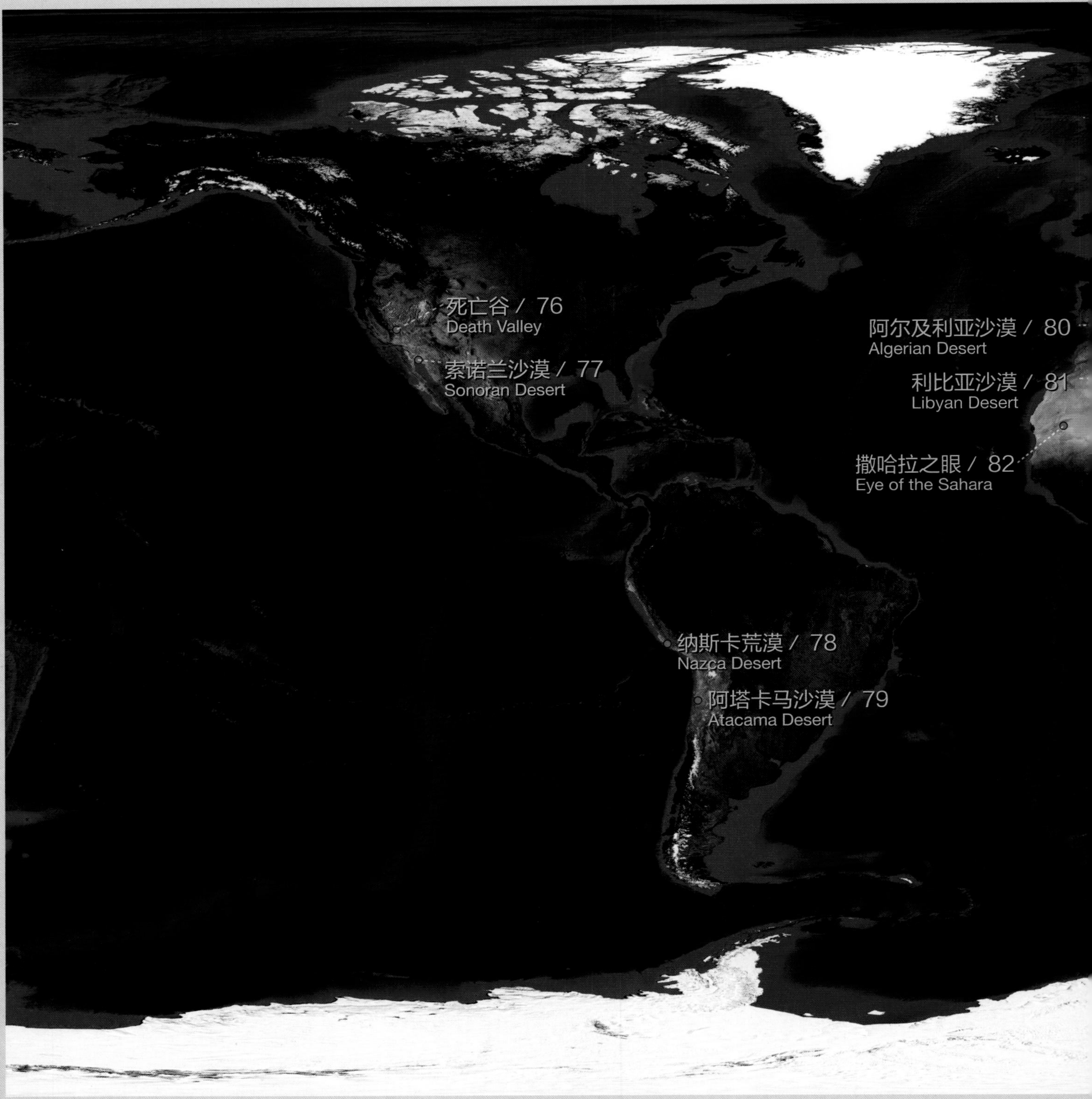

死亡谷 / 76
Death Valley

索诺兰沙漠 / 77
Sonoran Desert

阿尔及利亚沙漠 / 80
Algerian Desert

利比亚沙漠 / 81
Libyan Desert

撒哈拉之眼 / 82
Eye of the Sahara

纳斯卡荒漠 / 78
Nazca Desert

阿塔卡马沙漠 / 79
Atacama Desert

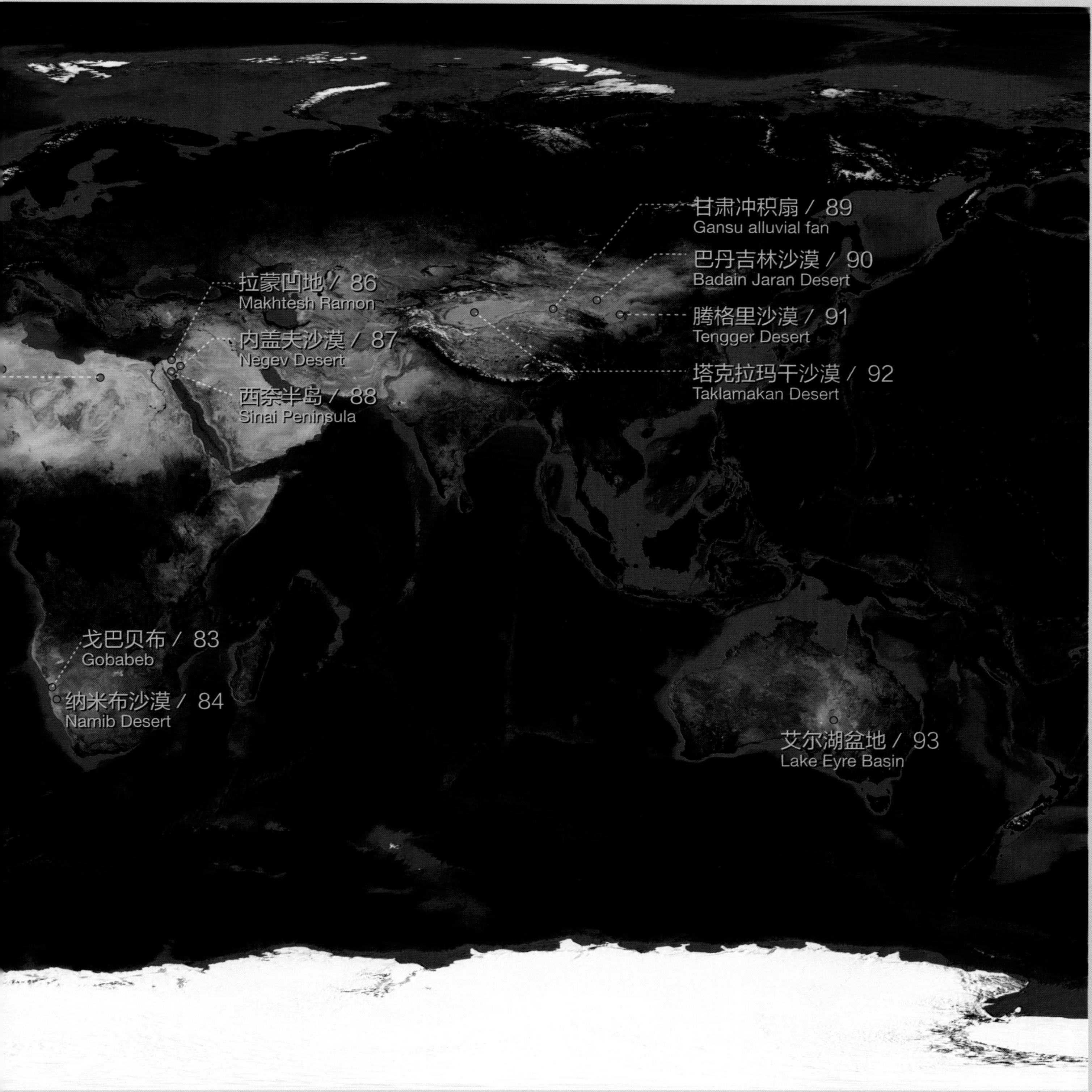

甘肃冲积扇 / 89
Gansu alluvial fan

巴丹吉林沙漠 / 90
Badain Jaran Desert

拉蒙凹地 / 86
Makhtesh Ramon

腾格里沙漠 / 91
Tengger Desert

内盖夫沙漠 / 87
Negev Desert

塔克拉玛干沙漠 / 92
Taklamakan Desert

西奈半岛 / 88
Sinai Peninsula

戈巴贝布 / 83
Gobabeb

纳米布沙漠 / 84
Namib Desert

艾尔湖盆地 / 93
Lake Eyre Basin

三波段热红外图像（波段组合：3-2-1） 成像时间：2023-09-26 日间
Three band thermal infrared image (band combination: 3-2-1)
Imaging time: 2023-09-26 Daytime

死亡谷
Death Valley

死亡谷是加利福尼亚州东部的一个沙漠山谷，是北美洲最低、最干旱的地区，被认为是夏季地球上最热的地方，最高气温高达 55℃。死亡谷里有一块会漂移的石块。此三波段合成的日间热红外影像有效地展示了死亡谷及周围地形的热量区别。

Death Valley is a desert valley in eastern California, which is the lowest and driest region in North America. It is considered the hottest place on Earth in summer, with a maximum temperature of up to 55°C. There is a drifting stone in Death Valley. The daytime thermal infrared image synthesized from three bands effectively displays the thermal differences between Death Valley and the surrounding terrain.

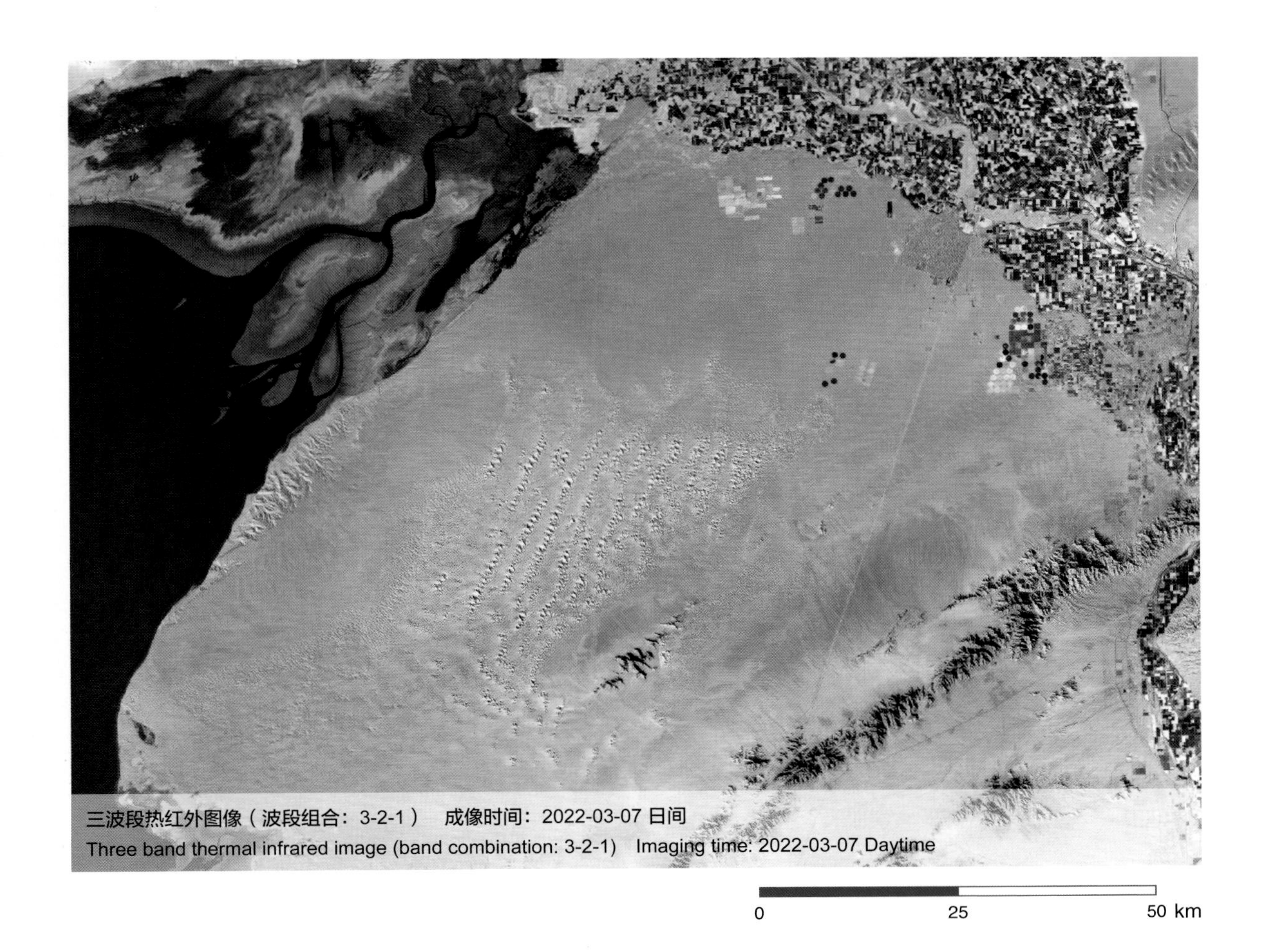
三波段热红外图像（波段组合：3-2-1） 成像时间：2022-03-07 日间
Three band thermal infrared image (band combination: 3-2-1) Imaging time: 2022-03-07 Daytime

索诺兰沙漠
Sonoran Desert

索诺兰沙漠位于墨西哥和美国边境附近，是一个大而平坦的半干旱地区，植被覆盖很少随季节性变化。索诺兰沙漠海拔 37m，水平能见度为 30 ~ 45km。索诺兰沙漠空间变化和方向效应较小，被广泛用于西半球上空的卫星校准。此三波段合成的日间热红外影像包括海水和沙漠，明显地看到海水和沙漠的分界线，以及均匀的沙漠场地。

The Sonoran Desert is located near the Mexico and United States border. It is a large flat semi-arid region with limited vegetation cover and known seasonal variations. The site is 37m above sea level and has horizontal visibilities between 30~45km. The site is widely used for satellites positioned over the Western Hemisphere for calibration purposes because of its small spatial variations and directional effects. The daytime thermal infrared image synthesized from three bands includes seawater and desert, with clear boundaries between seawater and desert, as well as a uniform desert site.

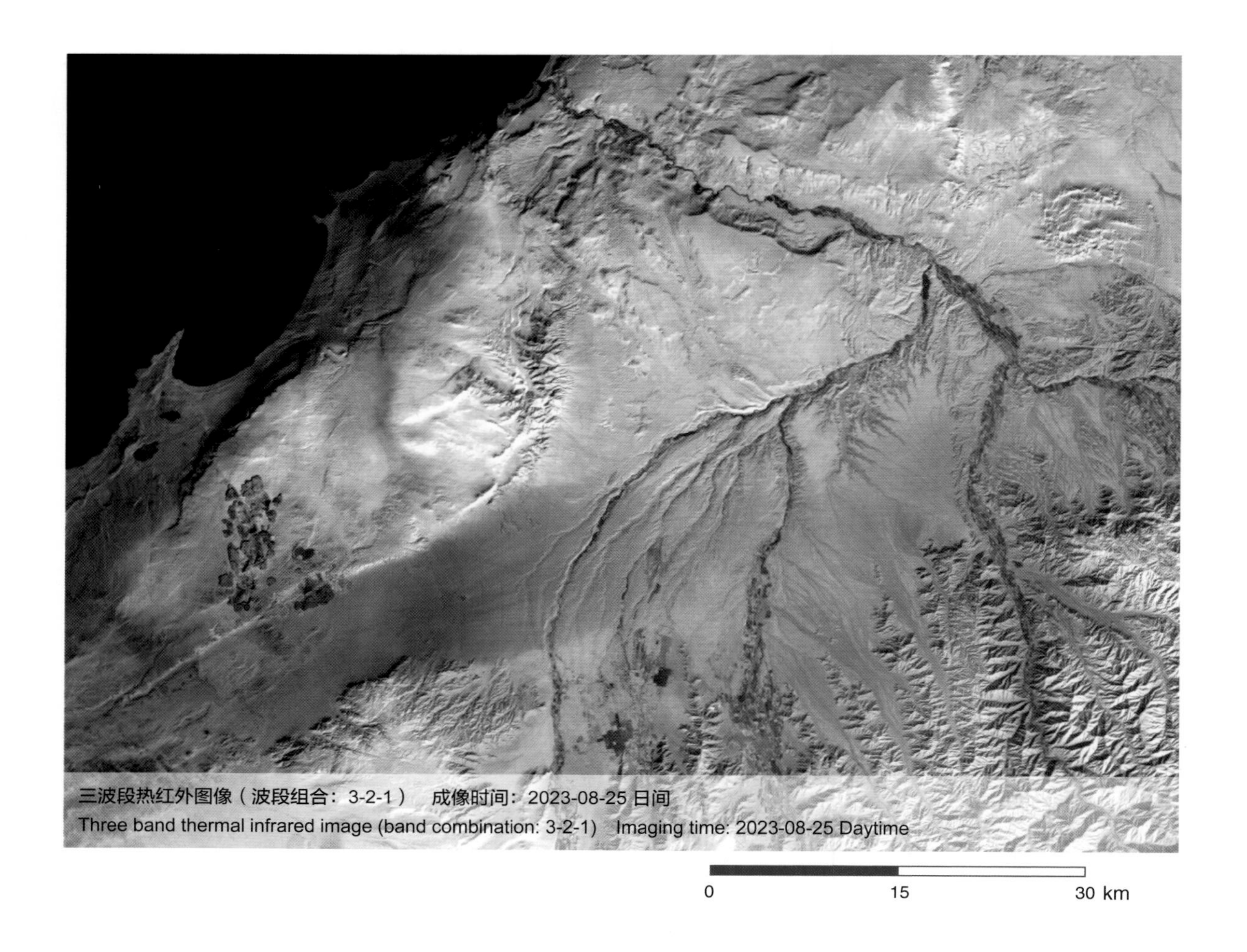

三波段热红外图像（波段组合：3-2-1） 成像时间：2023-08-25 日间
Three band thermal infrared image (band combination: 3-2-1) Imaging time: 2023-08-25 Daytime

纳斯卡荒漠
Nazca Desert

纳斯卡荒漠位于秘鲁海岸的干旱草原上。纳斯卡荒漠最为人所知的是其神秘的纳斯卡线条图。这些线条描绘了动物、植物和想象的形象。此三波段合成的日间热红外影像包括海洋、干涸的河流和沙漠，相较于海水，沙漠温度较高，产生了明显对比，有效展示了纳斯卡荒漠的热分布。

The Nazca Desert is located on arid grasslands along the coast of Peru. The most well-known part of the Nazca Desert is its mysterious Nazca Line. These lines depict the images of animals, plants and imagination. The daytime thermal infrared image synthesized from three bands includes the ocean, dry rivers, and desert. Compared with seawater, the desert has a higher temperature, which creates a clear contrast and effectively displays the thermal distribution of the Nazca Desert.

三波段热红外图像（波段组合：3-2-1） 成像时间：2022-09-14 日间
Three band thermal infrared image (band combination: 3-2-1) Imaging time: 2022-09-14 Daytime

阿塔卡马沙漠
Atacama Desert

阿塔卡马沙漠位于智利北部南美洲太平洋沿岸的沙漠高原。阿塔卡马沙漠是世界上最干燥的非极地沙漠，总体上第二干旱沙漠，是唯一一个降水量少于极地沙漠的炎热沙漠，也是世界上最大的雾沙漠。由于该地区与火星环境的相似性，该地区已被用作火星探险模拟的实验场所。此三波段合成的日间热红外影像包括海洋和沙漠，相较于沙漠，海洋温度较低，产生了明显对比，展示了阿塔卡马沙漠的热环境情况。

The Atacama Desert is a desert plateau located on the Pacific coast of South America, in the north of Chile. The Atacama Desert is the driest nonpolar desert in the world, and the second driest overall. It is the only hot true desert to receive less precipitation than polar deserts, and the largest fog desert in the world. The area has been used as an experimentation site for Mars expedition simulations due to its similarities to the Martian environment. The daytime thermal infrared image synthesized from three bands includes the ocean and desert. Compared with the desert, the ocean temperature is lower, resulting in a significant contrast, showcasing the thermal environment of Atacama Desert.

阿尔及利亚沙漠
Algerian Desert

阿尔及利亚沙漠是位于阿尔及利亚境内北非中部的沙漠，是撒哈拉沙漠的一部分。沙漠占阿尔及利亚总面积的 90% 以上。阿尔及利亚沙漠原本是一片石质沙漠，逐渐变成内陆的沙丘沙漠。阿尔及利亚沙漠每年降雨量不到 100mm。阿尔及利亚沙漠由金字塔形的沙丘组成，高 100m，随机分布，坡度从 15° 到 30° 不等。此三波段合成的夜间热红外影像有效地展示了夜间沙丘与沙漠平地的温度差，清晰地显示了夜间沙漠的形态。

The Algerian Desert is a desert located in central North Africa within Algeria, constituting part of the Sahara. The desert covers more than 90% of Algeria's total area. It was originally a stony desert, gradually changing into a sand dune desert inland. The annual rainfall in the Algerian desert is less than 100mm. The Algerian desert is composed of dunes having a pyramidal shape, 100m high, randomly disposed, with a slope ranging from 15 to 30 degrees. The nighttime thermal infrared image synthesized from three bands effectively displays the temperature difference between nighttime sand dunes and desert plains, and clearly displays the morphology of nighttime deserts.

三波段热红外图像（波段组合：3-2-1）　成像时间：2023-12-03 夜间
Three band thermal infrared image (band combination: 3-2-1)　Imaging time: 2023-12-03 Nighttime

利比亚沙漠
Libyan Desert

利比亚沙漠位于撒哈拉沙漠的东北部，从利比亚东部延伸到埃及西南部以及苏丹的西北部。此区域自然条件恶劣，干旱且荒凉，不适宜居住。利比亚沙漠主要由光秃秃的岩石高原和石质或沙质平原组成。利比亚沙漠是全球辐射定标场的选址区域。此三波段合成的日间热红外影像能清晰的刻画利比亚沙漠的表面特征——绵延起伏的沙丘。

Libyan Desert, northeastern portion of the Sahara, extending from eastern Libya through southwestern Egypt into the extreme northwest of Sudan. This area is harsh, arid, and inhospitable, making it unsuitable for living. The Libyan Desert is mainly composed of bare rocky plateaus and stony or sandy plains. The Libyan Desert is the site selection area for global radiometric calibration sites. The daytime thermal infrared image synthesized from three bands can clearly depict the surface features of the Libyan Desert – the undulating sand dunes.

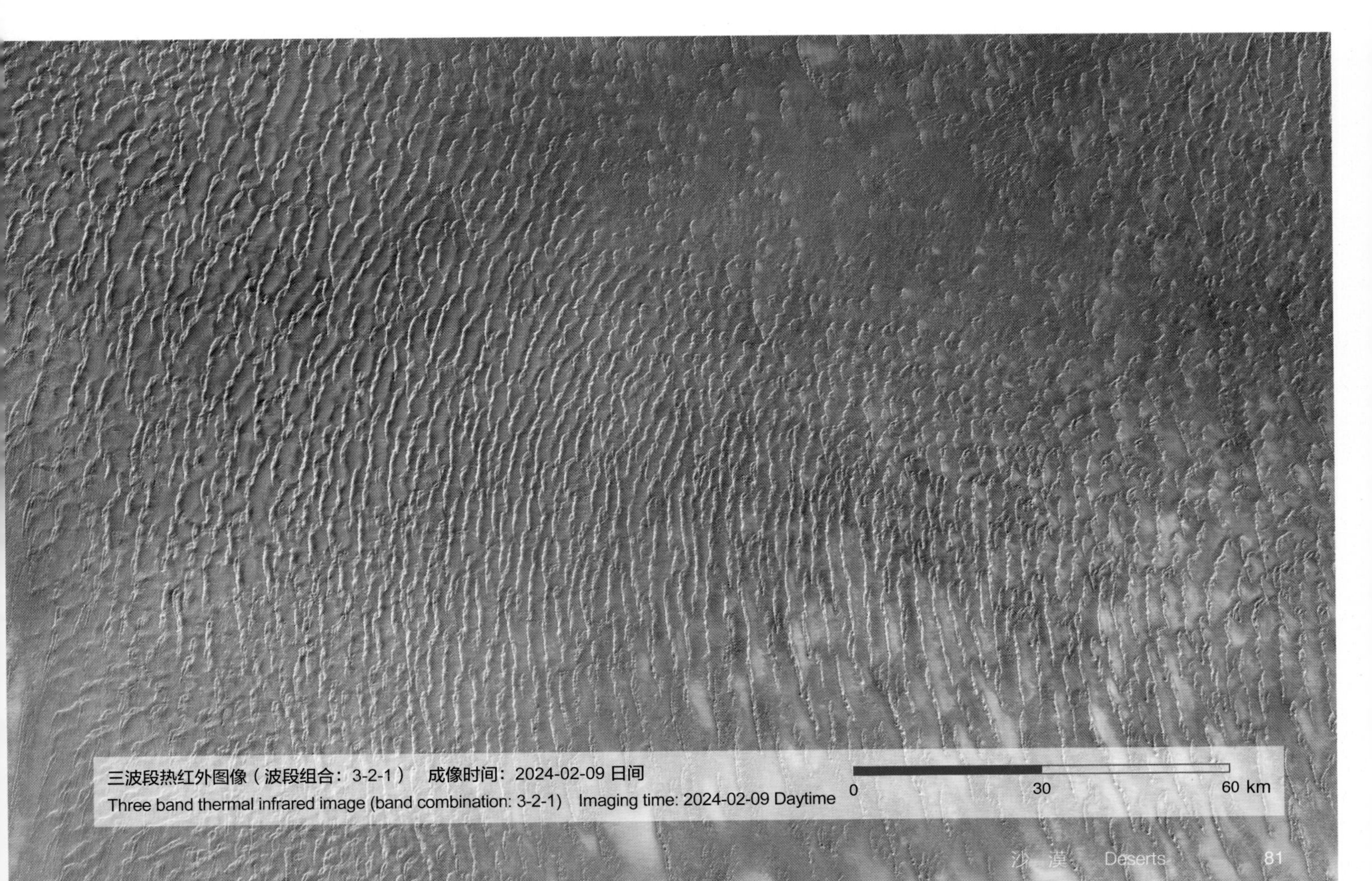

三波段热红外图像（波段组合：3-2-1）　成像时间：2024-02-09 日间
Three band thermal infrared image (band combination: 3-2-1)　Imaging time: 2024-02-09 Daytime

撒哈拉之眼
Eye of the Sahara

撒哈拉之眼位于毛里塔尼亚境内。撒哈拉之眼是一个巨大的同心圆，直径达到了48km，在太空上清晰可见。撒哈拉之眼海拔约400m，中心十分平坦，四周是一些矮山丘，远处是黄沙。此三波段合成的夜间热红外影像展示了夜间撒哈拉之眼与其他地方的热差异，清晰地刻画了夜间撒哈拉之眼的热分布。

Eye of the Sahara is located in Mauritania. Eye of the Sahara is a huge concentric circle with a diameter of 48km, clearly visible from space. Eye of the Sahara has an altitude of about 400m, with a very flat center surrounded by some low hills and yellow sand in the distance. The nighttime thermal infrared image synthesized from three bands displays the thermal differences between the Eye of the Sahara at night and other places, clearly depicting the thermal distribution of the Eye of the Sahara at night.

三波段热红外图像（波段组合：3-2-1） 成像时间：2023-11-18 夜间
Three band thermal infrared image (band combination: 3-2-1) Imaging time: 2023-11-18 Nighttime

戈巴贝布
Gobabeb

戈巴贝布坐落在极度干旱的纳米布沙漠的中心地带，年平均降雨量为 25mm。戈巴贝布具有丰富且高度多样化的暗甲虫。戈巴贝布具有稳定的空间特征以及极少的降雨量，被选为全球辐射定标场地。此三波段合成的日间热红外影像中，沙漠中起伏的沙丘能被精细地勾勒出。定标场位于沙丘和其余干旱地区分界线的上方。

The Gobabeb sits in the heart of the hyperarid Namib Desert, with an average annual rainfall of 25mm. The Gobabeb has a rich and highly diverse population of dark beetles. Gobabeb with its stable spatial characteristics and minimal rainfall, was chosen as a global radiometric calibration site. The daytime thermal infrared image synthesized from three bands shows that undulating sand dunes can be finely delineated, and the calibration site is located above the boundary between the dunes and the rest of the arid area.

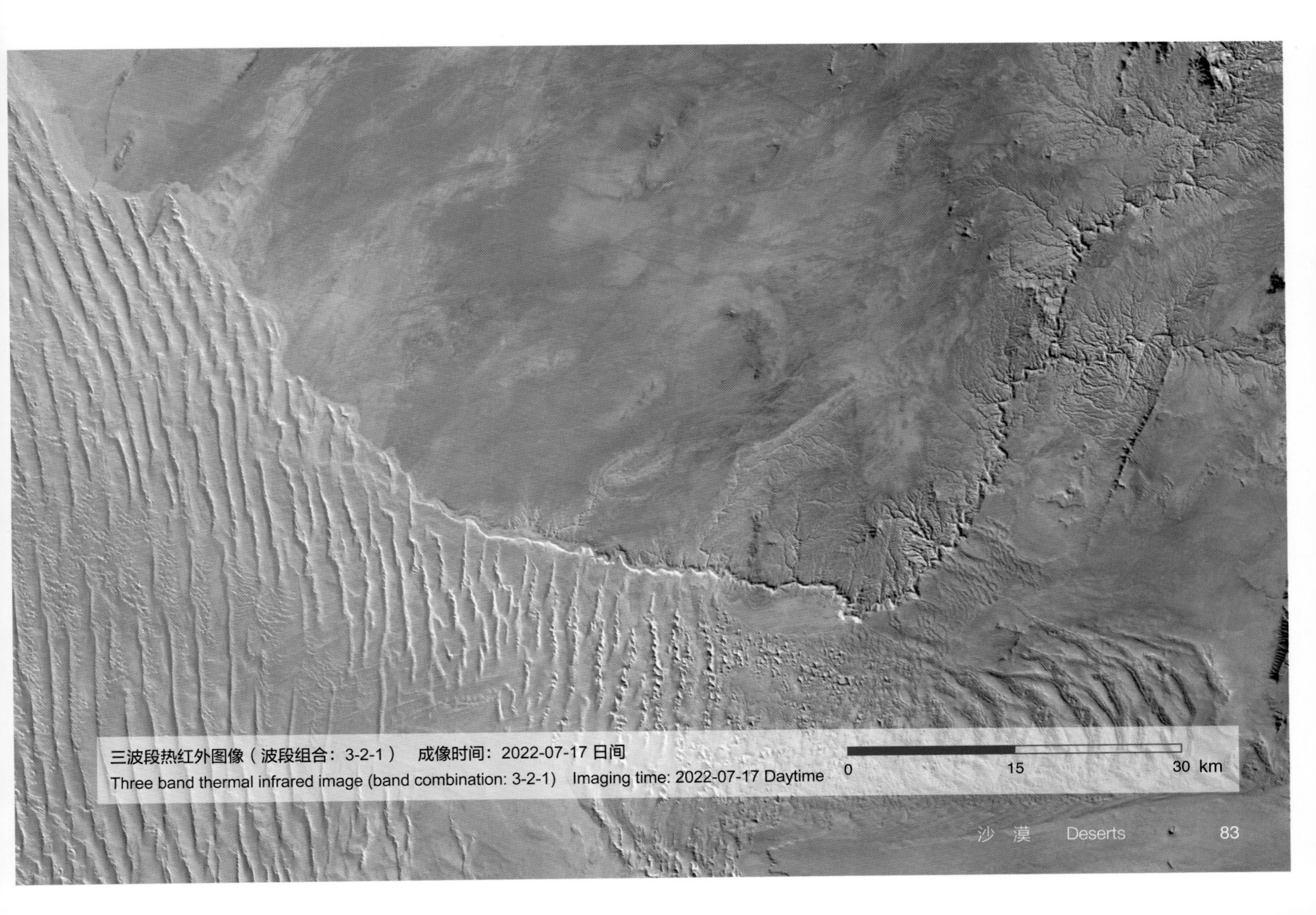

三波段热红外图像（波段组合：3-2-1）　成像时间：2022-07-17 日间
Three band thermal infrared image (band combination: 3-2-1)　Imaging time: 2022-07-17 Daytime

纳米布沙漠
Namib Desert

纳米布沙漠在非洲纳米比亚，西边紧邻大西洋，是世界上最古老、最干燥的沙漠之一。麻点和沙丘混合存在构成了纳米布沙漠，尤其在西北地区更为明显。该地区具有稳定的空间特征，被用来作为辐射定标场。此三波段合成的夜间热红外影像包括沙漠和海洋，相较于陆地，海洋夜间温度较高，产生了明显的对比，热红外影像在夜间也能清晰地展示地表状态。

The Namib Desert in Namibia, Africa, bordering the Atlantic Ocean to the west, is one of the oldest and driest deserts in the world. A mixture of pockmarks and dunes forms the Namib Desert, especially in the northwest. This region has stable spatial characteristics and is used as a radiometric calibration site. The nighttime thermal infrared images synthesized from three bands includes desert and ocean. Compared with the land, the sea has a higher night temperature, which produces an obvious contrast. The thermal infrared image can also clearly show the state of the land surface at night.

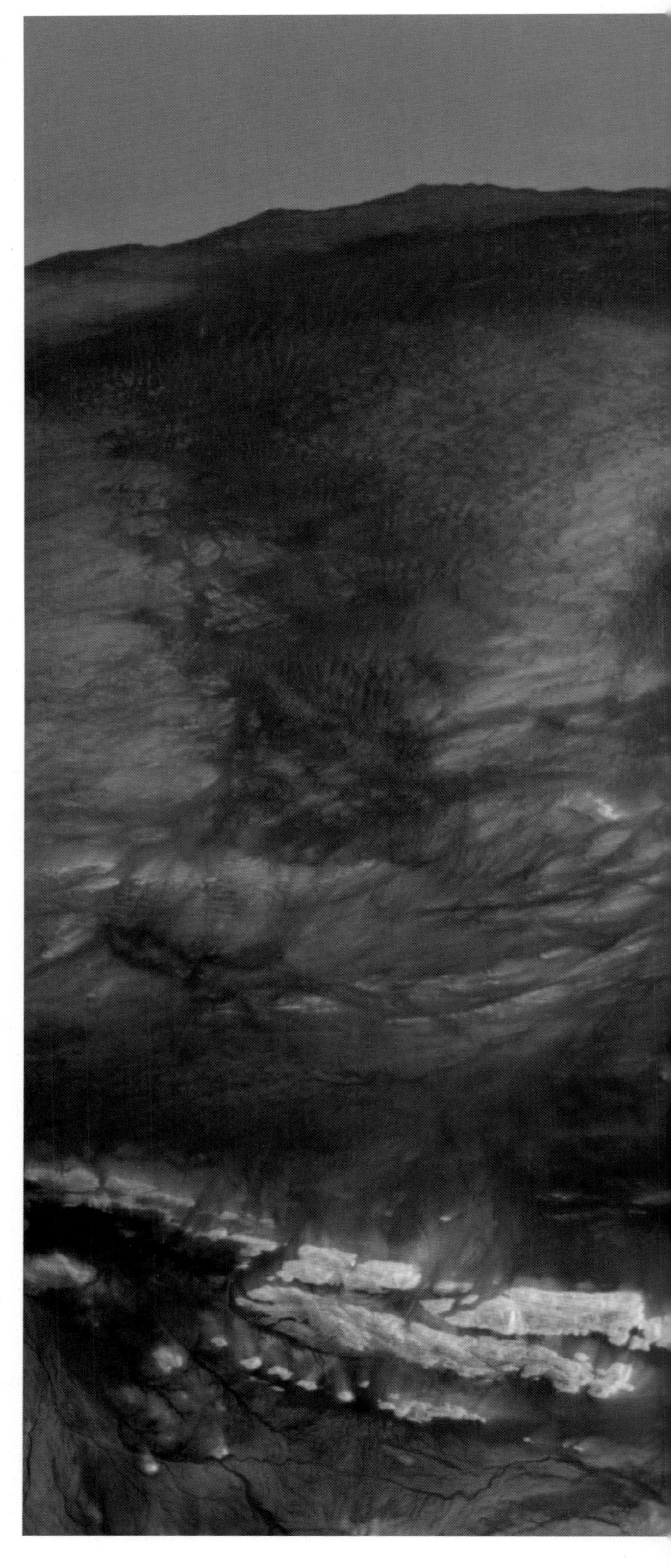

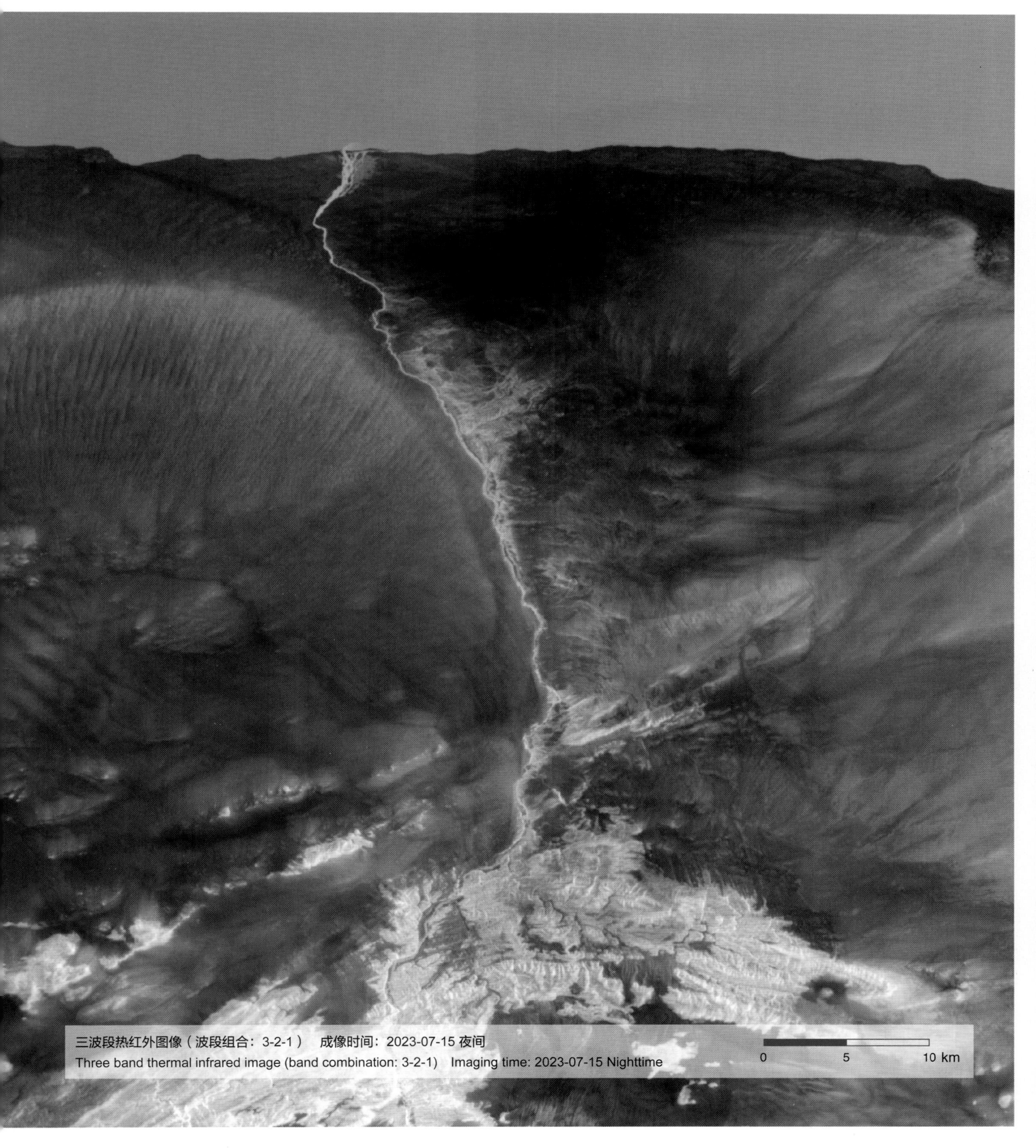

三波段热红外图像（波段组合：3-2-1）　成像时间：2023-07-15 夜间
Three band thermal infrared image (band combination: 3-2-1)　Imaging time: 2023-07-15 Nighttime

拉蒙凹地
Makhtesh Ramon

拉蒙凹地是一个四面以陡峭的壁为界的盆地，长 40km、宽 2 ~ 10km、深 500m，形状像一个细长的心脏。此区域年降水量为 25mm，气候干旱，植被和土壤覆盖受到限制。此三波段合成的日间热红外影像展示了盆地中地表与其他地物呈现的地表温度。

Makhtesh Ramon is a basin bounded by steep clifs — the formation is 40km long, 2~10km wide, and 500m deep, and is shaped like an elongated heart. The annual precipitation in this area is 25mm, with an arid conditions, vegetation and soil cover are limited. This 3-band daytime thermal infrared image shows clear contrast between different types of objects in the desert.

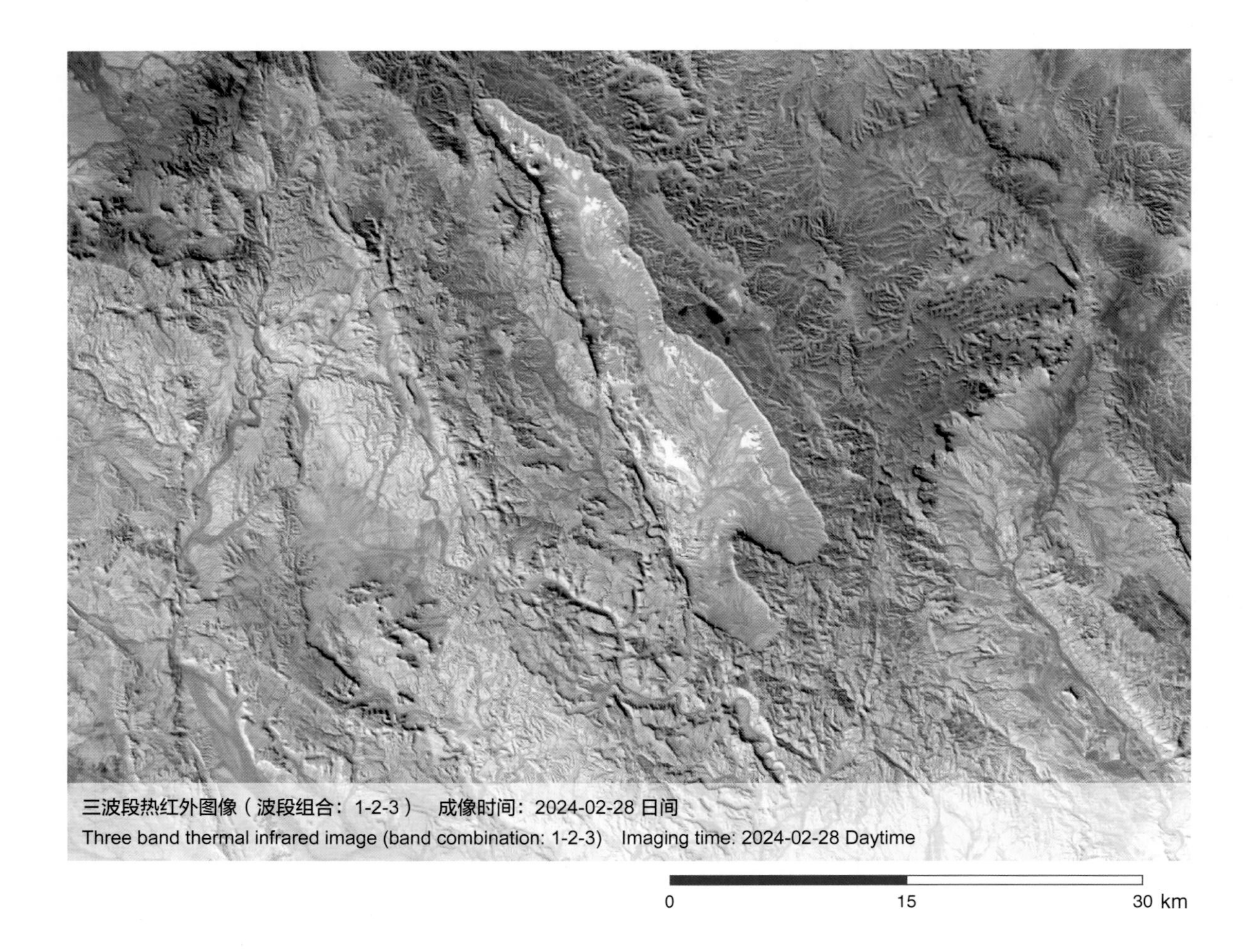

三波段热红外图像（波段组合：1-2-3） 成像时间：2024-02-28 日间
Three band thermal infrared image (band combination: 1-2-3) Imaging time: 2024-02-28 Daytime

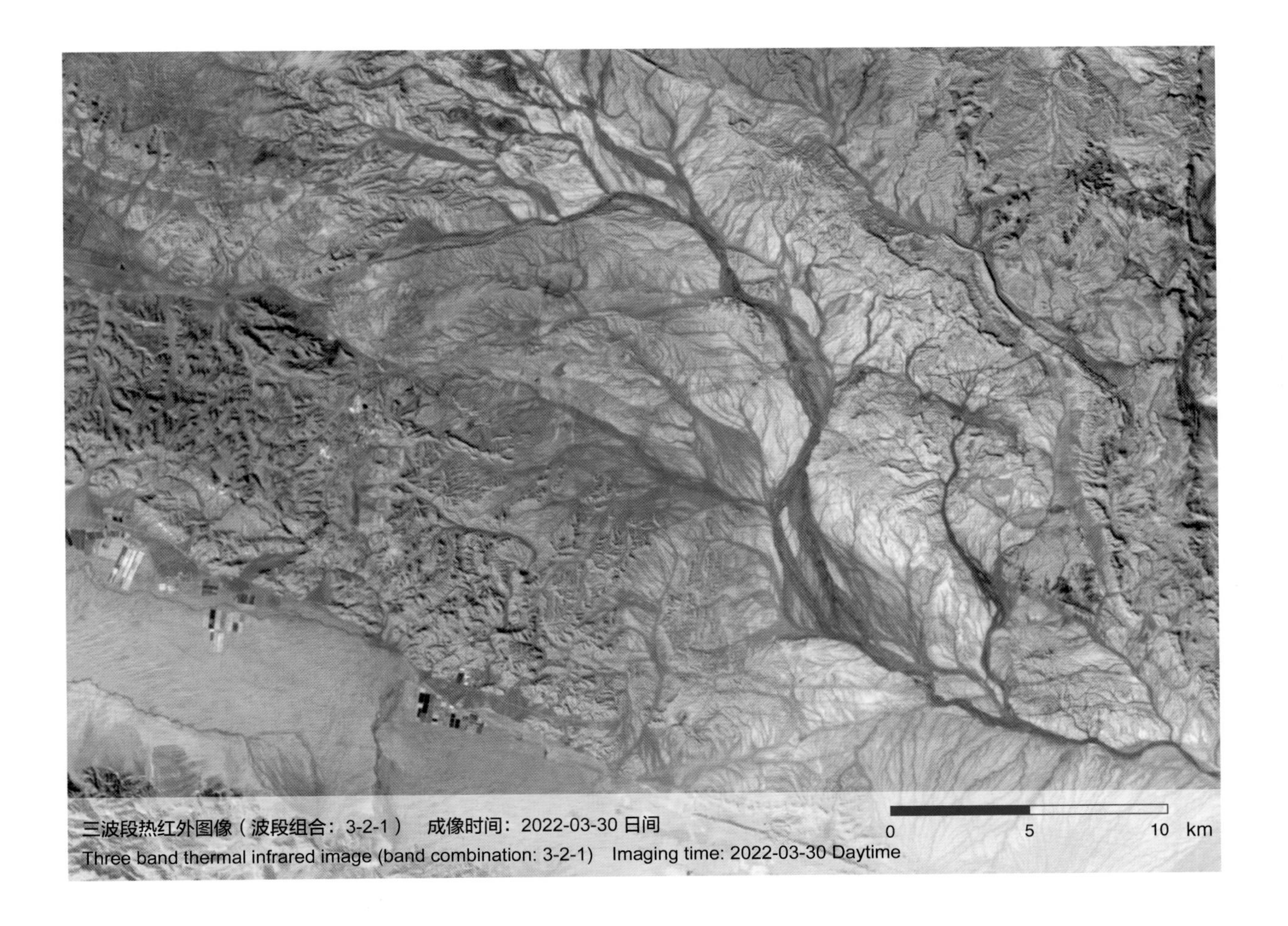

三波段热红外图像（波段组合：3-2-1） 成像时间：2022-03-30 日间
Three band thermal infrared image (band combination: 3-2-1) Imaging time: 2022-03-30 Daytime

内盖夫沙漠
Negev Desert

内盖夫沙漠是以色列南部的大片沙漠地区，沙丘高达 30m，是一片多岩石沙漠。该沙漠植被稀疏，每年的降雨约为 24mm，生长着一些特有的植物。全球遥感卫星的辐射定标场之一位于内盖夫沙漠。根据不同的地表温度，此三波段合成的日间热红外影像能清晰的分辨出水系和沙漠以及岩石。

Negev Desert is a large desert region in southern Israel and has sand dunes that reach heights of up to 30m, with a rocky desert. Vegetation in the Negev Desert is sparse, with annual rainfall of 24mm, but certain trees and plants thrive there. One of the radiometric calibration sites for global remote sensing satellites is located in Negev Desert. According to different surface temperatures, the daytime thermal infrared image synthesized from three bands can clearly distinguish water systems, deserts, and rocks.

西奈半岛
Sinai Peninsula

西奈半岛是连接非洲及亚洲的三角形半岛，气候炎热干燥，植被稀少，仅在干谷处生长有刺灌木和草本植物。半岛上广大的干燥地区称为西奈沙漠。此三波段合成的日间热红外影像有效展示了西奈沙漠热量分布情况。

The Sinai Peninsula is a triangular peninsula connecting Africa and Asia0. The climate of the Sinai Peninsula is hot and dry, with sparse vegetation, and only thorny shrubs and herbaceous plants grow in dry valleys. The vast dry areas on the peninsula are called the Sinai Desert. The daytime thermal infrared image synthesized from three bands effectively displays the distribution of heat in the Sinai Desert.

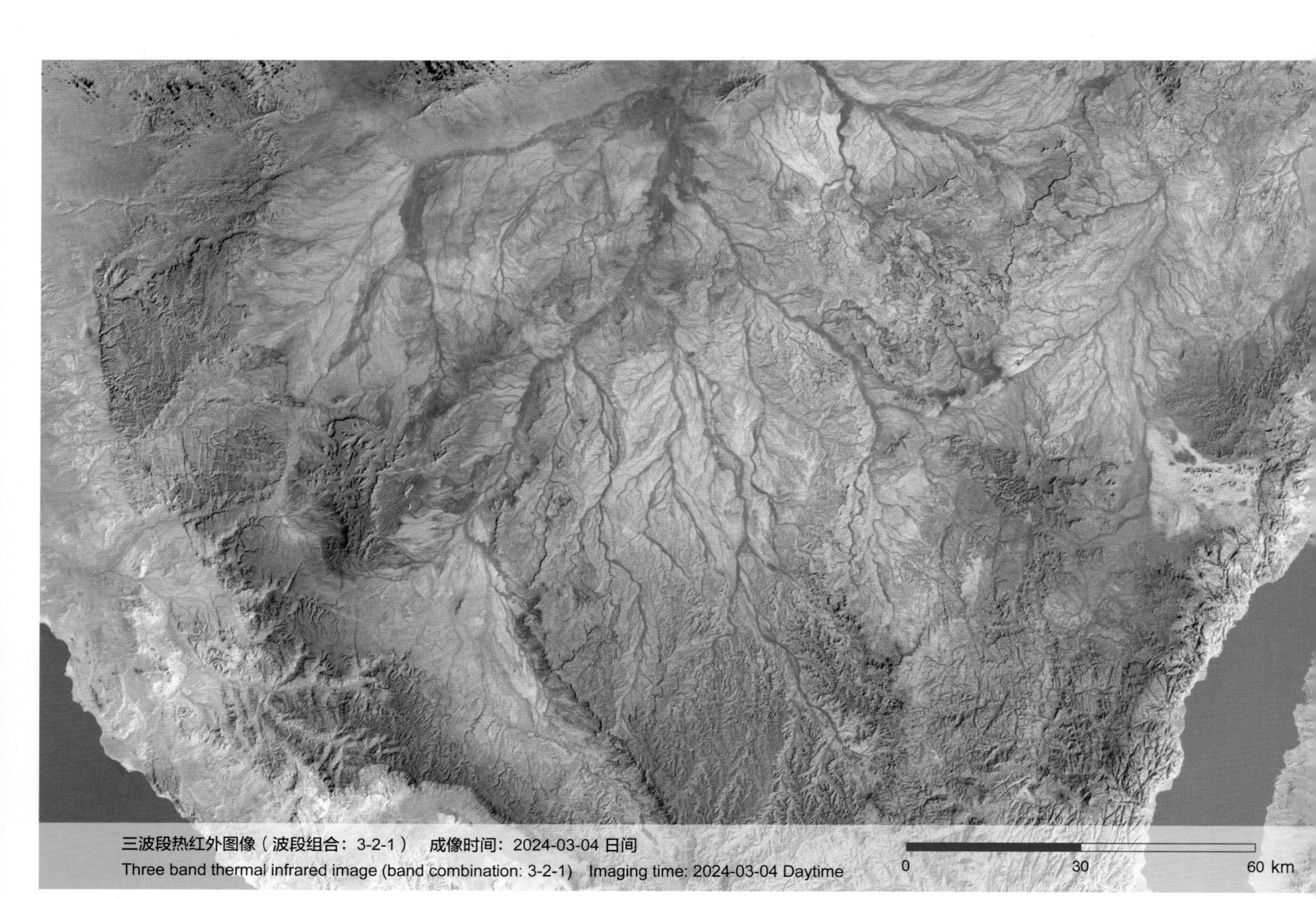

三波段热红外图像（波段组合：3-2-1）　成像时间：2024-03-04 日间
Three band thermal infrared image (band combination: 3-2-1)　Imaging time: 2024-03-04 Daytime

三波段热红外图像（波段组合：3-2-1） 成像时间：2022-01-09 夜间
Three band thermal infrared image (band combination: 3-2-1) Imaging time: 2022-01-09 Nighttime

甘肃冲积扇
Gansu alluvial fan

该冲积扇位于中国甘肃省境内，发源于祁连山脉。冲积扇是河流出山口处的扇形堆积体，当河流流出谷口时摆脱侧向约束，其携带物质便铺散沉积下来。平面上呈扇形，扇顶伸向谷口，立体上大致呈半埋藏的锥形。此三波段合成的日间热红外影像中涵盖了多个冲积扇区域，刻画了冲积扇的线条分布及河流流向，其中玉门疏勒河冲积扇属于内陆干旱环境大型冲积扇。

The alluvial fan is located in Gansu Province of China and originates in the Qilian Mountains. Alluvial fan is a fan-shaped deposit body at the mouth of a river. When the river flows out of the mouth of a valley and gets rid of the lateral constraint, the material it carries is dispersed and deposited. The plane is fan-shaped, the fan top reaches to the mouth of the valley, and the stereoscopic shape is roughly half-buried cone. The daytime thermal infrared image synthesized from three bands covers a number of alluvial fan regions, describing the line distribution and river flow direction of alluvial fans, among which the Shulehe alluvial fan in Yumen is a large alluvial fan in the inland arid environment.

巴丹吉林沙漠
Badain Jaran Desert

巴丹吉林沙漠位于中国内蒙古自治区西部的银额盆地底部。巴丹吉林沙漠的有些沙丘高度超过 500m，大多数沙丘的平均高度约为 200m。沙漠中还有 100 多个泉水湖泊，位于沙丘之间，俗称“沙漠千湖”。此三波段合成的夜间热红外影像包括湖泊和沙丘，由于湖泊和沙丘的温度不相同，沙丘中的蓝色或白色的区域就是湖泊，有效地展示了夜间沙漠的热分布情况。

The Badain Jaran Desert is located at the bottom of the Yin'e Basin in the western part of Inner Mongolia Autonomous Region, China. Some sand dunes in the Badain Jaran Desert reach a height of more than 500m, while most average around 200m. The desert also features over 100 spring-fed lakes that lie between the dunes, commonly known as "Thousand Lakes of Desert". The nighttime thermal infrared image synthesized from three bands includes lakes and sand dunes. Due to the different temperatures of lakes and sand dunes, the blue or white areas in the sand dunes are lakes, effectively displaying the thermal distribution of nighttime deserts.

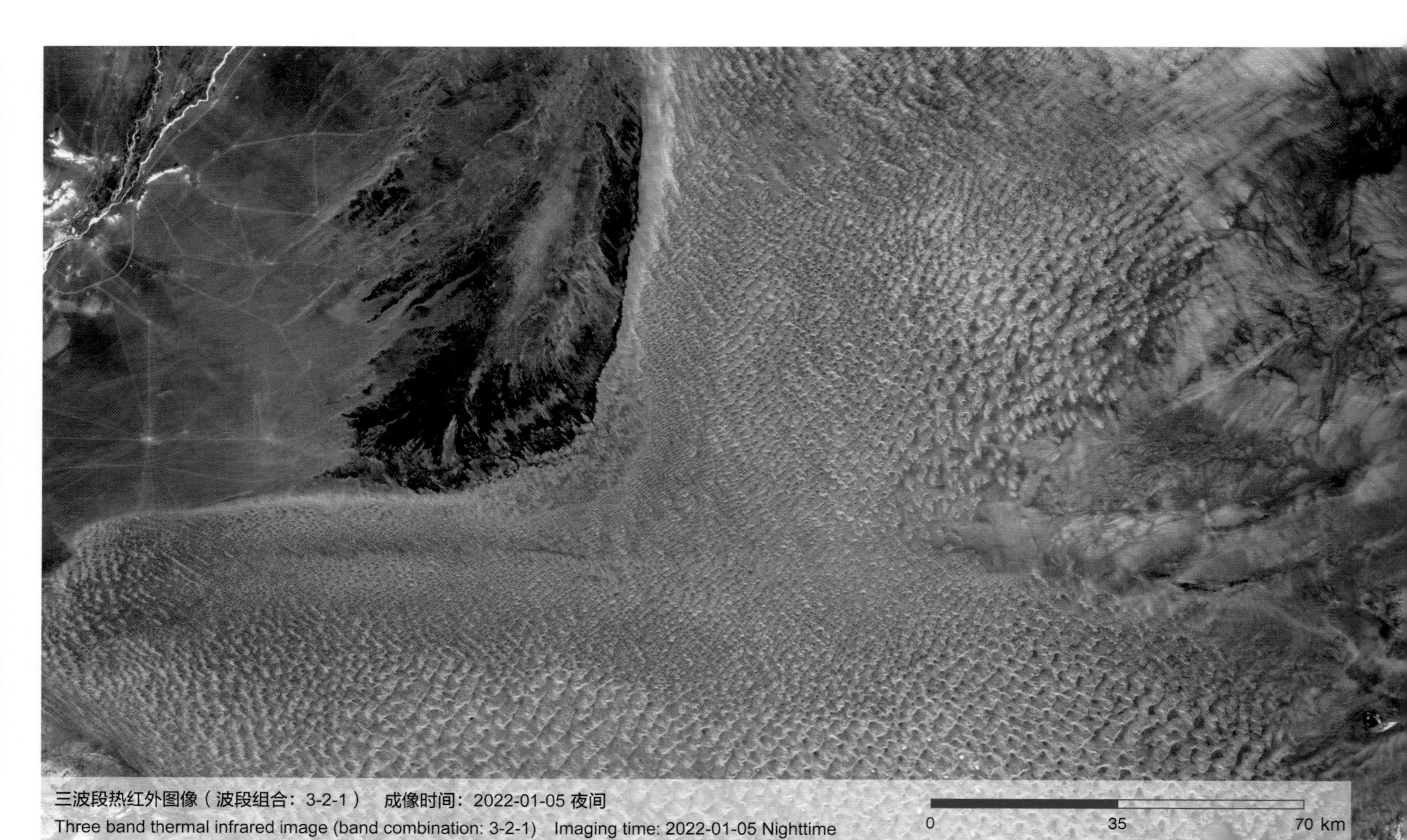

三波段热红外图像（波段组合：3-2-1）　成像时间：2022-01-05 夜间
Three band thermal infrared image (band combination: 3-2-1)　Imaging time: 2022-01-05 Nighttime

腾格里沙漠
Tengger Desert

腾格里沙漠是中国第四大沙漠，沙丘面积占71%，以流动沙丘为主，大多为格状沙丘链及新月形沙丘链，高度在 10 ~ 20m。此三波段合成的夜间热红外影像包括湖泊和沙丘，由于湖泊和沙丘的温度不相同，沙丘中的蓝色或白色的区域就是湖泊，有效地展示了夜间沙漠的热分布情况。

Tengger Desert is the fourth largest desert in China. The area of sand dunes in Tengger Desert accounts for 71%, which is dominated by mobile dune, most of which are lattice dune chains and crescent dune chains, with a height of 10~20m. The nighttime thermal infrared image synthesized from three bands includes lakes and sand dunes. Due to the different temperatures of lakes and sand dunes, the blue or white areas in the sand dunes are lakes, effectively displaying the thermal distribution of nighttime deserts.

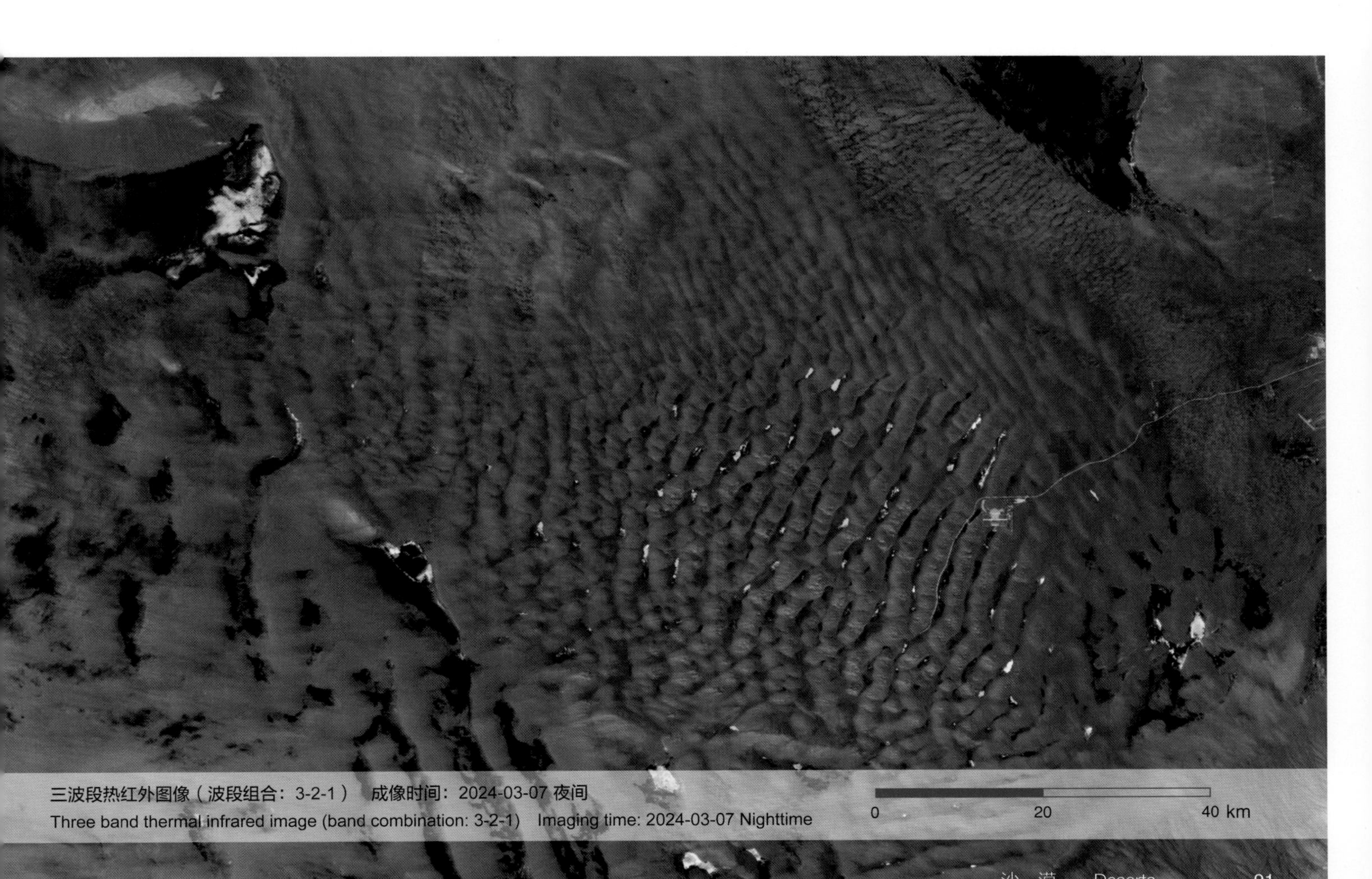

三波段热红外图像（波段组合：3-2-1） 成像时间：2024-03-07 夜间
Three band thermal infrared image (band combination: 3-2-1) Imaging time: 2024-03-07 Nighttime

塔克拉玛干沙漠
Taklamakan Desert

塔克拉玛干沙漠是中国西北地区新疆西南地区的一片沙漠，是世界上最大的沙漠之一。沙漠自然条件恶劣，风沙危害严重。该地区的特点是气候干燥、降雨量少、蒸发量大。此三波段合成的夜间热红外影像包括穿越沙漠的河流，相较于沙漠，河流的温度较低，产生了明显对比，有效展示了内陆沙漠的热环境情况。

The Taklamakan Desert is a desert in Southwest Xinjiang in Northwest China, one of the largest sandy deserts in the world. The desert's natural conditions are adverse, and there are serious hazards from wind-drift sand. The area is characterized by a dry climate, low rainfall and high evaporation. The nighttime thermal infrared image synthesized from three bands includes rivers crossing deserts. Compared with deserts, rivers have lower temperatures, creating a clear contrast and effectively displaying the thermal environment of inland deserts.

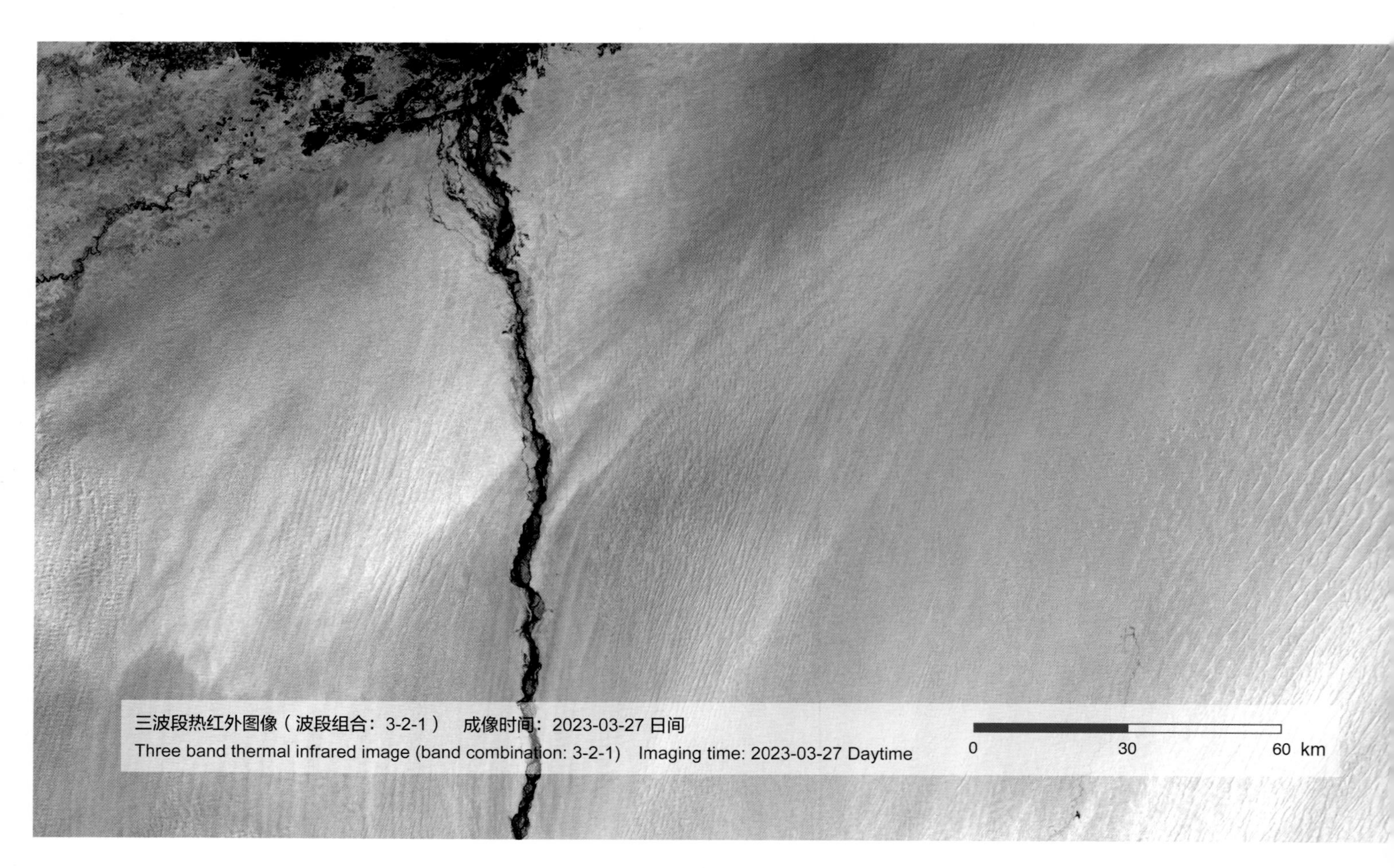

三波段热红外图像（波段组合：3-2-1）　成像时间：2023-03-27 日间
Three band thermal infrared image (band combination: 3-2-1)　Imaging time: 2023-03-27 Daytime

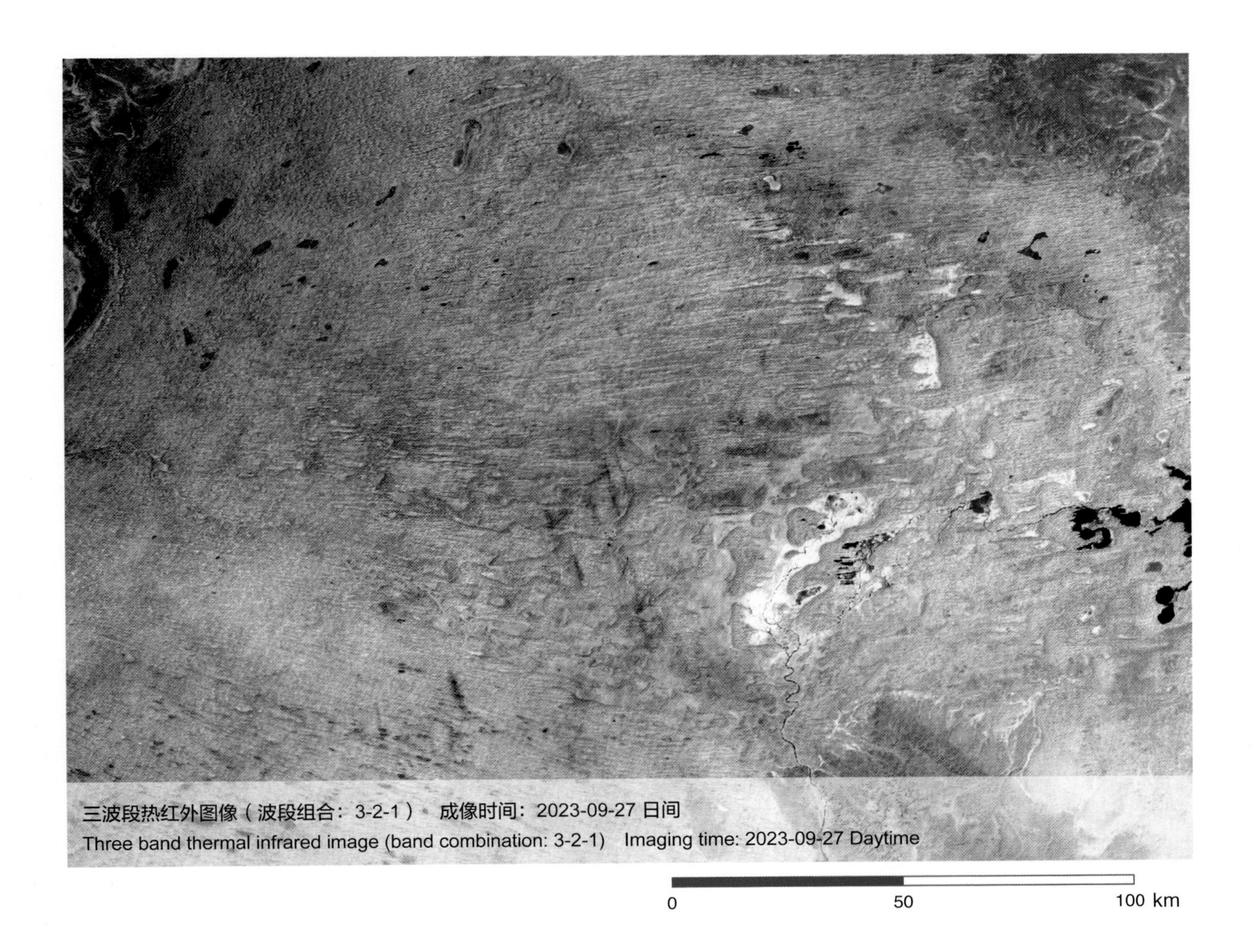

三波段热红外图像（波段组合：3-2-1） 成像时间：2023-09-27 日间
Three band thermal infrared image (band combination: 3-2-1) Imaging time: 2023-09-27 Daytime

艾尔湖盆地
Lake Eyre Basin

艾尔湖盆地位于昆士兰与南澳大利亚交界处，弗林德斯山脉的北部，盆地地表多为低矮的浅色沙丘，覆盖着干旱盐湖。地区生态环境独特且脆弱，有数种当地独有的物种。该地区具有空间特征的稳定性适合作为卫星定标场。此三波段合成的日间热红外影像包含艾尔湖盆地的部分沙漠地区，南北区域沙漠的温度不同产生了明显的对比，有效展示了沙漠中的热环境分布情况。

The Lake Eyre Basin is located on the border between Queensland and South Australia, north of the Flinders Ranges. The surface is mostly low, light-colored sand dunes covered by arid salt lakes. The ecology of the region is unique and fragile, with several species unique to the region. The stability of spatial characteristics of the area is suitable as a satellite calibration site. The daytime thermal infrared image synthesized from three bands contains part of the desert area in the Lake Eyre Basin, and the different temperatures of the deserts in the north and south regions produce obvious contrasts, effectively demonstrating the distribution of thermal environments in the desert.

Fire Incidents

火灾

• 西北地区（加拿大）野火 / 98
Wildfires in Northwest Territories (Canada)

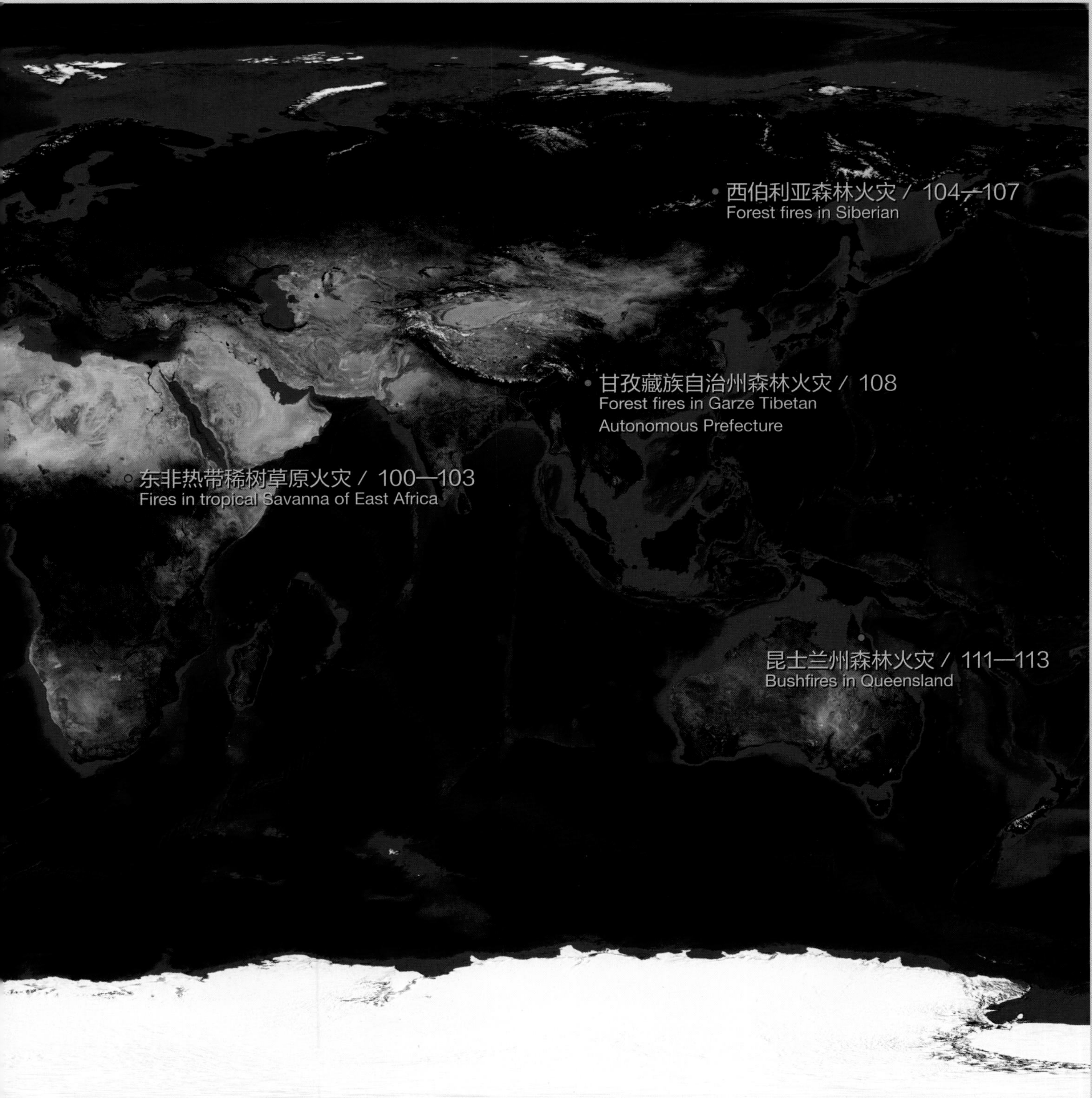

西伯利亚森林火灾 / 104—107
Forest fires in Siberian

甘孜藏族自治州森林火灾 / 108
Forest fires in Garze Tibetan
Autonomous Prefecture

东非热带稀树草原火灾 / 100—103
Fires in tropical Savanna of East Africa

昆士兰州森林火灾 / 111—113
Bushfires in Queensland

西北地区（加拿大）野火

Wildfires in Northwest Territories（Canada）

西北地区是加拿大第二大的地区，面积广阔，人烟稀少。该地区气候高寒，接近北极圈，冬季温度可达到零下 40℃。其首府为耶洛奈夫，区内有大熊湖、大奴湖、麦根士河等地标性景点。区域内野火常由闪电造成，对森林生态环境造成复杂影响。热红外图像中过火区域明显，可以协助野火监控。

The Northwest Territories is Canada's second largest region, vast and sparsely populated. The region has an alpine climate, close to the Arctic Circle, and winter temperatures can reach -40°C. Its capital is Yellowknife, and the region is home to landmarks such as Big Bear Lake, Big Slave Lake, and the McGuinness River. Wildfires in the region are often caused by lightning and have a complex impact on the forest ecosystem. Burned areas are evident in thermal infrared imagery and can assist in wildfire monitoring.

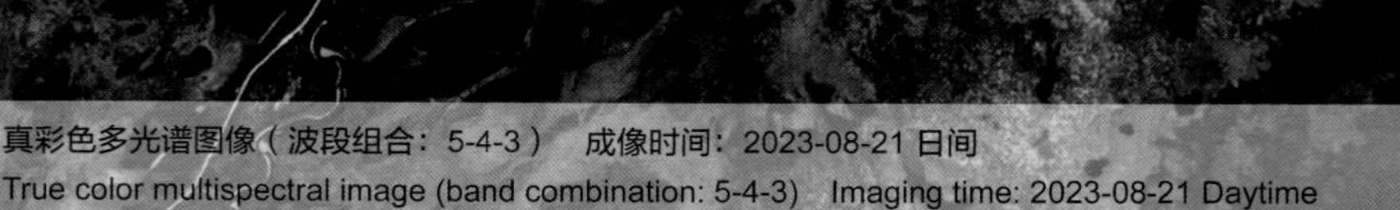

真彩色多光谱图像（波段组合：5-4-3）　成像时间：2023-08-21 日间
True color multispectral image (band combination: 5-4-3)　Imaging time: 2023-08-21 Daytime

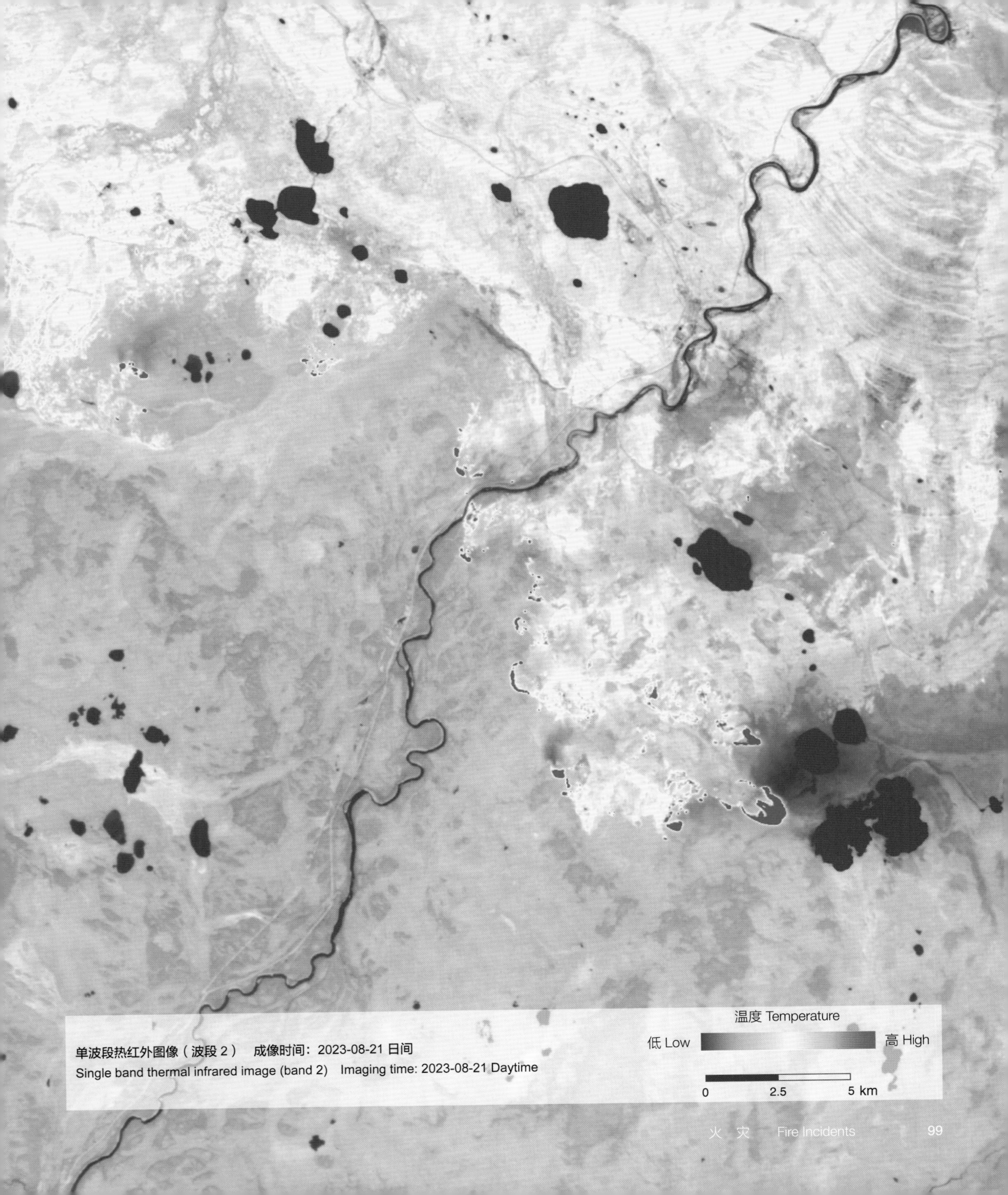

单波段热红外图像（波段 2）　成像时间：2023-08-21 日间
Single band thermal infrared image (band 2)　Imaging time: 2023-08-21 Daytime

真彩色多光谱图像（波段组合：5-4-3）　成像时间：2024-01-20 日间
True color multispectral image (band combination: 5-4-3)　Imaging time: 2024-01-20 Daytime

东非热带稀树草原火灾
Fires in tropical Savanna of East Africa

南苏丹共和国是东非的一个内陆国家，气候类型主要为热带草原气候，每年 5 ~ 10 月为雨季，11 月到次年 4 月为旱季。南苏丹与中非共和国的边界附近常有火点。大规模草原火灾对气候环境以及人体健康会造成影响。热红外图像中，火点明显。通过热红外数据，可以做到对火点的精准锁定，辅助相关政策制定与管理。

South Sudan is a landlocked country in East Africa with a predominantly savannah climate, with a rainy season from May to October and a dry season from November to April. Fires are common near South Sudan's border with the Central African Republic. Large-scale grassland fires can have an impact on the climate and environment as well as on human health. Fire points are evident in thermal infrared images. Thermal infrared data can be used to pinpoint fires and assist in policy making and management.

单波段热红外图像（波段 2）　成像时间：2024-01-20 日间
Single band thermal infrared image (band 2)　Imaging time: 2024-01-20 Daytime

温度 Temperature
低 Low　高 High
0　2　4 km

真彩色多光谱图像（波段组合：5-4-3）　成像时间：2024-01-25 日间
True color multispectral image (band combination: 5-4-3)　Imaging time: 2024-01-25 Daytime

东非热带稀树草原火灾
Fires in tropical Savanna of East Africa

单波段热红外图像（波段 2） 成像时间：2024-01-25 日间
Single band thermal infrared image (band 2) Imaging time: 2024-01-25 Daytime

温度 Temperature
低 Low 高 High
0 2 4 km

真彩色多光谱图像（波段组合：5-4-3） 成像时间：2023-07-02 日间
True color multispectral image (band combination: 5-4-3) Imaging time: 2023-07-02 Daytime

西伯利亚森林火灾
Forest fires in Siberian

西伯利亚森林，绝大部分地区属于亚寒带气候。地区植被种类最多的是针叶林，约占全区山地面积的六七成，树种主要为耐寒的落叶松、云杉等。西伯利亚是北半球冬季最寒冷的地方之一。热红外图像中的火点有明显高温。对于地广人稀的区域，热红外遥感图像可以精准锁定林火地点与过火面积，有效辅助林火监控与管理。

The vast majority of the Siberian forests have a subarctic climate. The largest variety of regional vegetation is coniferous forests, which account for about 60% to 70% of the region's mountainous area, and are dominated by larch and spruce. Siberia has one of the coldest winters in the Northern Hemisphere. The fire spots in the thermal infrared image have significantly higher temperatures. For the sparsely populated areas, thermal infrared remote sensing images can accurately target forest fire locations and burned areas, effectively assisting forest fire monitoring and management.

单波段热红外图像（波段 2）　成像时间：2023-07-02 日间
Single band thermal infrared image (band 2)　Imaging time: 2023-07-02 Daytime

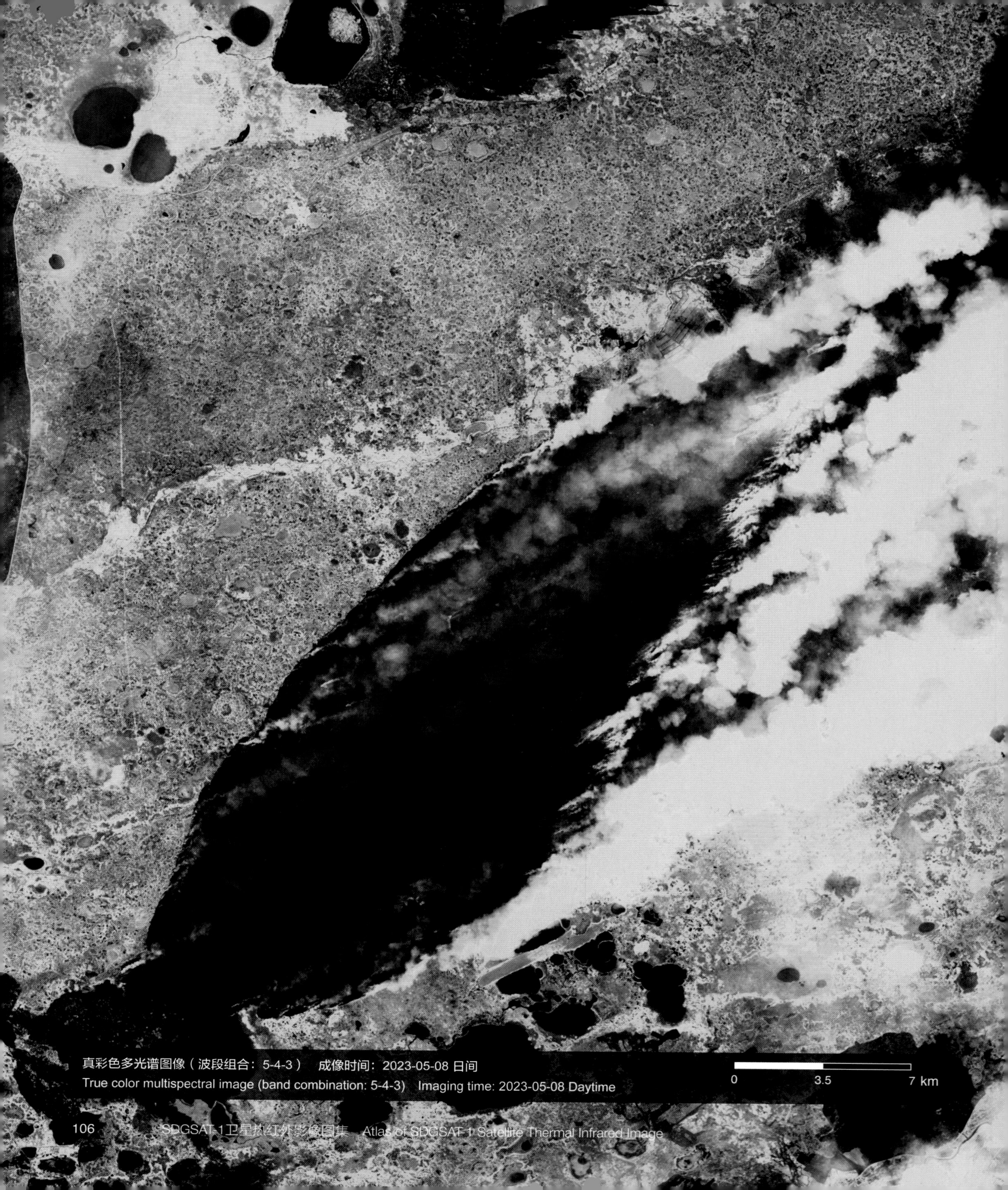

真彩色多光谱图像（波段组合：5-4-3）　成像时间：2023-05-08 日间
True color multispectral image (band combination: 5-4-3)　Imaging time: 2023-05-08 Daytime

西伯利亚森林火灾

Forest fires in Siberian

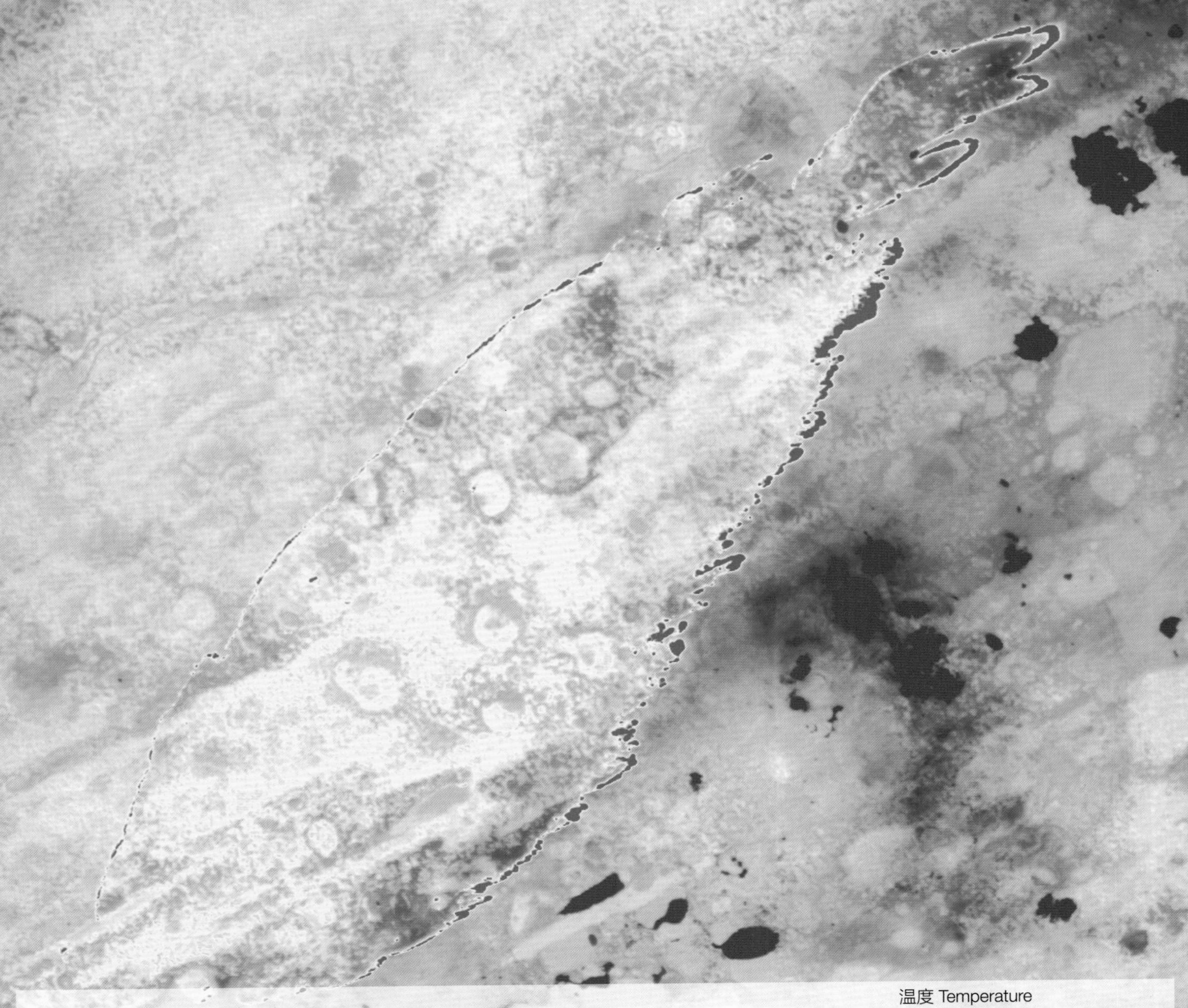

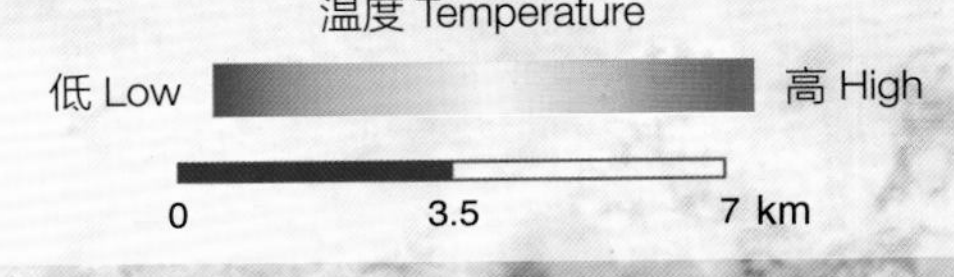

单波段热红外图像（波段 2） 成像时间：2023-05-08 日间

Single band thermal infrared image (band 2) Imaging time: 2023-05-08 Daytime

甘孜藏族自治州森林火灾
Forest fires in Garze Tibetan Autonomous Prefecture

四川省甘孜藏族自治州位于中国西南腹地，青藏高原边缘，海拔较高、地形复杂、森林覆盖广。2024 年 3 月，甘孜藏族自治州南部发生森林火灾。热红外图像显示，山脉上有高温分布，为正在燃烧的火点。同时相的微光影像显示大火正熊熊燃烧。通过热红外影像，可明确火点分布位置以及过火面积，辅助救灾力量完成对火灾的控制。

Garze Tibetan Autonomous Prefecture in Sichuan Province is located in the hinterland of southwestern China, on the edge of the Tibetan Plateau, with high elevations and complex terrain with extensive forest cover.In March 2024, a forest fire broke out in the southern part of Ganzi Prefecture. Thermal infrared imagery shows high temperatures distributed over the mountain range as burning fires. Simultaneous microlight images show that the fire is burning. The thermal infrared image can clarify the distribution location of the fire point and the fire area, which can assist the rescue force to contain the spread of the fire.

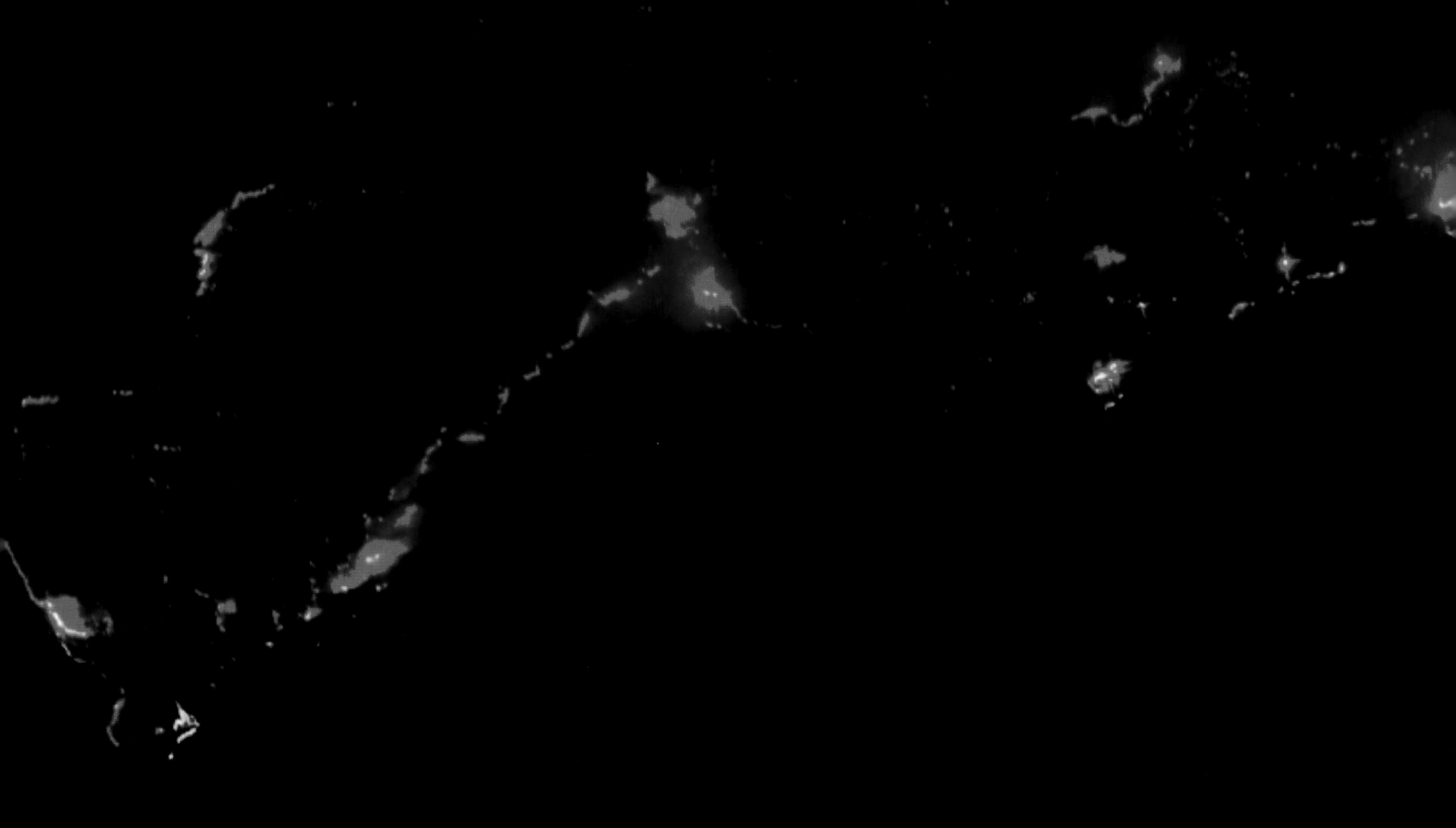

真彩色微光图像（波段组合：3-2-1） 成像时间：2024-03-16 夜间
True color glimmer image (band combination: 3-2-1) Imaging time: 2024-03-16 Nighttime
0 3.25 6.5 km

单波段热红外图像（波段 2）　成像时间：2022-03-16 日间
Single band thermal infrared image (band 2)　Imaging time: 2022-03-16 Daytime

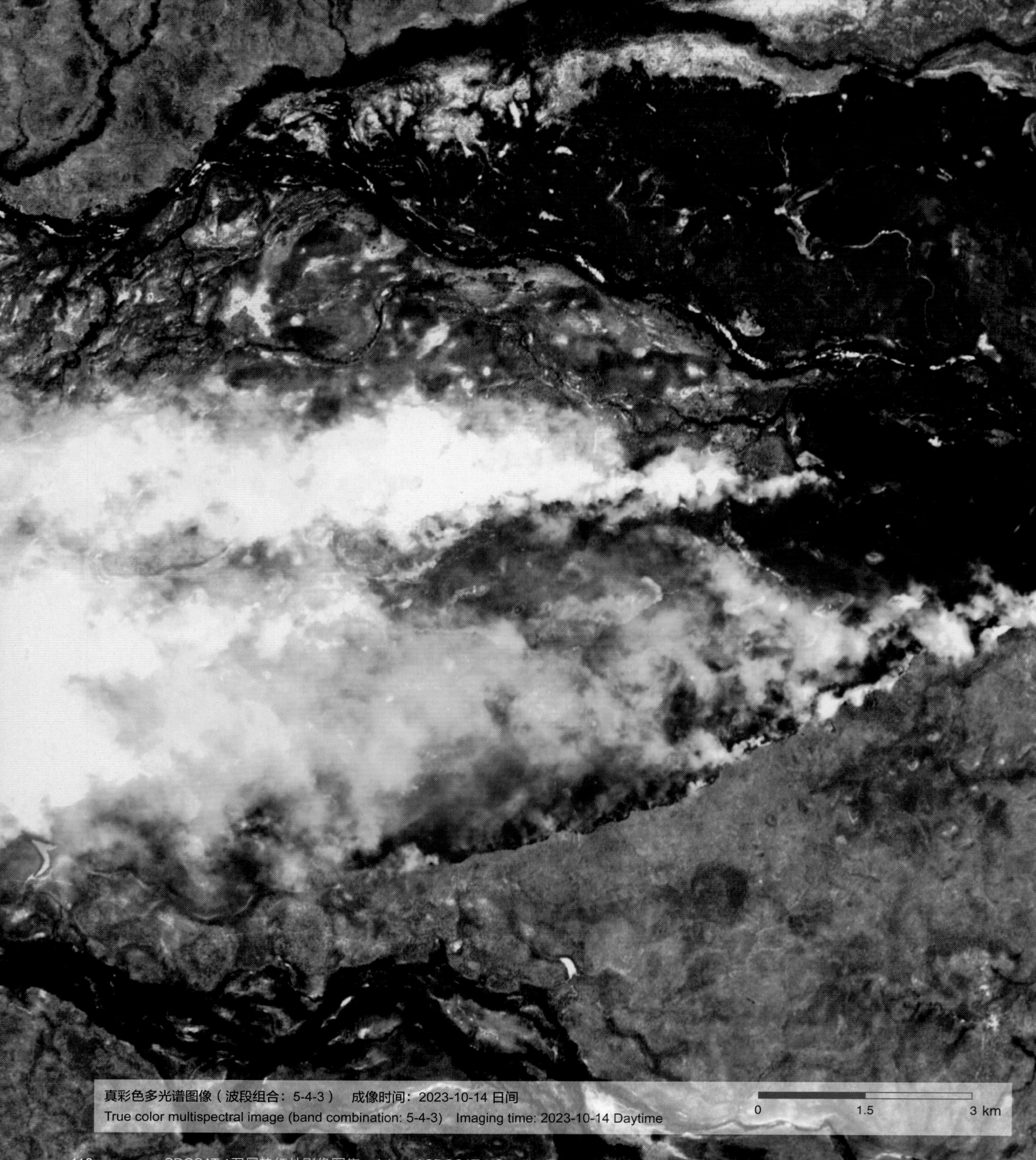

真彩色多光谱图像（波段组合：5-4-3） 成像时间：2023-10-14 日间
True color multispectral image (band combination: 5-4-3) Imaging time: 2023-10-14 Daytime

昆士兰州森林火灾
Bushfires in Queensland

昆士兰州位于澳大利亚东北部，三面环海，气候条件多变。每年 5 月到 10 月湿度较低，晴天居多，多发山火，导致人员伤亡和财产损失。重大森林火灾常常对区域动植物生存造成复杂的长期影响。热红外图像中地面高温异常点明显，与多烟雾覆盖的多光谱图像对比，可以更好地提供火点位置信息，协助山火研究与管理。

Queensland is located in the north-east of Australia and is surrounded by the sea on three sides, with variable weather conditions. The low humidity and predominance of sunny days from May to October each year result in a high incidence of bushfires, leading to casualties and property damage. Major bushfires often have a complex and long-term impact on the regional flora and fauna. The high ground temperature anomalies evident in thermal infrared imagery, in contrast to multispectral imagery where smoke coverage is thick, can provide better information on the location of fires and assist in bushfires research and management.

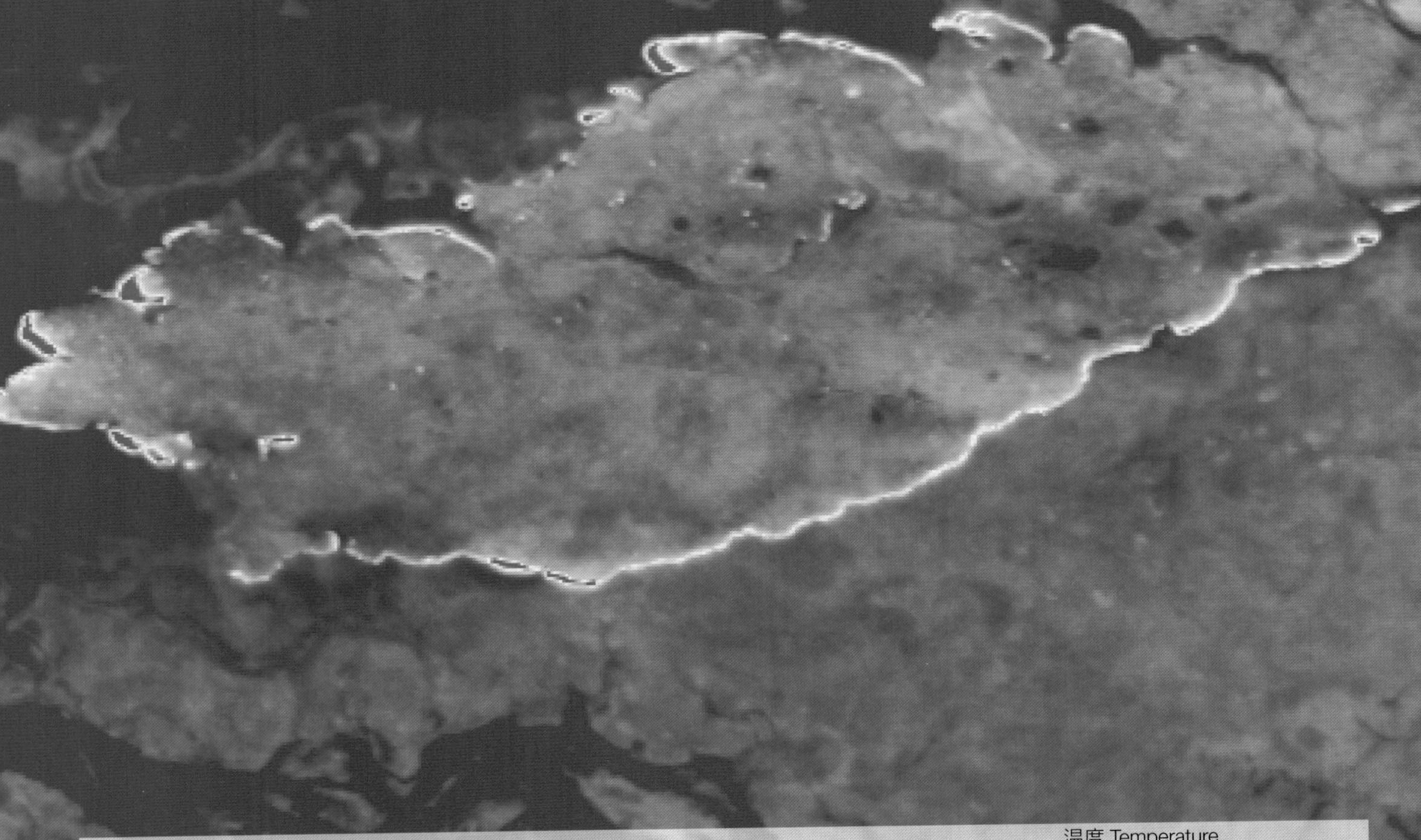

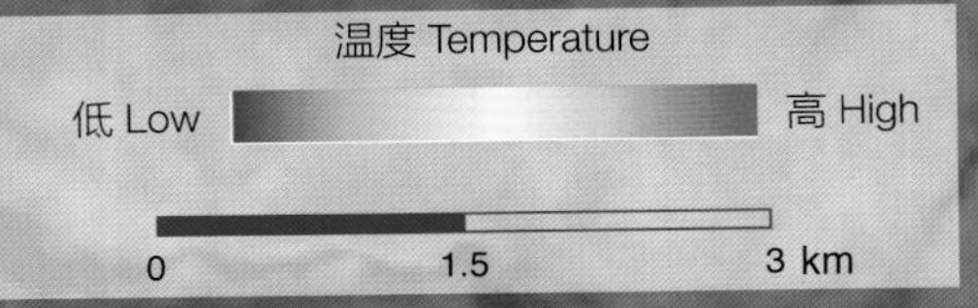

单波段热红外图像（波段 2） 成像时间：2023-10-14 日间
Single band thermal infrared image (band 2) Imaging time: 2023-10-14 Daytime

真彩色多光谱图像（波段组合：5-4-3） 成像时间：2023-10-14 日间
True color multispectral image (band combination: 5-4-3) Imaging time: 2023-10-14 Daytime

昆士兰州森林火灾
Bushfires in Queensland

单波段热红外图像（波段 2） 成像时间：2023-10-14 日间

Single band thermal infrared image (band 2) Imaging time: 2023-10-14 Daytime

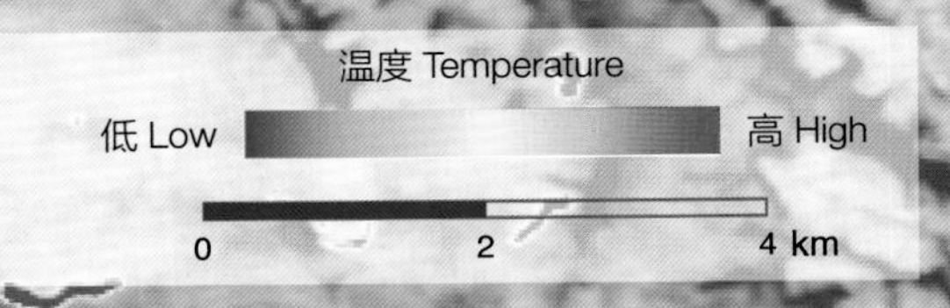

海冰

Sea Ice

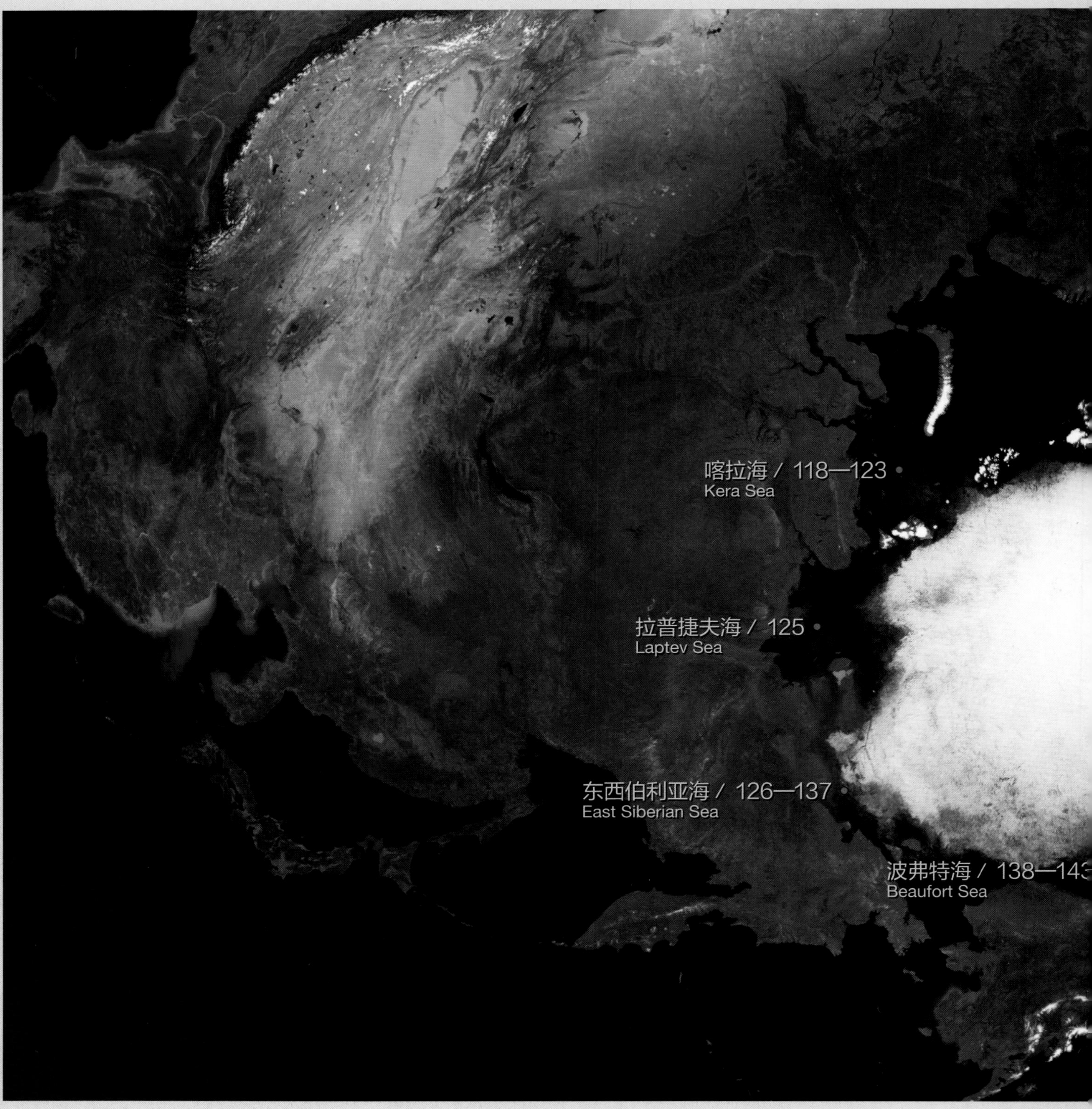

喀拉海 / 118—123
Kera Sea

拉普捷夫海 / 125
Laptev Sea

东西伯利亚海 / 126—137
East Siberian Sea

波弗特海 / 138—143
Beaufort Sea

巴芬湾 / 144
Baffin Bay

伊丽莎白女王群岛海岸 / 147
Coast of the Queen Elizabeth Islands

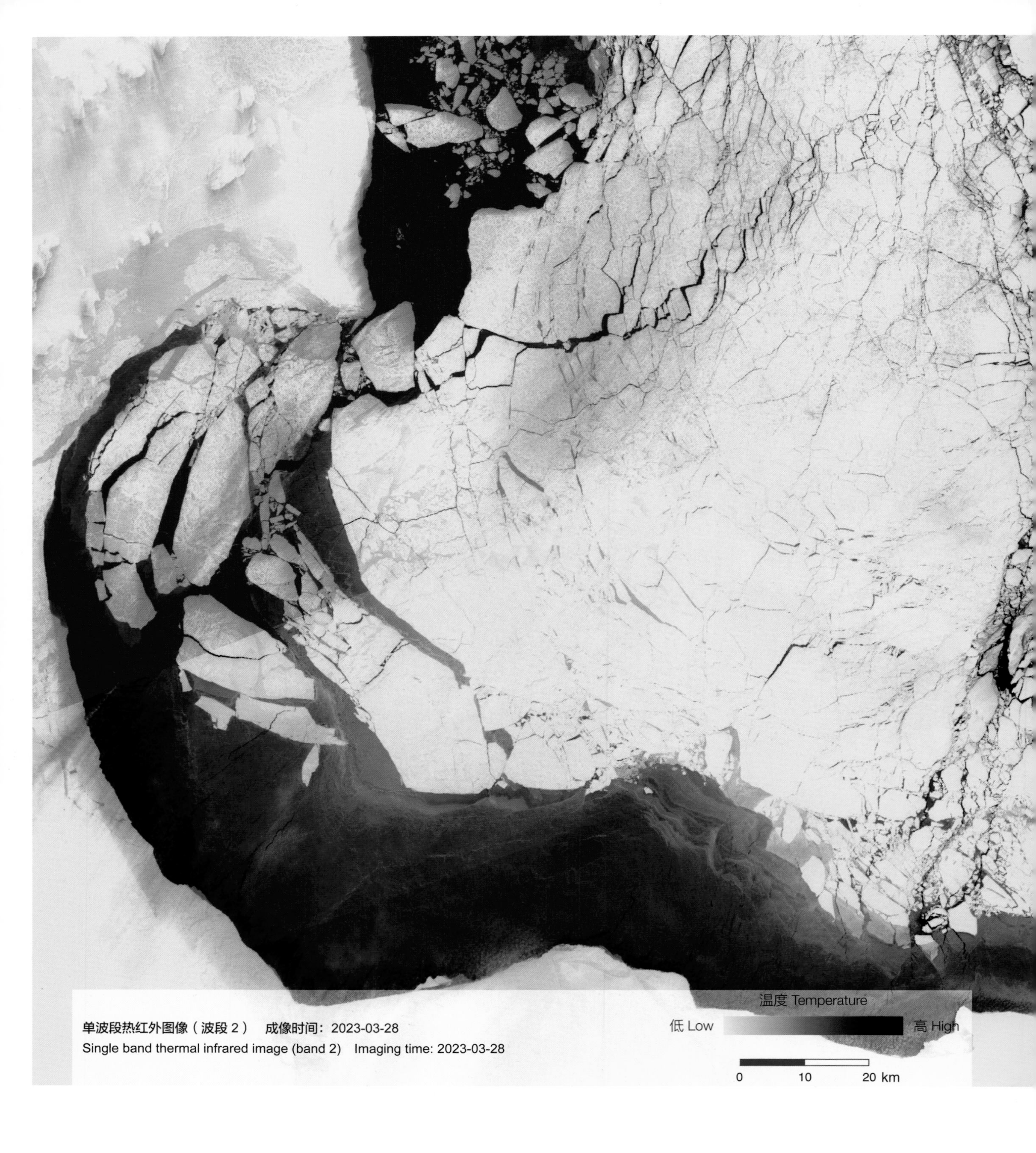

单波段热红外图像（波段 2）　成像时间：2023-03-28

Single band thermal infrared image (band 2)　Imaging time: 2023-03-28

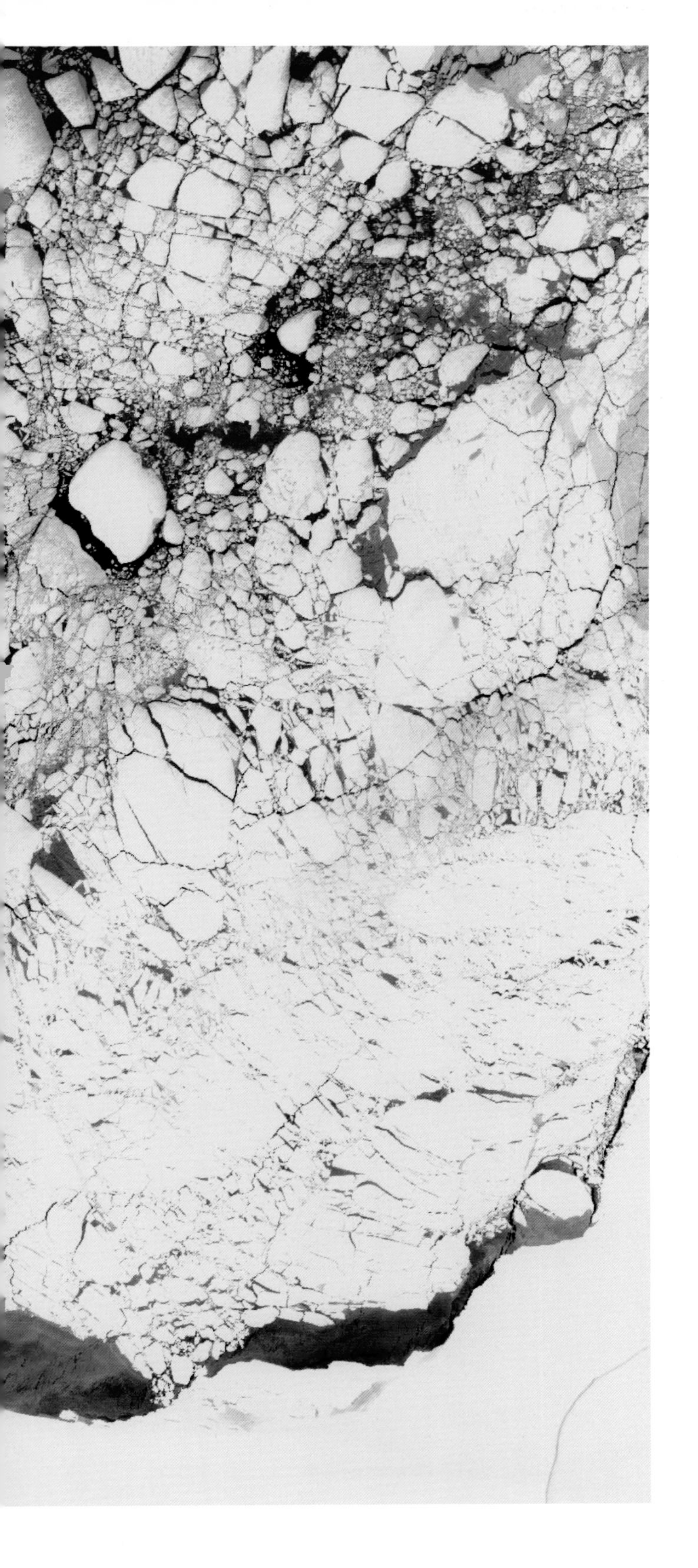

喀拉海（北地群岛沿岸）

Kara Sea off the coast of the Northland Islands

喀拉海是北冰洋的边缘海，位于亚洲大陆西北部沿岸和新地岛、北地群岛之间。西通巴伦支海，东连拉普捷夫海，北接北冰洋。喀拉海区气候异常寒冷，几乎终年冰封。北地群岛位于喀拉海东部，将喀拉海与拉普捷夫海分隔开。群岛气候严寒，解冻期只有两个半月。植被主要是地衣和低矮灌木，动物稀少。

此图为 2023 年 3 月 28 日在北极喀拉海东部北地群岛沿岸获取的 SDGSAT-1 热红外数据单波段（10.3 ~ 11.3μm）伪彩色影像，以浅蓝到深蓝的冷暖渐变表示海冰与海水的温度变化。大片深蓝色带状区域是一个尺度较大的沿岸冰间湖，其间暴露出温度较高的海水以及薄冰，与较厚的浮冰产生鲜明的温度对比。在 30m 分辨率下，海冰中细小的裂隙（冰间水道）也清晰可见，展现出丰富的细节特征。

The Kara Sea is a marginal sea of the Arctic Ocean, located between the northwestern coast of the Asian continent and the islands of Novaya Zemlya and the Northland Islands. It is bordered by the Barents Sea to the west, the Laptev Sea to the east, and the Arctic Ocean to the north. The climate in the Kara Sea region is exceptionally cold, with nearly year-round ice cover. The Severnaya Zemlya archipelago is situated in the eastern part of the Kara Sea, separating it from the Laptev Sea. The archipelago experiences extremely cold climates, with a thawing period lasting only two and a half months. Vegetation mainly consists of lichens and low shrubs, while animal life is scarce.

This image, captured on March 28, 2023, along the coast of the Severnaya Zemlya Archipelago in the eastern Kara Sea, depicts single-band (10.3~11.3μm) false-color thermal infrared data from the SDGSAT-1. The gradient from light blue to deep blue represents temperature variations between sea ice and seawater. The extensive dark blue band represents a large-scale coastal polynya, revealing areas of relatively warm seawater and thin ice, contrasting sharply with thicker pack ice. At a resolution of 30m, fine fractures (sea ice leads) within the sea ice cover are visible, showcasing rich detail features.

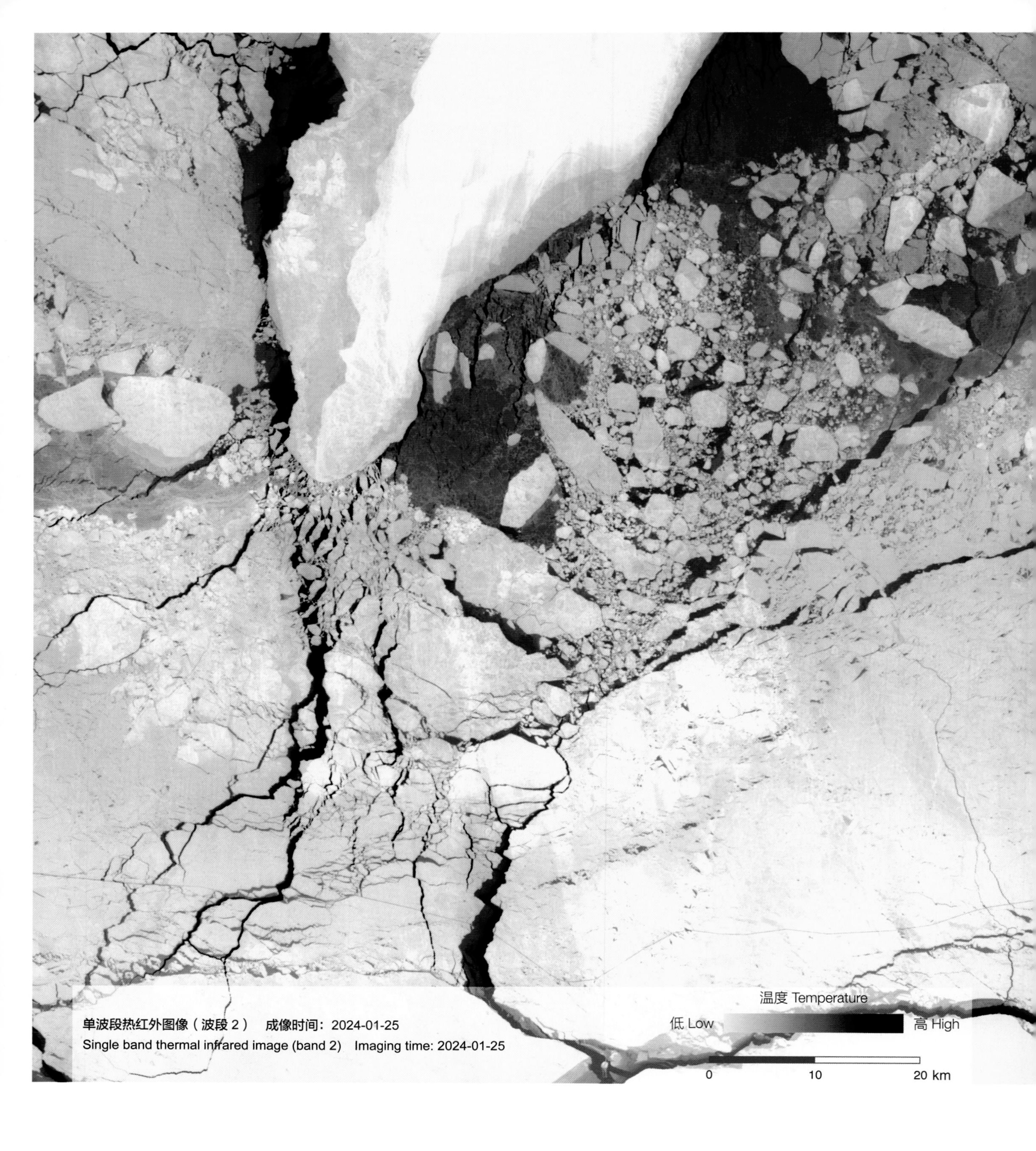

单波段热红外图像（波段 2） 成像时间：2024-01-25
Single band thermal infrared image (band 2) Imaging time: 2024-01-25

喀拉海
（泰梅尔半岛沿岸）
Kera Sea off the coast of Taimyr Peninsula

泰梅尔半岛位于喀拉海北缘，它是亚洲最北的半岛，其北部沿海为狭长平原；海岸线曲折，多峡湾。

本图为 2024 年 1 月 25 日在喀拉海泰梅尔半岛沿岸获取的 SDGSAT-1 卫星热红外数据单波段（10.3 ~ 11.3μm）伪彩色图，以浅蓝到深蓝的冷暖渐变表示海冰与海水的温度变化。此时处于海冰的冻结期，仅沿岸边断裂，尺寸不同的浮冰群向海域内聚集、堆积，形成相互交错的裂隙，暴露出部分温度较高的冰下海面以及薄冰，与温度较低的海冰形成鲜明的温度差异，展示出沿岸海冰的分布细节。

The Taymyr Peninsula lies to the north of the Kara Sea, making it the northernmost peninsula in Asia. Its northern coastal area consists of a narrow plain, with a winding coastline and numerous bays.

This image, captured on January 25, 2024, along the coast of the Taimyr Peninsula in the Kara Sea, depicts single-band (10.3~11.3μm) false-color thermal infrared data from the SDGSAT-1. The color gradient from light blue to deep blue represents temperature variations between sea ice and seawater. During the freezing period of sea ice, fragmented sea ice floes accumulate and pile up towards the Kera Sea, forming intersecting sea ice leads. This exposes portions of relatively warmer underwater ice surfaces and thin ice, creating distinct temperature differences with the higher sea ice, and showcasing detailed distributions of coastal sea ice.

喀拉海（泰梅尔半岛沿岸）
Kera Sea off the coast of Taimyr Peninsula

本图为 2024 年 2 月 4 日在喀拉海泰梅尔半岛沿岸获取的 SDGSAT-1 卫星热红外数据单波段（10.3 ~ 11.3μm）伪彩色图，以浅蓝到深蓝的冷暖渐变表示海冰与海水的温度变化。此时处于海冰的冻结期，图中两个冰封的小岛改变了海冰分布的连续性，小岛周围堆积着尺度不同的浮冰群，形成了方向不一、相互交错的海冰裂隙，暴露出冰下温度较高的海水，与浮冰产生鲜明温度差异，有效展示出细节丰富的沿岸海冰分布。

This image, captured on February 4, 2024, along the coast of the Taimyr Peninsula in the Kara Sea, depicts single-band (10.3~11.3μm) false-color thermal infrared data from the SDGSAT-1. The color gradient from light blue to deep blue represents temperature variations between sea ice and seawater. During the freezing period of sea ice, the two frozen islands depicted in the image disrupt the continuity of the sea ice. Around these islands, ice floes of varying scales accumulate, forming intersecting sea ice leads in different directions. This exposes relatively warmer seawater beneath the ice, creating distinct temperature differences with the ice floes, effectively illustrating the detailed distribution of coastal sea ice.

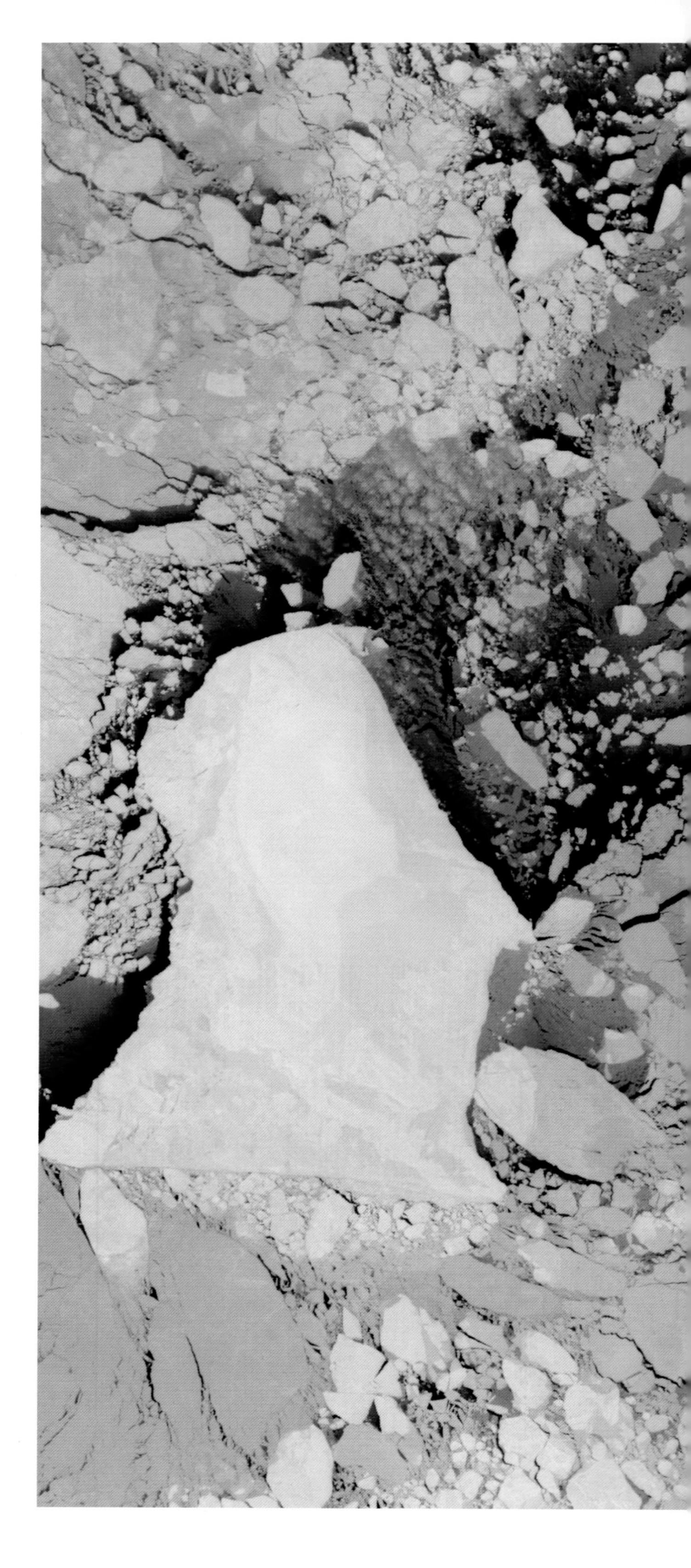

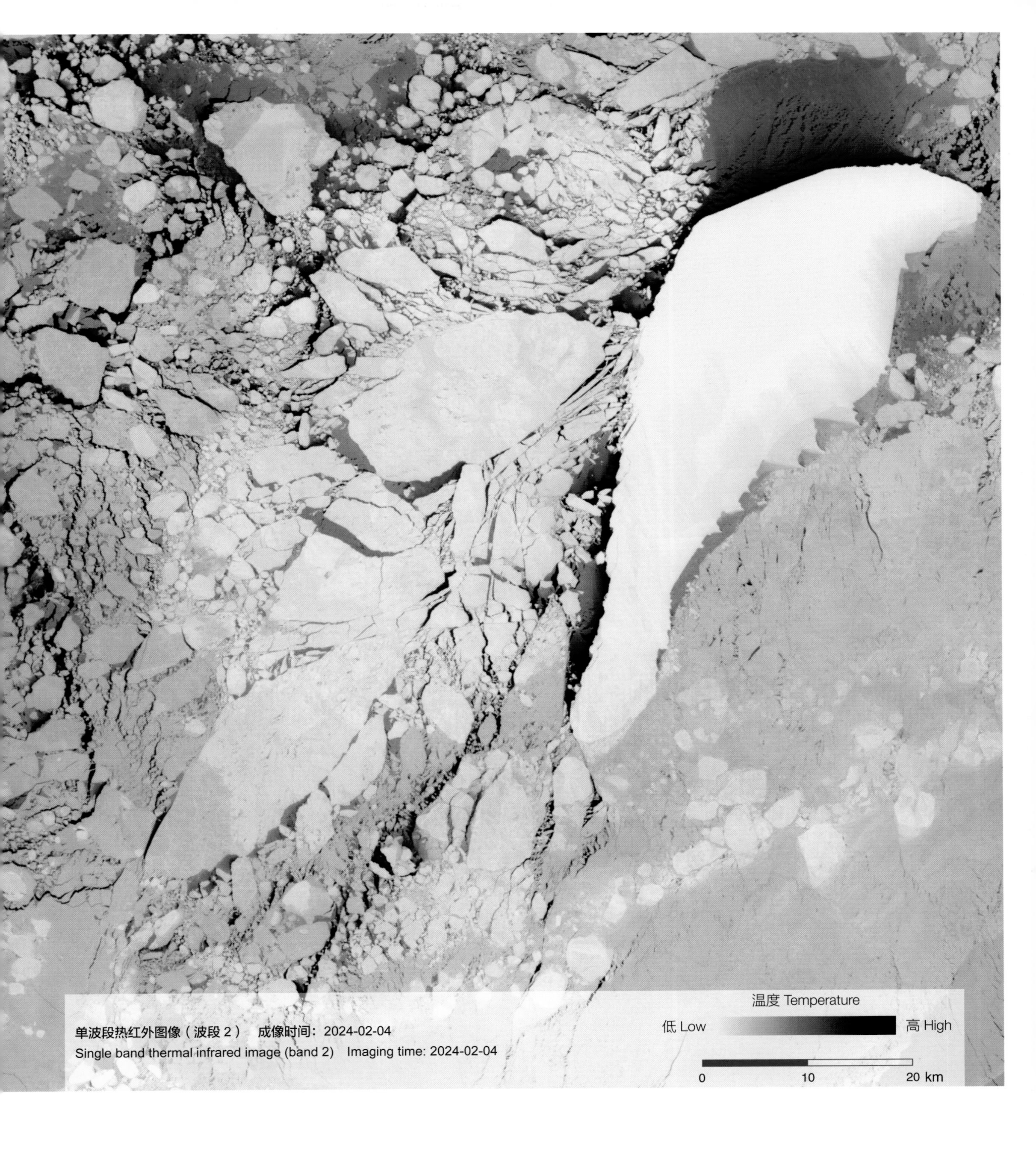

单波段热红外图像（波段 2）　成像时间：2024-02-04
Single band thermal infrared image (band 2)　Imaging time: 2024-02-04

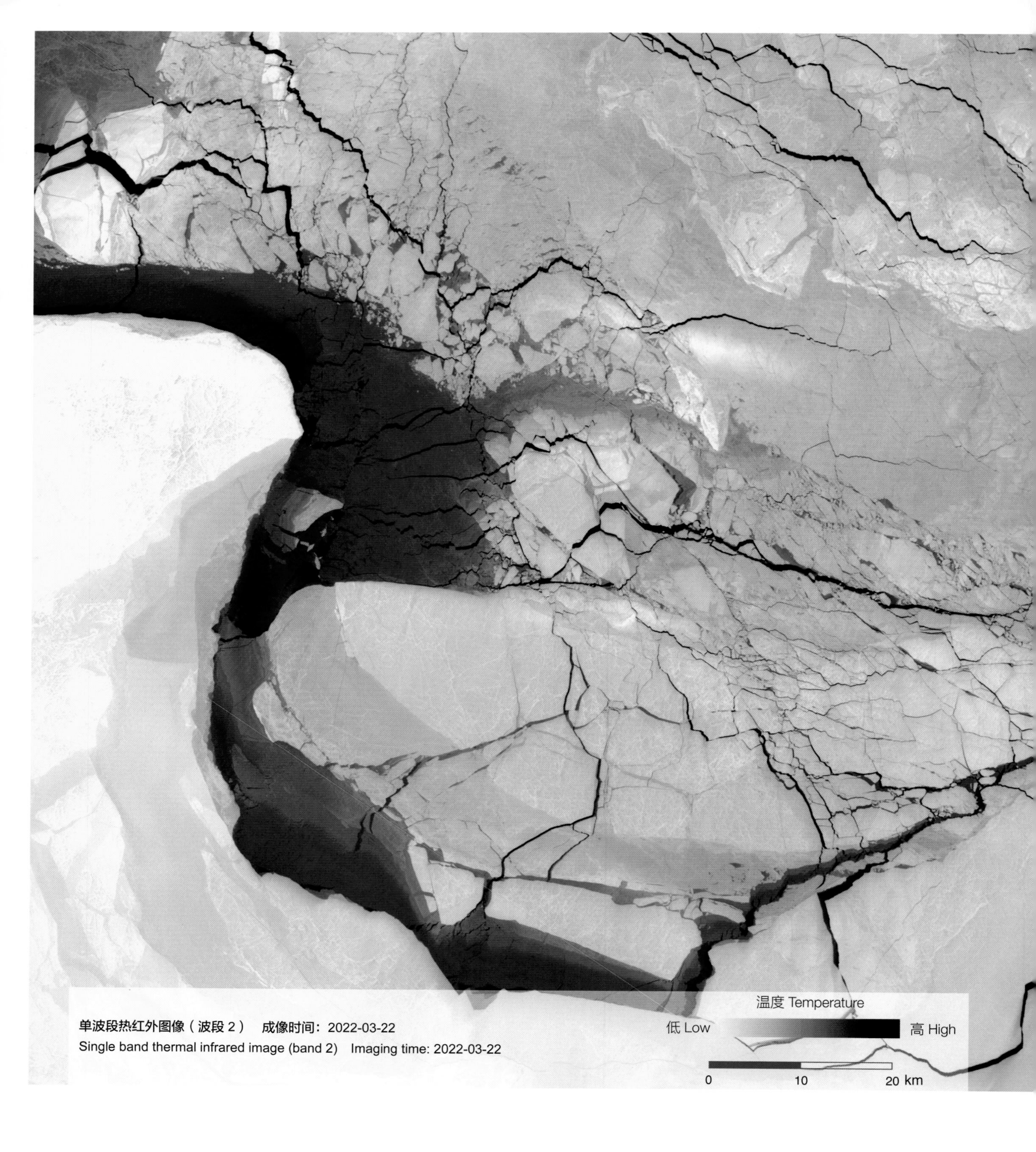

单波段热红外图像（波段 2）　成像时间：2022-03-22

Single band thermal infrared image (band 2)　Imaging time: 2022-03-22

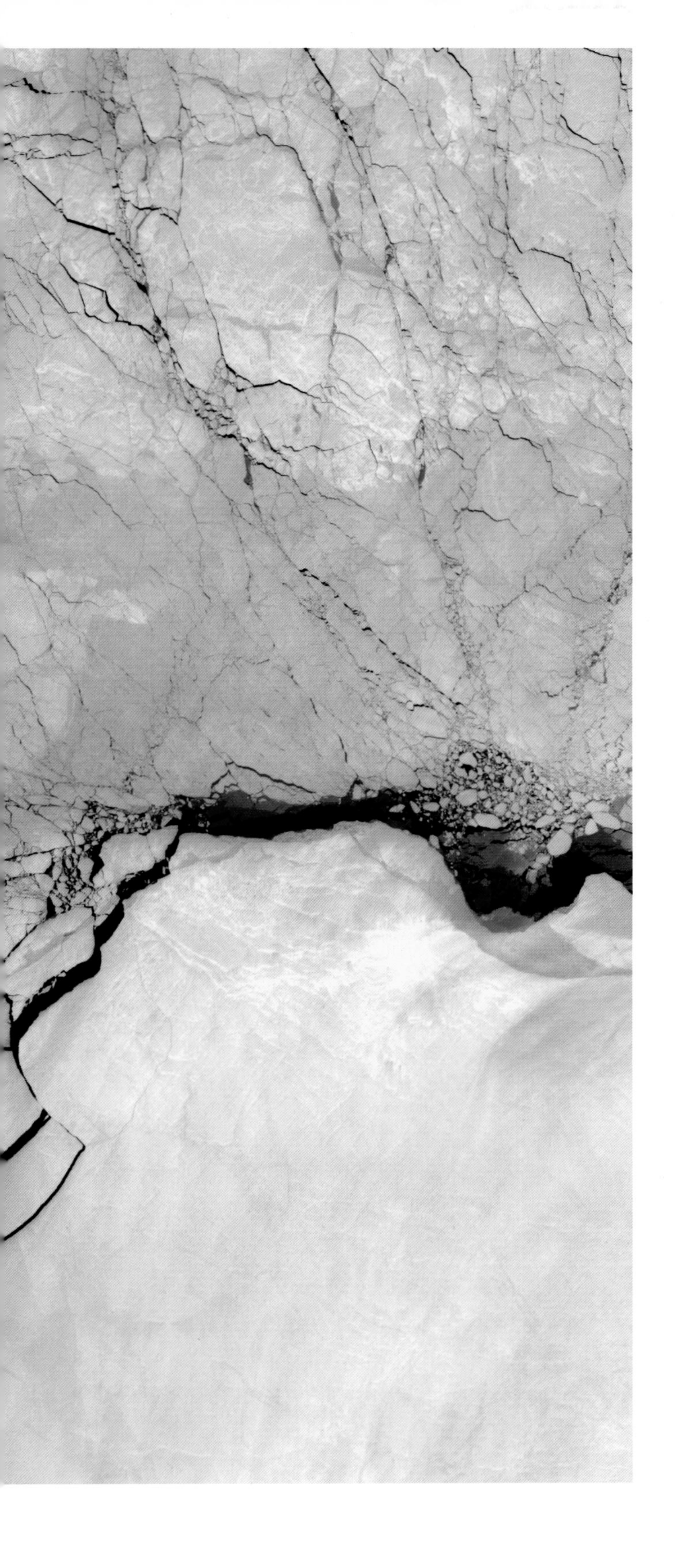

拉普捷夫海
Laptev Sea

拉普捷夫海是北冰洋的边缘海之一，位于西伯利亚东海岸、泰梅尔半岛、北地群岛和新西伯利亚群岛之间。拉普捷夫海气候严峻，南部地区极夜持续 3 个月，北部极夜长达 5 个月；每年的 8 月、9 月可通航。

此图为 2022 年 3 月 22 日在北极拉普捷夫海获取的 SDGSAT-1 热红外数据单波段（10.3 ～ 11.3μm）伪彩色影像，以浅蓝到深蓝的冷暖渐变表示海冰与海水的温度变化。此时正处春季，海冰沿陆缘断裂，蜿蜒的裂隙（冰间水道）逐渐向海域内延伸，暴露出冰下温度较高的海水以及薄冰，与厚冰形成鲜明的温度差异，在 30m 的空间分辨下，有效展示出丰富的海冰细节特征。

The Laptev Sea is one of the marginal seas of the Arctic Ocean, situated between the eastern coast of Siberia, the Taymyr Peninsula, the Severnaya Zemlya archipelago, and the New Siberian Islands. The climate of the Laptev Sea is severe, with the southern regions experiencing polar night for three months and the northern regions enduring polar night for up to five months. Navigation is possible only during August and September each year.

This image, captured on March 22, 2022, in the Arctic Laptev Sea, depicts single-band (10.3~11.3μm) false-color thermal infrared data from the SDGSAT-1. The color gradient from light blue to deep blue represents temperature variations between sea ice and seawater. In the spring season, sea ice is fracturing along the coast, with sinuous fractures (sea ice leads) gradually extending into the sea. This exposes relatively warmer seawater beneath the ice and thin ice, creating distinct temperature differences with the thicker ice. With a spatial resolution of 30m, it effectively showcases rich details of sea ice features.

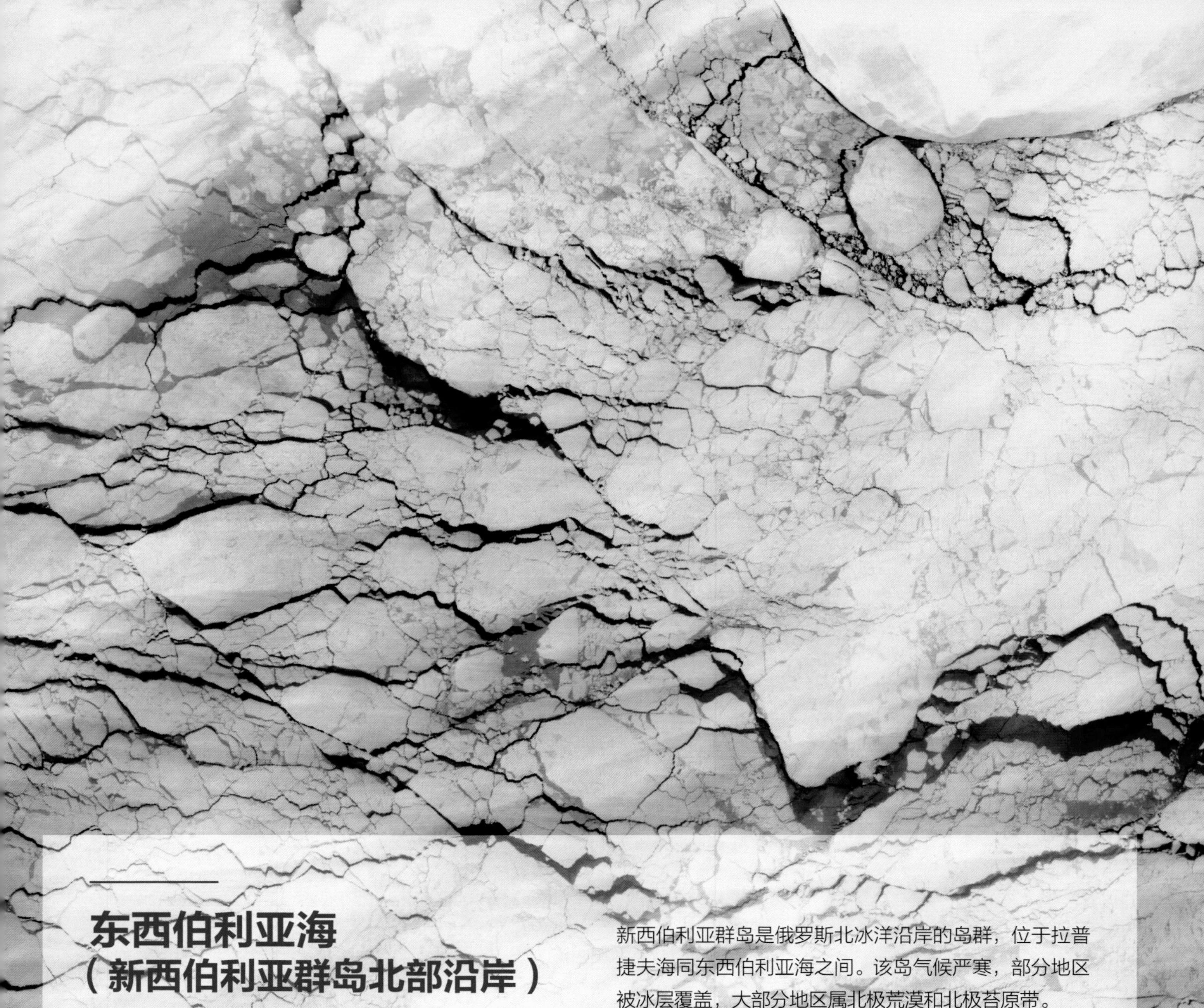

东西伯利亚海（新西伯利亚群岛北部沿岸）

East Siberian Sea off the northern coast of the New Siberian Islands

东西伯利亚海是北冰洋的边缘海，位于北面的北极角和南面的西伯利亚之间。东面隔弗兰格尔岛和楚科奇海相邻、西侧则以新西伯利亚群岛与拉普捷夫海分隔开来。东西伯利亚海长期受极地东风带影响，气候非常严寒，结冰期长达9个月，航行困难，仅限于8月和9月可通航。

新西伯利亚群岛是俄罗斯北冰洋沿岸的岛群，位于拉普捷夫海同东西伯利亚海之间。该岛气候严寒，部分地区被冰层覆盖，大部分地区属北极荒漠和北极苔原带。

本图为2024年2月4日在新西伯利亚群岛北部的东西伯利亚海区域获取的SDGSAT-1卫星热红外数据单波段（10.3 ~ 11.3μm）伪彩色图，以浅蓝到深蓝的冷暖渐变表示海冰与海水的温度变化。此时正处于海冰的冻结期，海面上覆盖着厚厚的海冰，图中两个冰封的小岛改变了海冰的连续性，以小岛为中心向四周辐射状分散着细小的海冰裂隙，暴露出冰下温度较高海水，与海冰形成温度差异，从而展现出冻结期海冰的分布特征。

The East Siberian Sea is a marginal sea of the Arctic Ocean located between the North Pole to the north and Siberia to the south. It is bordered by the neighboring Chukchi Sea to the east across Wrangel Island and by the Laptev Sea to the west across the New Siberian Islands. The East Siberian Sea is heavily influenced by the polar easterlies, resulting in extremely cold climates with a freezing period lasting up to nine months, making navigation difficult. Navigability is limited to August and September. The New Siberian Islands are an archipelago along the Russian Arctic coast, situated between the Laptev Sea and the East Siberian Sea. The islands experience severe cold climates, with some areas covered by ice layers, while most belong to the Arctic desert and tundra zones.

This image, captured on February 4, 2024, in the northern part of the New Siberian Islands in the East Siberian Sea, depicts single-band (10.3~11.3μm) false-color thermal infrared data from the SDGSAT-1. The color gradient from light blue to deep blue represents temperature variations between sea ice and seawater. During the freezing period of sea ice, although the sea surface is covered with thick ice, the presence of two frozen islands disrupts the continuity of the sea ice. Narrow sea ice leads radiate outward from the islands, exposing relatively warmer seawater beneath the ice and creating temperature differences with the sea ice, effectively illustrating the distribution characteristics of sea ice during the freezing period.

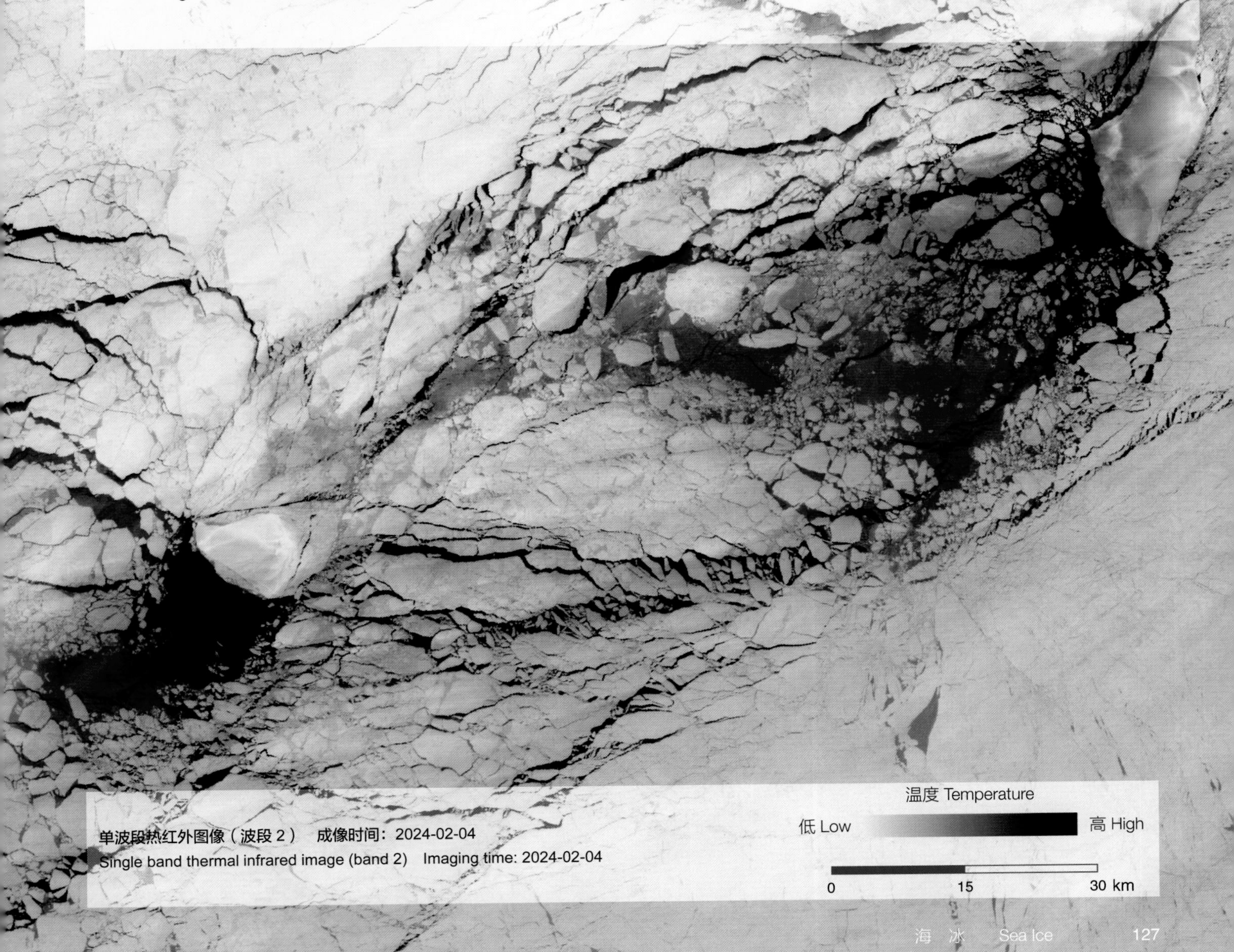

单波段热红外图像（波段 2） 成像时间：2024-02-04

Single band thermal infrared image (band 2) Imaging time: 2024-02-04

东西伯利亚海（新西伯利亚群岛北部沿岸）

East Siberian Sea off the northern coast of the New Siberian Islands

本图为 2024 年 2 月 20 日在新西伯利亚北部海域获取的 SDGSAT-1 卫星热红外数据单波段（10.3 ~ 11.3μm）伪彩色图，以浅蓝到深蓝的冷暖渐变表示海冰与海水的温度变化。随着冻结期温度持续降低，海面上覆盖着厚厚的海冰，冰间水道主要分布在岛屿四周，并向海域内延伸，暴露出温度较高的冰下海水以及薄冰，与厚冰形成温度差异，有效展现出冻结期海冰的细节特征。

This image, captured on February 20, 2024, in the northern waters of the New Siberian Islands, depicts single-band (10.3~11.3μm) false-color thermal infrared data from the SDGSAT-1. The color gradient from light blue to deep blue represents temperature variations between sea ice and seawater. As the temperature continues to drop during the freezing period, the sea surface is covered with thick sea ice. The sea ice leads are primarily distributed around the islands and extend into the East Siberian Sea, exposing relatively warmer seawater beneath the ice and thin ice. This creates temperature differences with the thicker ice, effectively illustrating detailed characteristics of sea ice during the freezing period.

单波段热红外图像（波段 2） 成像时间：2024-02-20

Single band thermal infrared image (band 2) Imaging time: 2024-02-20

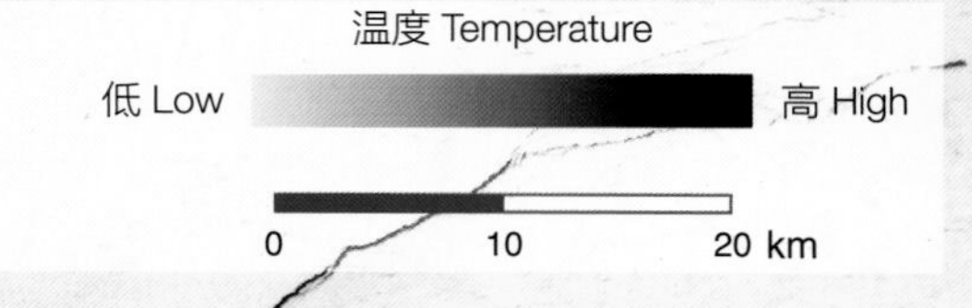

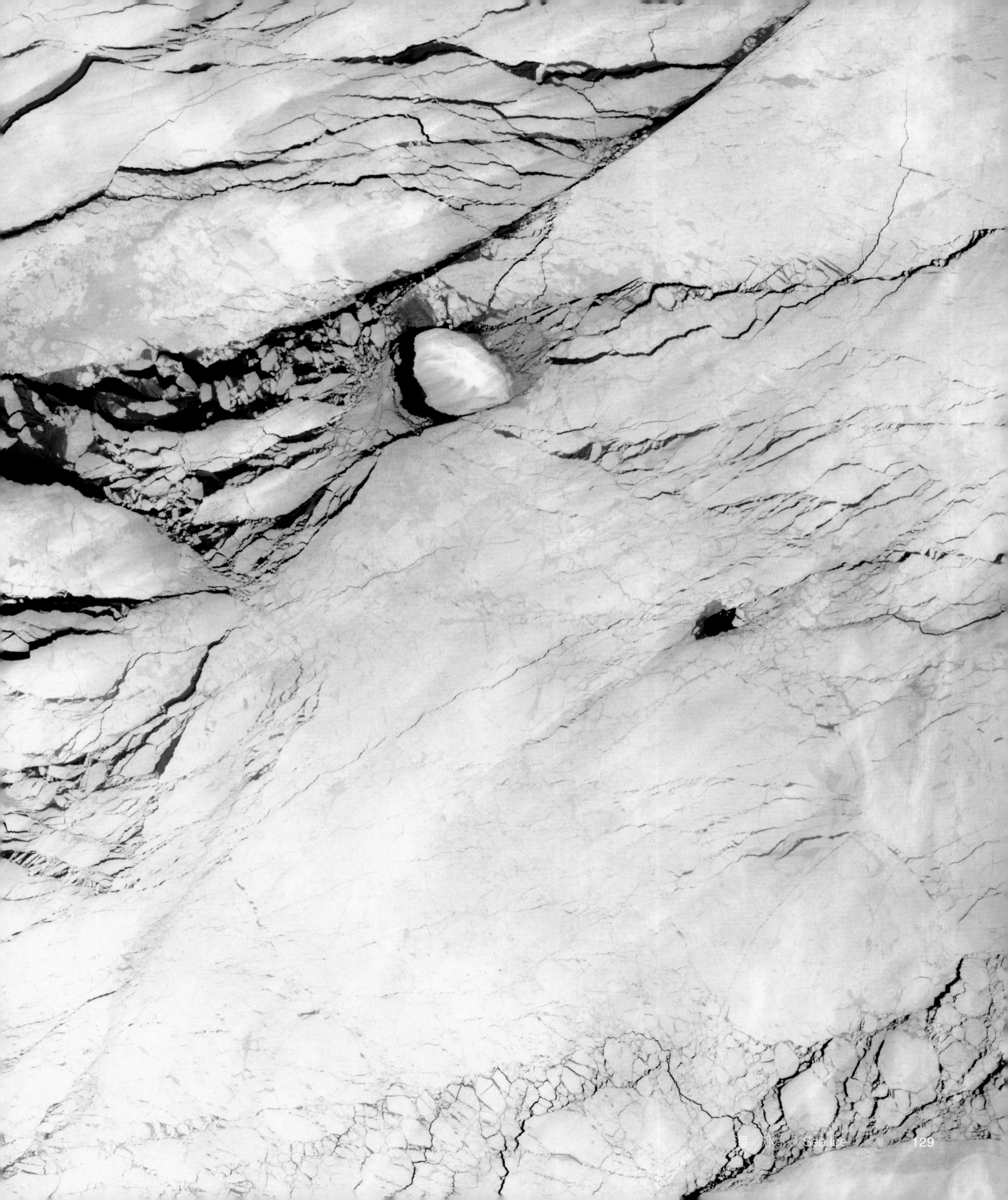

东西伯利亚海（新西伯利亚群岛沿岸）

East Siberian Sea off the coast of the New Siberian Islands

本图为 2024 年 2 月 19 日在新西伯利亚群岛沿岸获取的 SDGSAT-1 卫星热红外数据单波段（10.3 ~ 11.3μm）伪彩色图，以浅蓝到深蓝的冷暖渐变表示海冰与海水的温度变化。此时正处于冻结期，图中大片的白色区域为几乎全部冰封的群岛。海冰沿岸边产生断裂，并在冰面上分布着细小、蜿蜒的裂隙，暴露出温度较高的冰下海水以及薄冰，与较厚的浮冰形成温度差异，从而有效展现出沿岸海冰分布的细节特征。

This image, captured on February 19, 2024, along the coast of the New Siberian Islands, depicts single-band (10.3~11.3μm) false-color thermal infrared data from the SDGSAT-1. The color gradient from light blue to deep blue represents temperature variations between sea ice and seawater. During the freezing period, the islands are almost entirely ice-bound. Fractures occur along the coastal edges of the islands, and narrow, sinuous sea ice leads are distributed across the ice surface, exposing relatively warmer seawater beneath the ice and thin ice. This creates temperature differences with the thicker ice floes, effectively illustrating detailed characteristics of coastal sea ice distribution.

单波段热红外图像（波段 2）　成像时间：2024-02-19
Single band thermal infrared image (band 2)　Imaging time: 2024-02-19

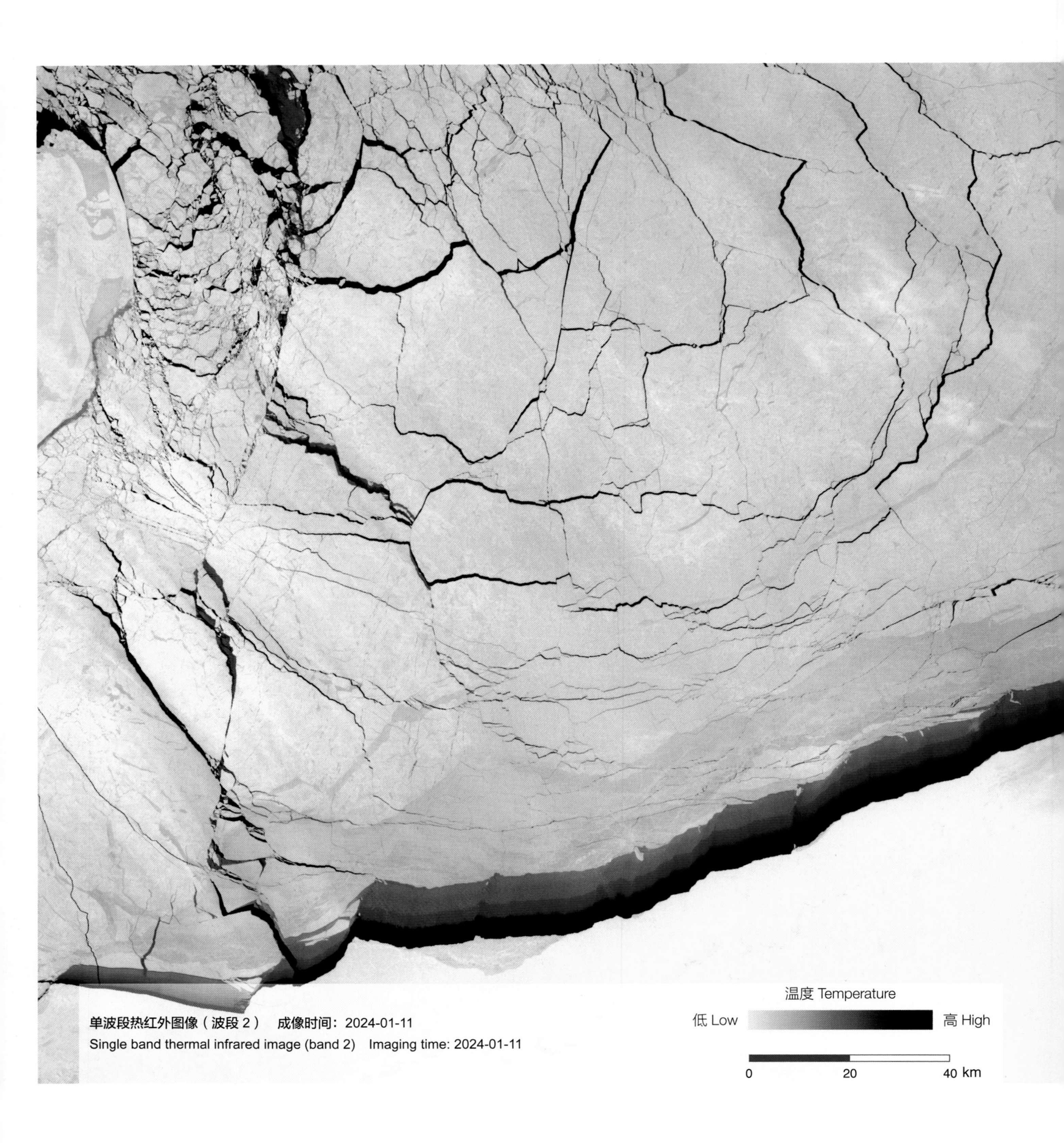

单波段热红外图像（波段 2） 成像时间：2024-01-11

Single band thermal infrared image (band 2) Imaging time: 2024-01-11

东西伯利亚海（新西伯利亚群岛沿岸）

East Siberian Sea off the coast of the New Siberian Islands

本图为 2024 年 1 月 11 日在新西伯利亚群岛沿岸获取的 SDGSAT-1 卫星热红外数据单波段（10.3 ~ 11.3μm）伪彩色图，以浅蓝到深蓝的冷暖渐变表示海冰与海水的温度变化。此时正处于冻结期，图中大片的白色区域为几乎全部冰封的群岛，海冰沿岸边产生断裂，冰面上分布着细小、网状的裂隙，并向海域内延伸，有效展现出冻结期海冰分布的细节特征。

This image, captured on January 11, 2024, along the coast of the New Siberian Islands, depicts single-band (10.3~11.3μm) false-color thermal infrared data from the SDGSAT-1. The color gradient from light blue to deep blue represents temperature variations between sea ice and seawater. During the freezing period, the islands are almost entirely ice-bound. Fractures occur along the coastal edges of the islands, and small, net-like sea ice leads are distributed across the ice surface, extending into the East Siberian Sea, effectively illustrating detailed characteristics of sea ice distribution during the freezing period.

东西伯利亚海
（西伯利亚大陆沿岸）
East Siberian Sea off the coast of the Siberian

本图为 2023 年 12 月 27 日在西伯利亚大陆沿岸东西伯利亚海获取的 SDGSAT-1 卫星热红外数据单波段（10.3 ~ 11.3μm）伪彩色图，以浅蓝到深蓝的冷暖渐变表示海冰与海水的温度变化。处于冻结季节的海面覆盖着厚厚的冰层，仅沿岸边断裂形成冰间裂隙并逐渐向海域内延伸，暴露出温度较高的冰下海水以及薄冰，与厚冰产生鲜明的温度差异，从而有效展现出海冰分布的细节特征。

This image, captured on December 27, 2023, in the East Siberian Sea along the coast of the Siberian, depicts single-band (10.3~11.3μm) false-color thermal infrared data from the SDGSAT-1. The color gradient from light blue to deep blue represents temperature variations between sea ice and seawater. During the freezing season, the sea surface is covered by thick ice. Fractures along the coastal edges form sea ice leads, gradually extending into the East Siberian Sea, exposing relatively warmer seawater beneath the ice and thin ice. This creates distinct temperature differences with the thicker ice, effectively illustrating detailed characteristics of sea ice distribution.

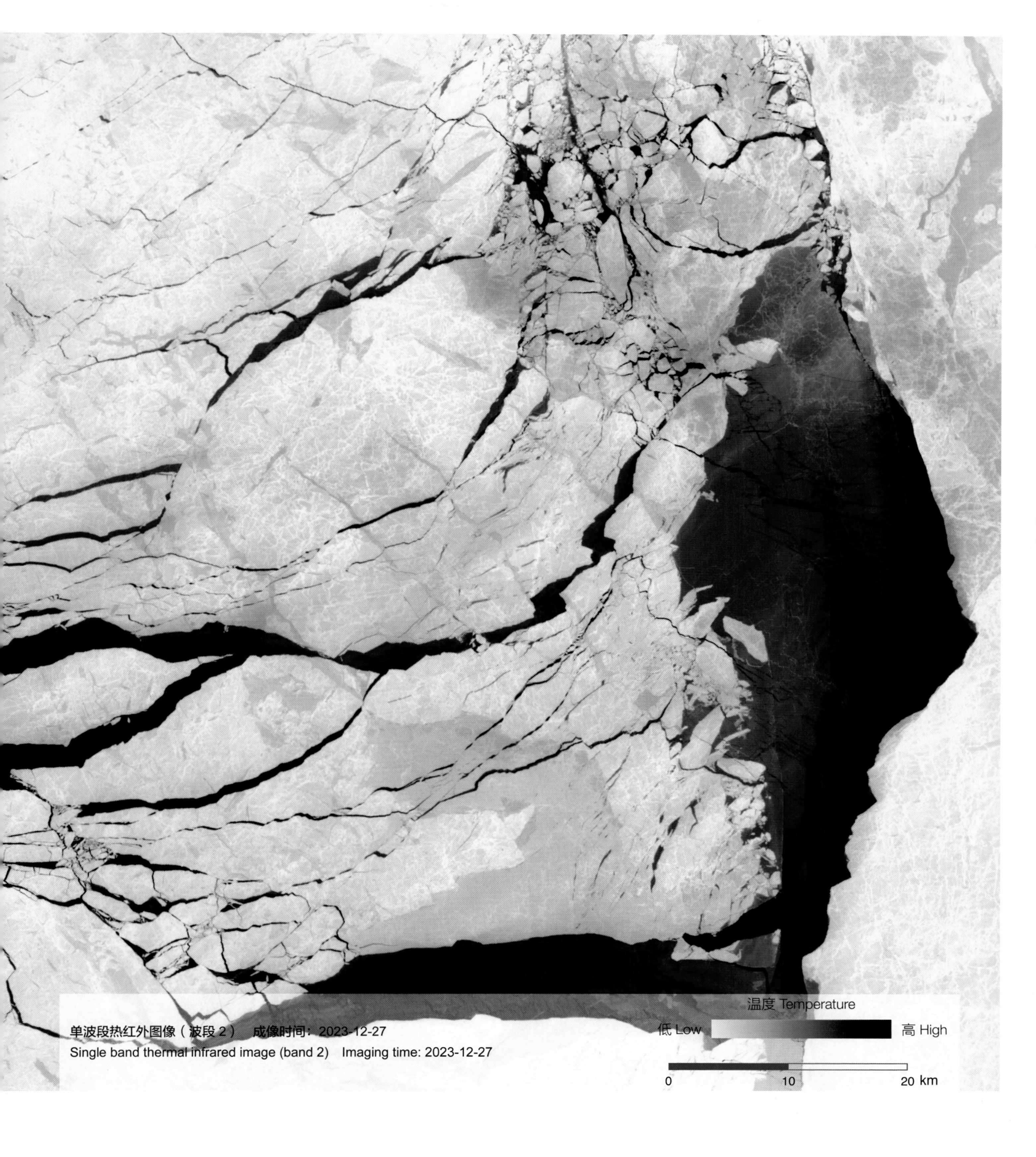

单波段热红外图像（波段 2） 成像时间：2023-12-27

Single band thermal infrared image (band 2) Imaging time: 2023-12-27

东西伯利亚海
（西伯利亚大陆沿岸）
East Siberian Sea off the coast of the Siberian

本图为 2024 年 1 月 24 日在西伯利亚大陆沿岸东西伯利亚海获取的 SDGSAT-1 卫星热红外数据单波段（10.3 ~ 11.3μm）伪彩色图，以浅蓝到深蓝的冷暖渐变表示海冰与海水的温度变化。冻结期海冰持续生长，形成厚厚的冰层，冰面上仅分布着细小的裂隙，其中主要为薄冰，产生温度差异，展现出冻结期海冰分布的细节特征。

This image, captured on January 24, 2024, in the East Siberian Sea along the coast of the Siberian, depicts single-band (10.3~11.3μm) false-color thermal infrared data from the SDGSAT-1. The color gradient from light blue to deep blue represents temperature variations between sea ice and seawater. During the freezing season, sea ice continues to grow and thicken. Only narrow sea ice leads, primarily consisting of thin ice, are observed on the ice surface, creating temperature differences and illustrating detailed characteristics of sea ice distribution during the freezing period.

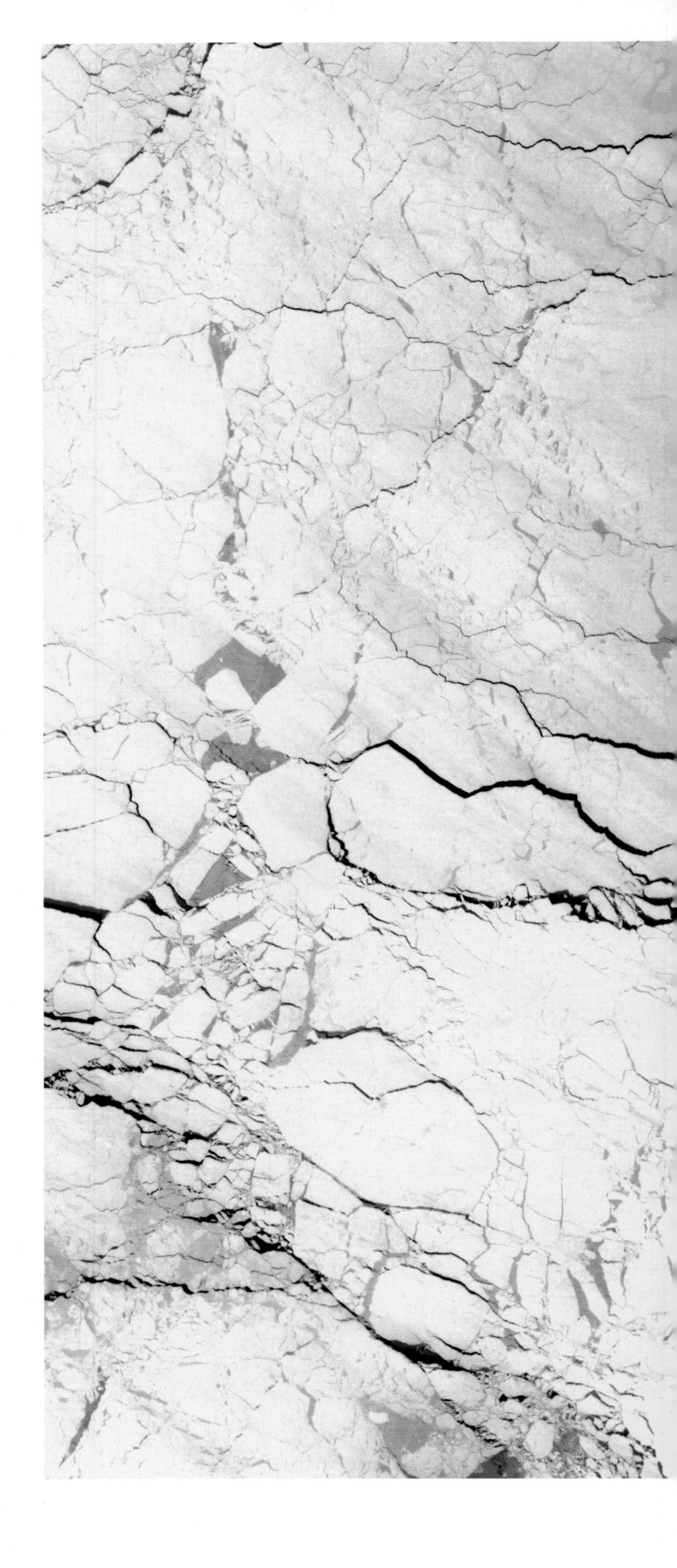

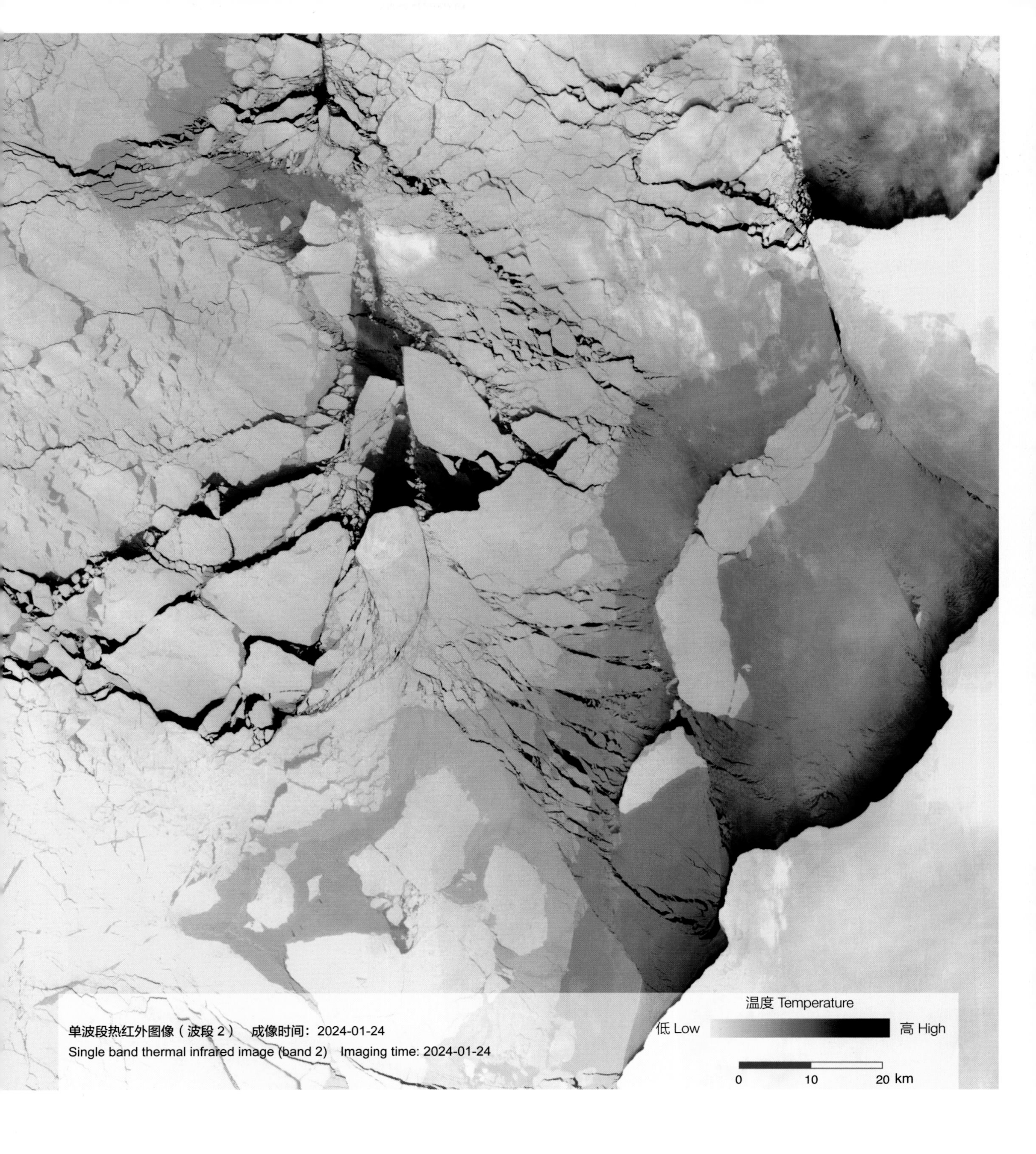
单波段热红外图像（波段 2）　成像时间：2024-01-24
Single band thermal infrared image (band 2)　Imaging time: 2024-01-24
温度 Temperature
低 Low
高 High
0
10
20 km

波弗特海
Beaufort Sea

波弗特海是北冰洋的边缘海，位于美国阿拉斯加州北部海岸和加拿大西北部海岸之间，向北延伸至北极群岛以西的班克斯岛，向东延伸至楚科奇湾。该海域气候严寒，海面几乎全年冰封，仅在 8 ～ 9 月沿岸出现狭窄的无冰海面，可供通航。

此图为 2022 年 7 月 7 日在北极波弗特海获取的单波段（10.3 ～ 11.3μm）伪彩色影像，以浅蓝到深蓝的冷暖渐变表示海冰与海水的温度变化。此时正处于融化季节，随着温度升高，海冰消融，在海面上形成不同尺度的浮冰群，间有松散浮冰和裂隙（冰间水道）分布，暴露出冰下温度较高的海水，与浮冰产生鲜明的温度差异，有效展现出浮冰分布的细节特征。

The Beaufort Sea is a marginal sea of the Arctic Ocean, situated between the northern coast of Alaska, USA, and the northwestern coast of Canada, extending northward to Banks Island, west of the Arctic Archipelago, and eastward to the Chukchi Sea. The climate in this area is extremely cold, with the sea surface being nearly ice-covered throughout the year. Only narrow ice-free areas appear along the coast during August to September, allowing for navigation.

This image, captured on July 7, 2022, within the Beaufort Sea, depicts false-color single-band (10.3~11.3μm) thermal infrared data. The color gradient from light blue to deep blue represents temperature fluctuations between sea ice and seawater. During the melting season, the elevated temperatures lead to ice melt and the formation of varying sizes of ice floes across the sea surface. The pack ice and fractures (sea ice leads) reveal warmer seawater beneath the ice, creating pronounced temperature differentials between the ice and water, effectively highlighting detailed characteristics of sea ice distribution.

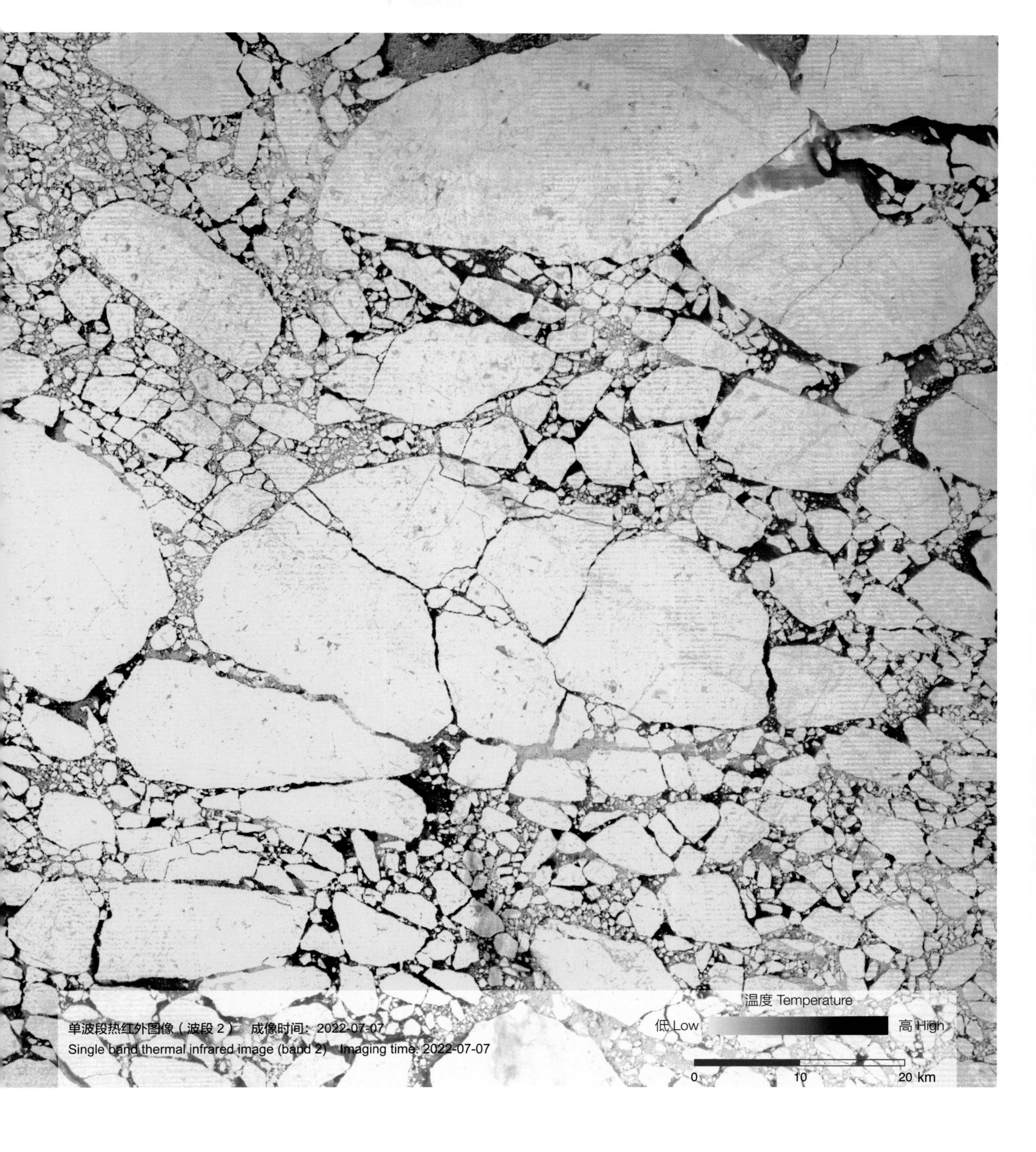

单波段热红外图像（波段 2） 成像时间：2022-07-07
Single band thermal infrared image (band 2) Imaging time: 2022-07-07

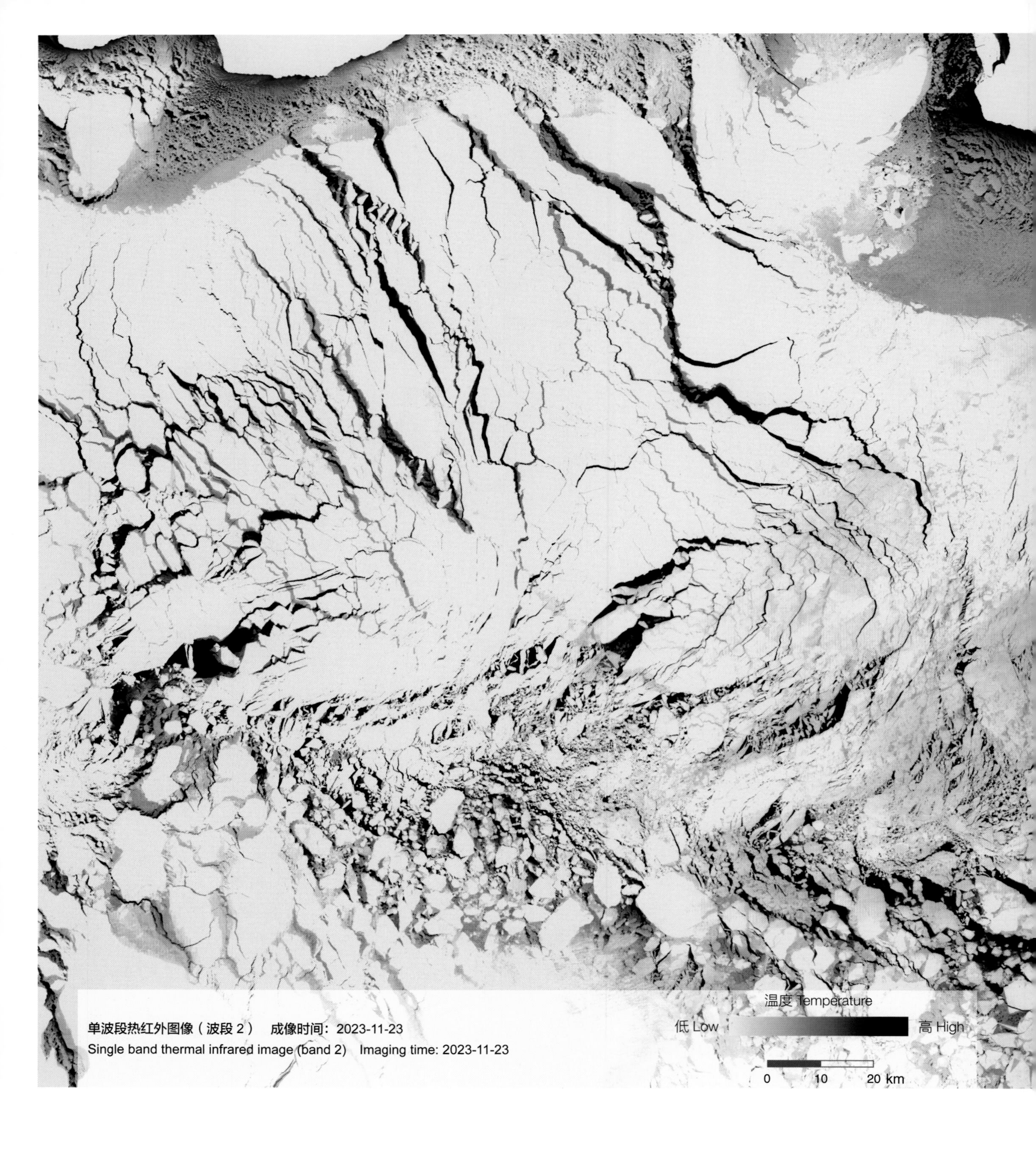

单波段热红外图像（波段 2） 成像时间：2023-11-23
Single band thermal infrared image (band 2) Imaging time: 2023-11-23

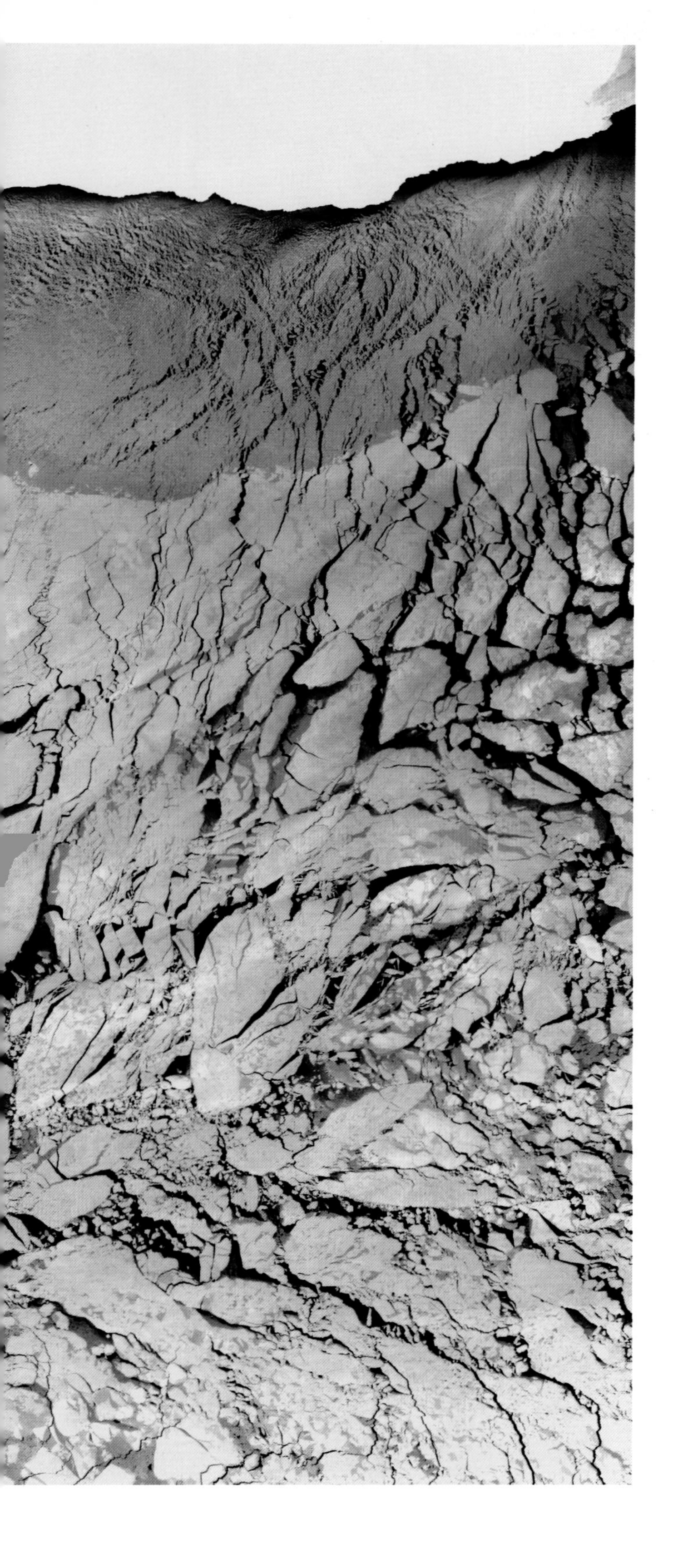

波弗特海
Beaufort Sea

本图为2023年11月23日在波弗特海获取的SDGSAT-1卫星热红外数据单波段（10.3 ~ 11.3μm）伪彩色图，以浅蓝到深蓝的冷暖渐变表示海冰与海水的温度变化。此时处于冻结期，图像包含部分阿拉斯加大陆，已经完全冰封。沿岸的浮冰群间分布着细小、蜿蜒的海冰裂隙，暴露出冰下温度较高的海面，与海冰产生鲜明的温度差异，有效展示出沿岸海冰的分布细节。

This image, acquired on November 23, 2023, within the Beaufort Sea, depicts single-band (10.3~11.3μm) false-color thermal infrared data from the SDGSAT-1. The color gradient from light blue to deep blue represents temperature variations between sea ice and seawater. During the freezing period, the image encompasses parts of the Alaskan mainland, which are completely ice-covered. Along the coast, scattered ice floes exhibit intricate and sinuous sea ice leads, exposing relatively warmer seawater beneath the ice, creating distinct temperature disparities between the sea ice and the seawater. This effectively showcases detailed distributions of coastal sea ice.

波弗特海（巴罗角沿岸）
The Beaufort Sea off the coast of Barrow Point

巴罗角是美国阿拉斯加州北冰洋海岸上的一个岬角，为美国最北点，以西为楚科奇海，以东为波弗特海。作为一个突出的岬角，巴罗角沿岸分布较多的海冰裂隙（冰间水道），展现出复杂的冰情。

此图为 2023 年 11 月 21 日在巴罗角东侧的波弗特海区域内获取的 SDGSAT-1 热红外数据单波段（10.3 ~ 11.3μm）伪彩色影像，以浅蓝到深蓝的冷暖渐变表示海冰与海水的温度变化。影像获取时间正处于海冰的冻结期，小尺度的浮冰在海面聚集、堆积。浮冰群中广泛分布着形态不一的海冰裂隙，暴露出冰下温度较高的海水，与浮冰群产生鲜明的温度差异，展现出细节丰富的海冰分布场景。

Barrow Point is a cape on the coast of the Arctic Ocean in the U.S. state of Alaska. It marks the northernmost point of the United States and is bordered by the Chukchi Sea to the west and the Beaufort Sea to the east. As a prominent cape, Barrow Point features prevailing sea ice leads (cracks in sea ice cover) along its coast, showcasing complex ice conditions.

This image, captured on November 21, 2023, within the Beaufort Sea area to the east of Barrow Point, depicts single-band (10.3~11.3μm) false-color thermal infrared data from the SDGSAT-1. The gradient from light blue to deep blue represents temperature variations between sea ice and seawater. During the freezing period of sea ice, small-scale ice floes congregate and accumulate on the sea surface. Within these ice floes, a wide distribution of differently shaped sea ice leads is evident, exposing relatively warmer seawater beneath the ice surface and creating pronounced temperature discrepancies between the ice floes, effectively illustrating a detailed scene of sea ice distribution.

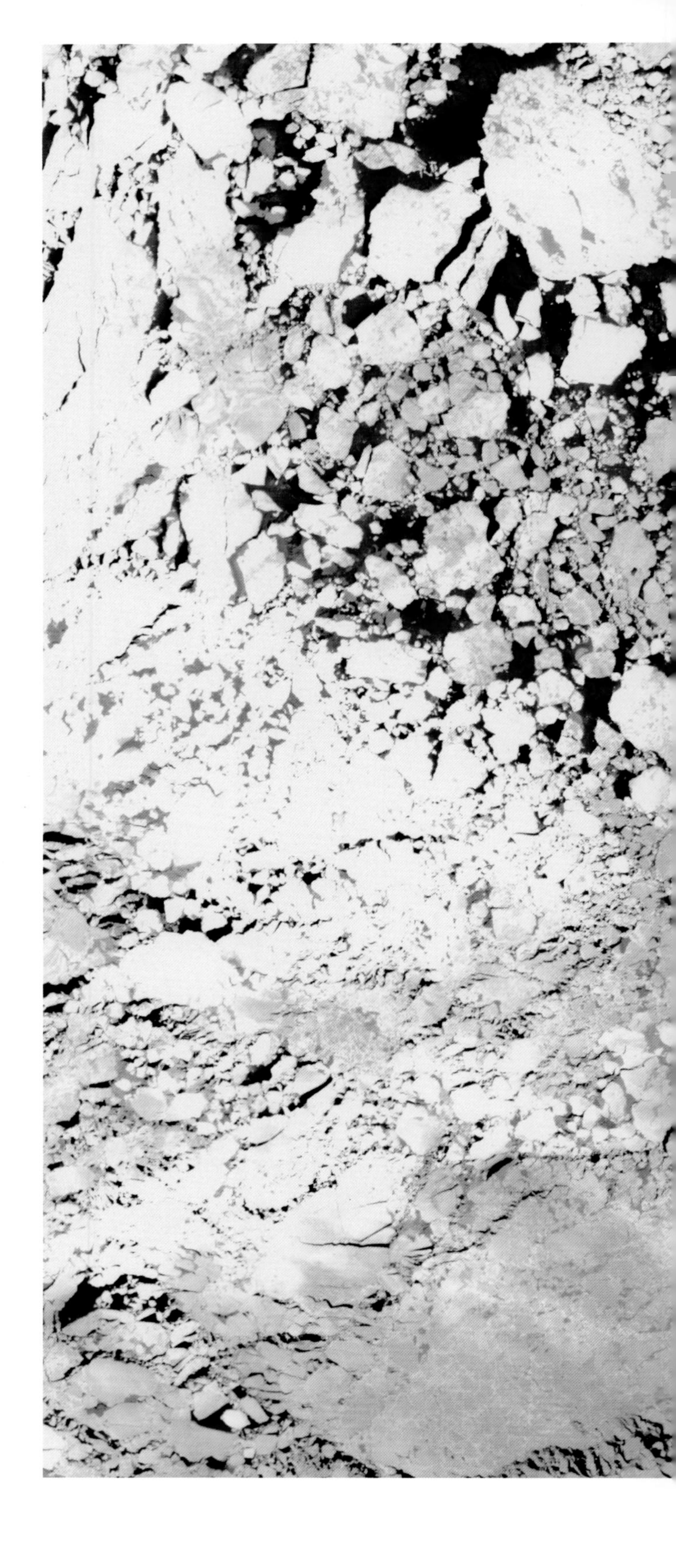

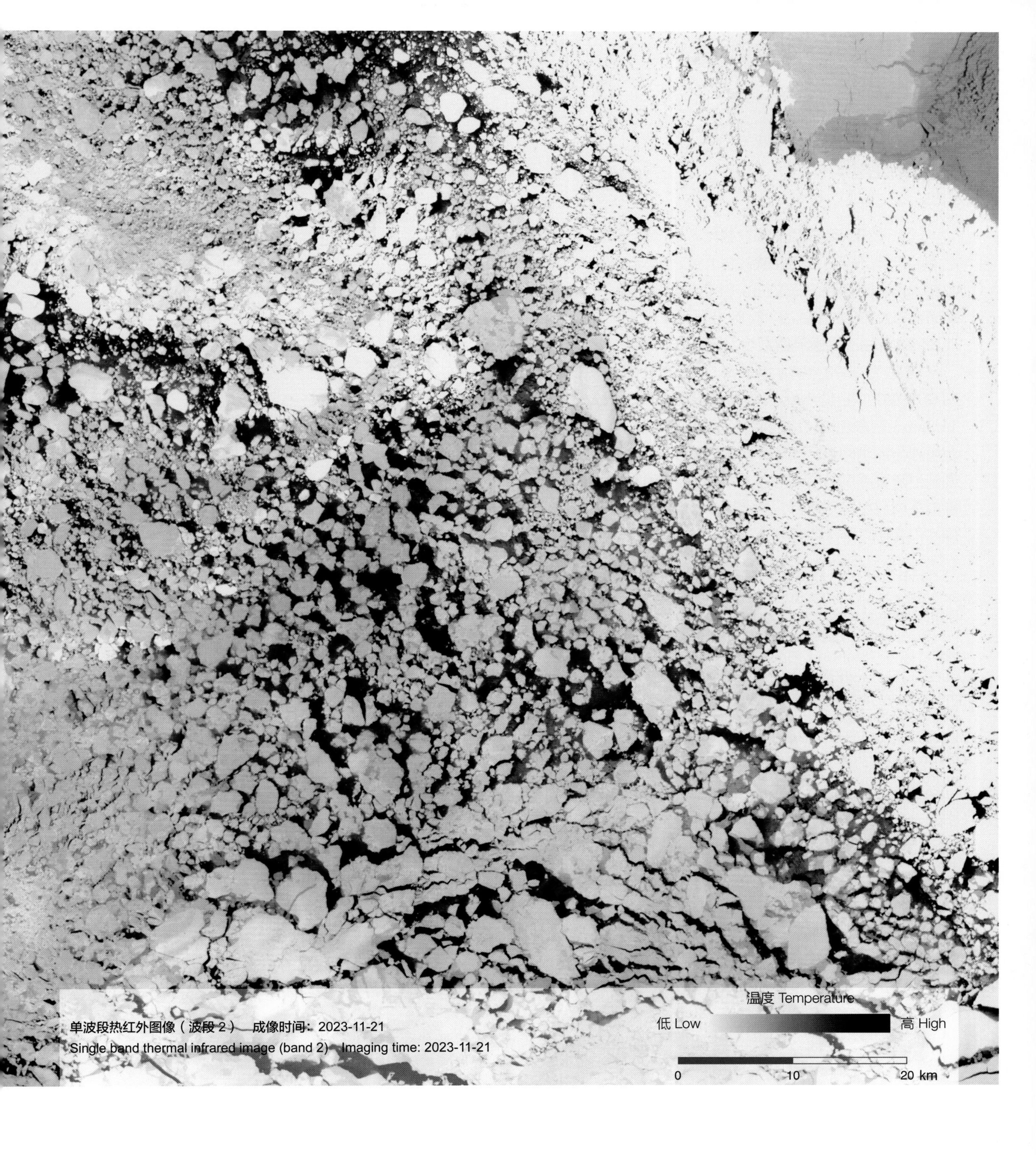

单波段热红外图像（波段 2） 成像时间：2023-11-21
Single band thermal infrared image (band 2) Imaging time: 2023-11-21

巴芬湾
Baffin Bay

巴芬湾是大西洋西北部在格陵兰岛与巴芬岛之间的延伸部分，位于北美洲东北部巴芬岛、埃尔斯米尔岛与格陵兰岛之间。巴芬湾气候严寒，海湾全年大部分时间封冰，仅8、9月可融冰通航；而海湾北部由于受暖流的影响，长年不封冻。

本图为2023年5月27日在巴芬湾获取的SDGSAT-1卫星热红外数据单波段（10.3～11.3μm）伪彩色图，以浅蓝到深蓝的冷暖渐变表示海冰与海水的温度变化。此时正处于海冰融化季节，随着温度升高，海冰逐渐消融，在海面上形成不同尺度的浮冰群，暴露出冰下温度较高的海水和清晰的浮冰边缘。

Baffin Bay is the northwestern extension of the Atlantic Ocean situated between Greenland and Baffin Island, located in the northeastern part of North America between Baffin Island, Ellesmere Island and Greenland. Baffin Bay has an extremely cold climate, with most of the bay covered by ice for the majority of the year, with only a brief period of ice melt allowing navigation in August and September; however, the northern part of the bay remains ice-free year-round due to the influence of warm currents.

This image, captured on May 27, 2023, in Baffin Bay, depicts single-band (10.3~11.3μm) false-color thermal infrared data from the SDGSAT-1. The color gradient from light blue to deep blue represents temperature variations between sea ice and seawater. During the sea ice melting season, as temperatures rise, the sea ice gradually forms floes of various sizes on the sea surface. This exposes relatively warmer seawater beneath the ice and clear edges of the ice floes, effectively illustrating detailed features of the sea ice distribution.

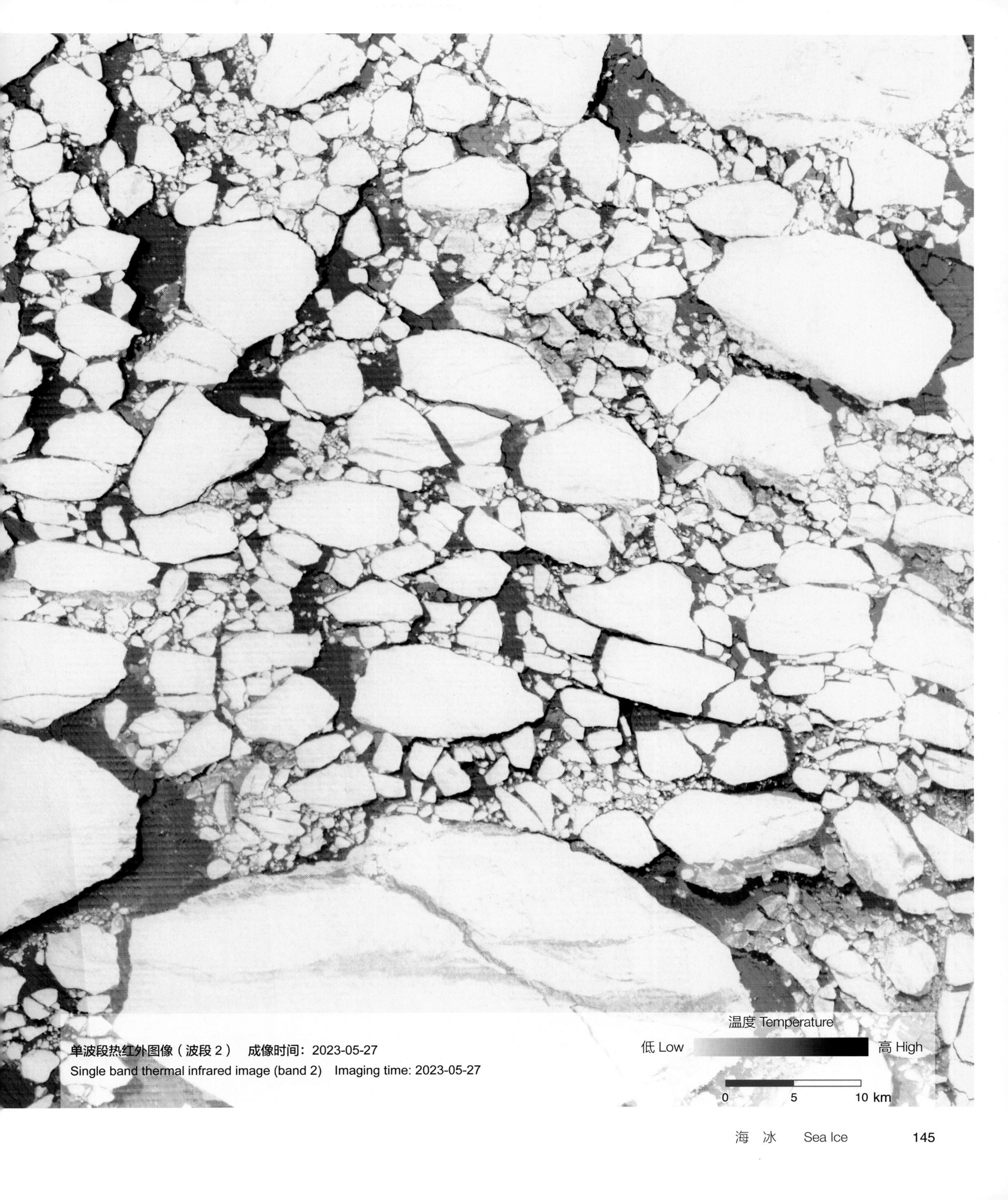

单波段热红外图像（波段 2） 成像时间：2023-05-27
Single band thermal infrared image (band 2) Imaging time: 2023-05-27

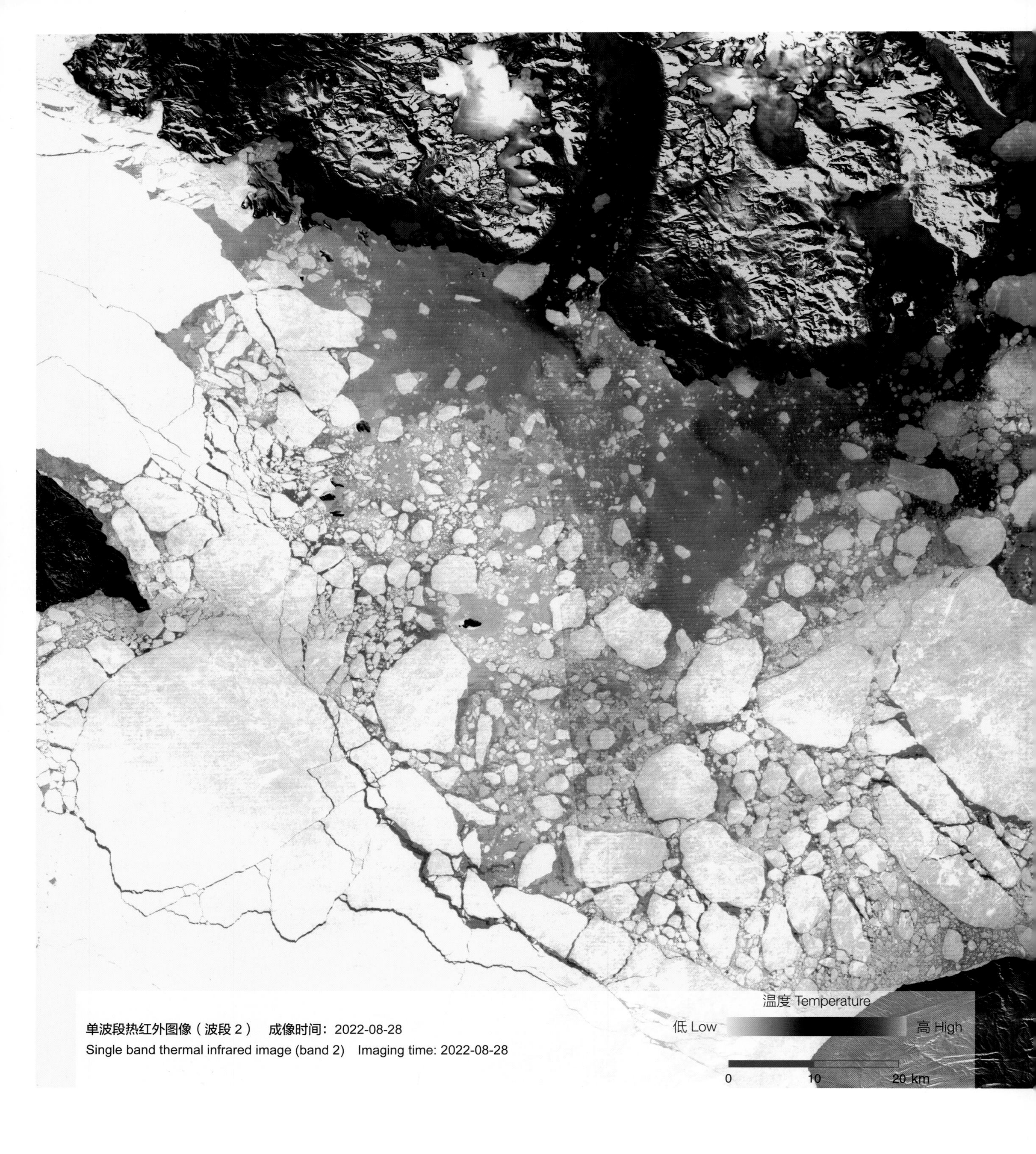

单波段热红外图像（波段 2）　成像时间：2022-08-28

Single band thermal infrared image (band 2)　Imaging time: 2022-08-28

伊丽莎白女王群岛海岸

Coast of the Queen Elizabeth Islands

伊丽莎白女王群岛位于加拿大北极群岛最北端，群岛属极地苔原气候，冬季严寒漫长，夏季温凉短暂，主要植被为苔藓、地衣。北极群岛位于北美洲加拿大最北端。群岛中的巴芬岛、埃尔斯米尔岛和维多利亚岛都是世界上面积最大的岛屿之一。群岛东与格陵兰岛相望，南隔哈得孙湾与加拿大本土相望。

本影像为 2022 年 8 月 28 日在伊丽莎白女王群岛获取的单波段（10.3 ~ 11.3μm）伪彩色影像。图像黄棕色表征温度较高的陆地，随着夏季温度升高，陆上冰雪逐渐融化，暴露出温度较高的陆地。图像以浅蓝到深蓝的冷暖渐变表示海冰与海水的温度变化，浮冰群与冰下海水之间的温度对比鲜明，从而有效展示出融化季海冰的特征。

The Queen Elizabeth Islands are situated at the northernmost tip of the Canadian Arctic Archipelago. The islands feature a polar tundra climate, characterized by long, severe winters and brief, cool summers. The predominant vegetation consists of mosses and lichens. The Arctic Archipelago is located at the northernmost point of North America, within Canada. Within the archipelago, islands like Baffin Island, Ellesmere Island, and Victoria Island rank among the largest islands globally. To the east, the archipelago faces Greenland, while to the south, it is separated from mainland Canada by Hudson Bay.

This image, acquired on August 28, 2022, over the Queen Elizabeth Islands, depicts single-band (10.3~11.3μm) false-color thermal infrared from the SDGSAT-1. The yellow-brown hues flanking the image denote elevated land surface temperatures, indicative of seasonal ice and snow melt during the summer months. The central portion of the image exhibits a gradation from light blue to deep blue, reflecting temperature differentials between sea ice and seawater. This stark thermal contrast between floating ice masses and the underlying seawater effectively illustrates the characteristics of seasonal sea ice melt.

工业热源

Industrial Heat Sources

亚利桑那核电站 / 152
Nuclear power plant in Arizona
密西西比河沿岸化工厂 / 155
Chemical plant by Mississippi River
夸察夸尔科斯河沿岸石化厂 / 157
Petrochemical plant by Coatzacoalcos River

布良斯克州发电站 / 158
Power station in Bryansk Oblast

扎波罗热核电站 / 159
Nuclear power plant in Zaporizhzhia Oblast

京津冀地区工业热源 / 169
Industrial heat sources in Beijing-Tianjin-Hebei region

波斯湾炼油厂 / 161
Oil refinery in Persian Gulf

拉斯拉凡炼油厂 / 162
Oil refinery in Ras Laffan

惠州市工业热源 / 170
Industrial heat sources in Huizong city

防城港区域工业热源 / 172
Industrial heat sources in Fangchenggang region

泰米尔纳德邦发电站 / 175
Power station in Tamil Nadu region

波洛夸内炼油厂 / 165
Oil refinery in Polokwane

约翰内斯堡矿井 / 167
Mine in Johannesburg

澳大利亚西部氧化铝精炼厂 / 176
Alumina refinery in western Australia

亚利桑那州核电站

Nuclear power plant in Arizona

该核电站位于亚利桑那州，是世界上唯一一个不靠近大型水域的核电站，也是美国唯一一个使用市政废水来进行冷却的核电站。该核电站是亚利桑那州南部和南加利福尼亚州人口密集地区的主要电力来源。在秋季夜间热红外影像中，植被覆盖区域表征为冷色调（蓝色或土黄色）。其中，核电站热排放区域表征出极热现象（红色）。

假彩色多光谱图像（波段组合：7-5-4）　成像时间：2023-11-04 日间

Pseudo color multispectral image (band combination: 7-5-4)　Imaging time: 2023-11-04 Daytime

0　1　2 km

Located in Arizona, this generating station is the only nuclear power plant in the world that is not located near a large body of water and the only one in the United States that uses municipal wastewater for cooling. It is the primary source of electricity for densely populated areas of southern Arizona and Southern California. In the fall nighttime thermal infrared imagery, areas of vegetation cover are characterized as cool (blue or earthy yellow). In particular, areas of thermal discharge from the nuclear power plant are characterized as extremely hot (red).

单波段热红外图像（波段 2）　成像时间：2023-10-27 夜间
Single band thermal infrared image (band 2)　Imaging time: 2023-10-27 Nighttime

假彩色多光谱图像（波段组合：7-5-4）　成像时间：2023-11-04 日间
Pseudo color multispectral image (band combination: 7-5-4)　Imaging time: 2023-11-04 Daytime

密西西比河沿岸化工厂

Chemical plant by Mississippi River

化工厂毗邻路易斯安那州的密西西比河，两岸工厂林立，形成了一系列的“钢铁走廊”和“化工走廊”等工业走廊，美国 80% 以上的钢铁厂、90% 以上的冶炼厂都在河流两岸，使密西西比河成为美国工业的“生命通道”。在冬季夜间热红外影像中，植被覆盖区域表征为冷色调（蓝色或土黄色），水体表征为暖色调（橙色）。其中，核电站热排放区域表征出极热现象（红色），核电站水循环路径刻画清晰（红色）。

Chemical plants adjacent to the Mississippi River in Louisiana, factories on both sides of the river, forming a series of “steel corridor” and “chemical corridor” and other industrial corridors, more than 80% of the steel mills, and more than 90% of the smelters in the United States are on both sides of the river, so that the Mississippi River has become the “lifeblood” of American industry. The Mississippi River has become the “lifeline” of American industry. In the winter nighttime thermal infrared image, the vegetation cover area is characterized by cool tones (blue or earthy yellow), and the water body is characterized by warm tones (orange). In particular, areas of thermal discharge from nuclear power plants are characterized by extreme heat (red), and the pathways of the nuclear power plant water cycle are clearly depicted (red).

单波段热红外图像（波段 2） 成像时间：2023-12-31 夜间

Single band thermal infrared image (band 2) Imaging time: 2023-12-31 Nighttime

假彩色多光谱图像（波段组合：7-5-4） 成像时间：2023-03-26 日间
Pseudo color multispectral image (band combination: 7-5-4) Imaging time: 2023-03-26 Daytime

夸察夸尔科斯河沿岸石化厂

Petrochemical plant by Coatzacoalcos River

该石化厂毗邻墨西哥的夸察夸尔科斯河，沿河岸建起的工业走廊中有大约 65 家石化厂。来自工业生产过程和广大居民区的废水，已经大大改变了这一重要江河流域的环境面貌。在冬季夜间热红外影像中，植被覆盖区域表征为冷色调（蓝色），水体表征为暖色调（黄绿色）。其中，石化厂热排放区域表征出极热现象（红色）。

These petrochemical plants are located adjacent to Mexico's Coatzacoalcos River, and there are about 65 petrochemical plants in an industrial corridor along the riverbank. Wastewater from the industrial processes and from the extensive residential areas has significantly altered the environmental profile of this important river basin. In the winter nighttime thermal infrared imagery, the vegetated areas are characterized by cool tones (blue) and the water bodies are characterized by warm tones (yellow-green). In particular, the area of thermal discharge from the petrochemical plant is characterized by extreme heat (red).

单波段热红外图像（波段 2）　成像时间：2024-02-28 夜间

Single band thermal infrared image (band 2)　Imaging time: 2024-02-28 Nighttime

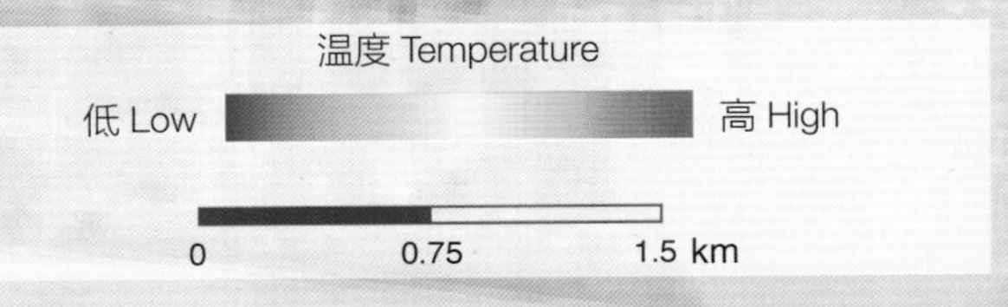

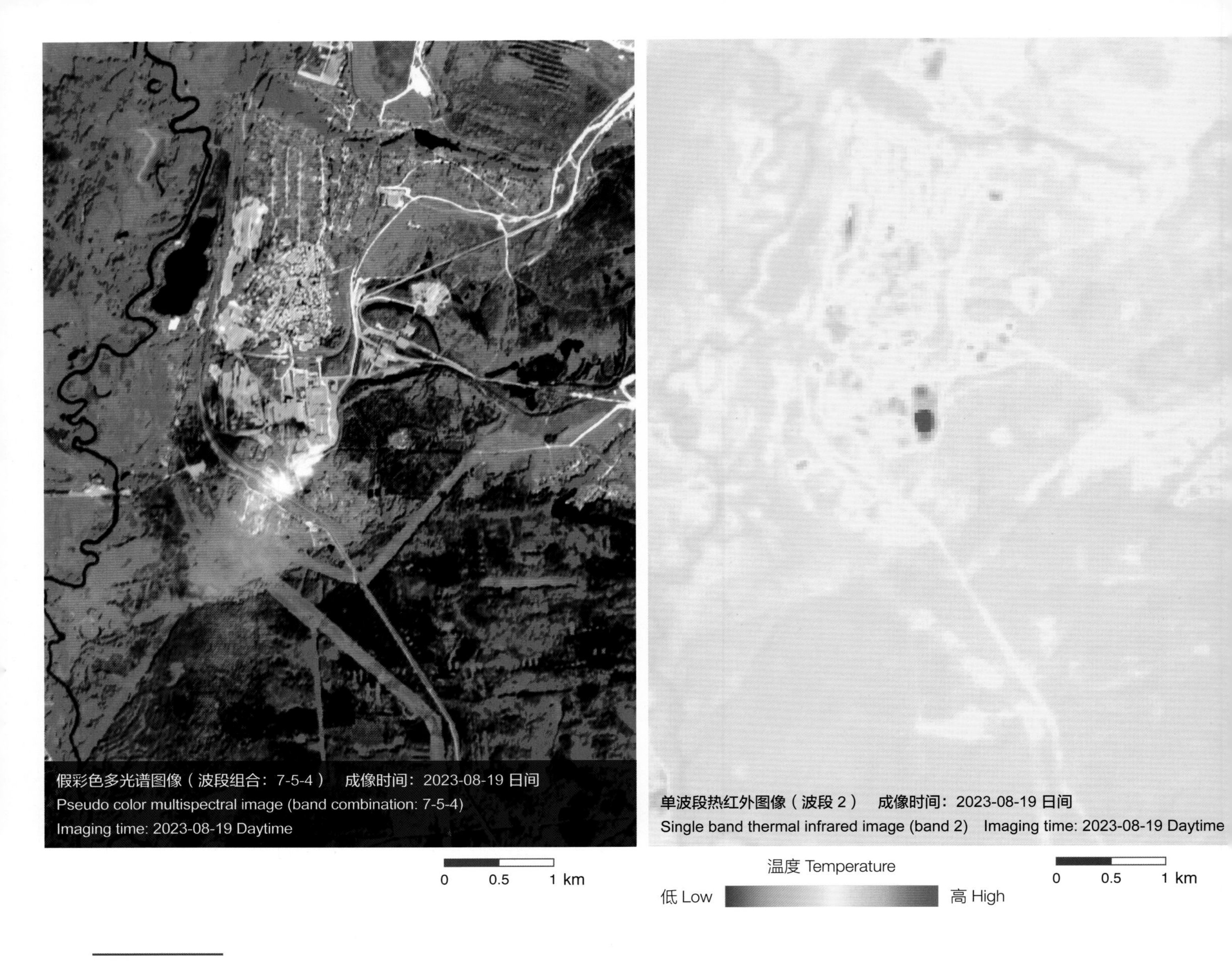

假彩色多光谱图像（波段组合：7-5-4） 成像时间：2023-08-19 日间
Pseudo color multispectral image (band combination: 7-5-4)
Imaging time: 2023-08-19 Daytime

单波段热红外图像（波段 2） 成像时间：2023-08-19 日间
Single band thermal infrared image (band 2) Imaging time: 2023-08-19 Daytime

布良斯克州发电站
Power station in Bryansk Oblast

该发电厂位于布良斯克州，该州是俄罗斯联邦主体之一，属中央联邦管区。与乌克兰和白俄罗斯接壤。在夏季夜间热红外影像中，植被覆盖区域表征为冷色调（浅绿色），建筑用地表征为暖色调（浅橙色）。其中，发电厂热排放区域表征出极热现象（红色），热排放区域周围表征为暖色调（浅红色）。

The power plant is located in the Bryansk Oblast, which is one of the constituent entities of the Russian Federation and is part of the Central Federal District. It shares borders with Ukraine and Belarus. In the summer nighttime thermal infrared imagery, vegetation-covered areas are characterized by cool tones (light green), and built-up land is characterized by warm tones (light orange). In particular, the area of thermal emissions from the power plant is characterized by extreme heat (red), and the area around the thermal emissions is characterized by warm tones (light red).

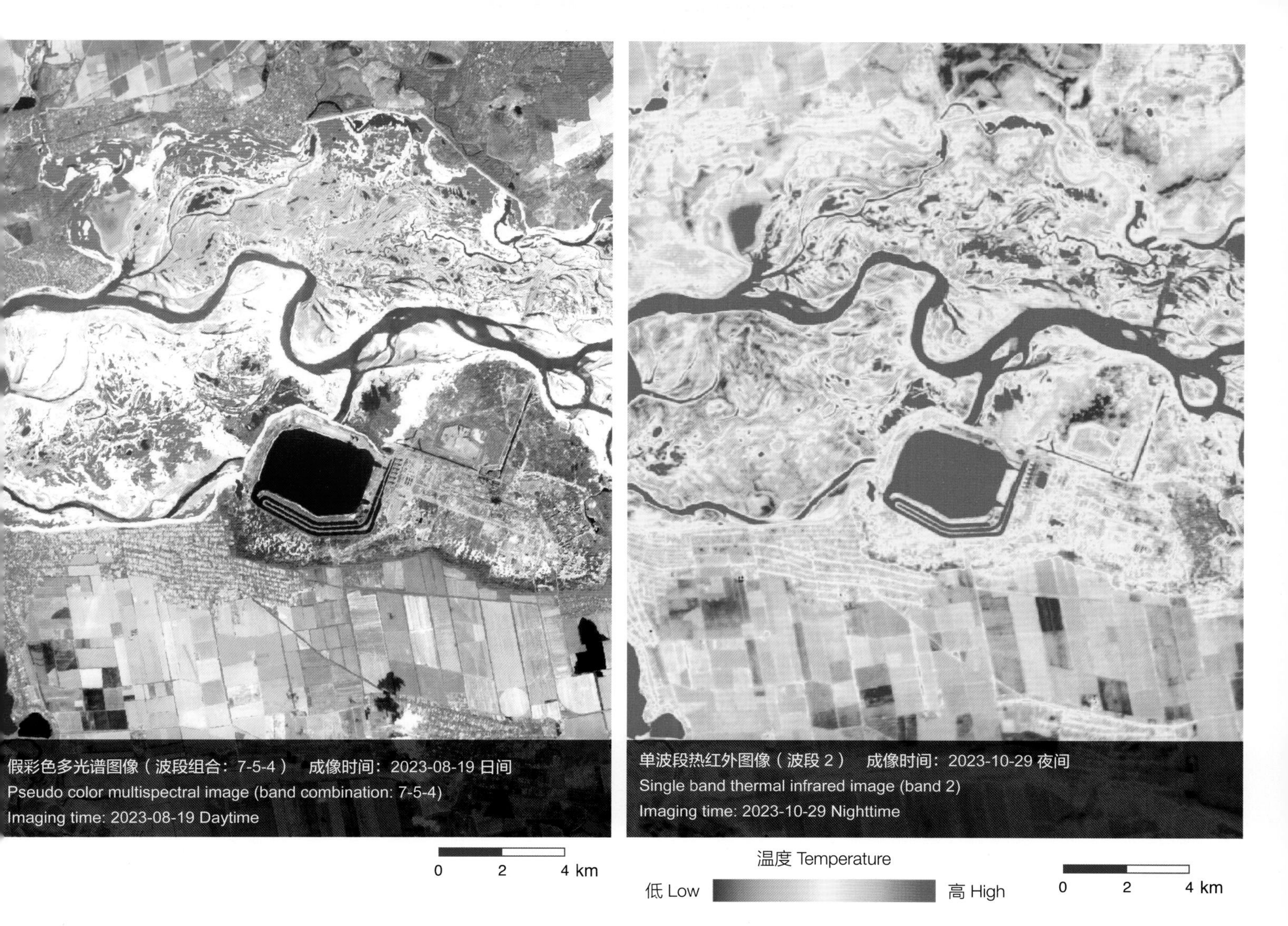

扎波罗热州核电站

Nuclear power plant in Zaporizhzhia Oblast

该核电站位于乌克兰第聂伯河河畔，是乌克兰最大的核电站。在秋季夜间热红外影像中，植被覆盖区域表征为冷色调（蓝色或土黄色），水体表征为暖色调（红色），核电站水循环路径刻画清晰（红色）。

This nuclear power plant is located on the banks of the Dnieper River in Ukraine and is the largest nuclear power plant in the country. In the autumn nighttime thermal infrared imagery, vegetation-covered areas are represented in cool tones (blue or earthy yellow), water bodies are represented in warm tones (red), and the paths of the nuclear power plant's water cycle are clearly delineated (red).

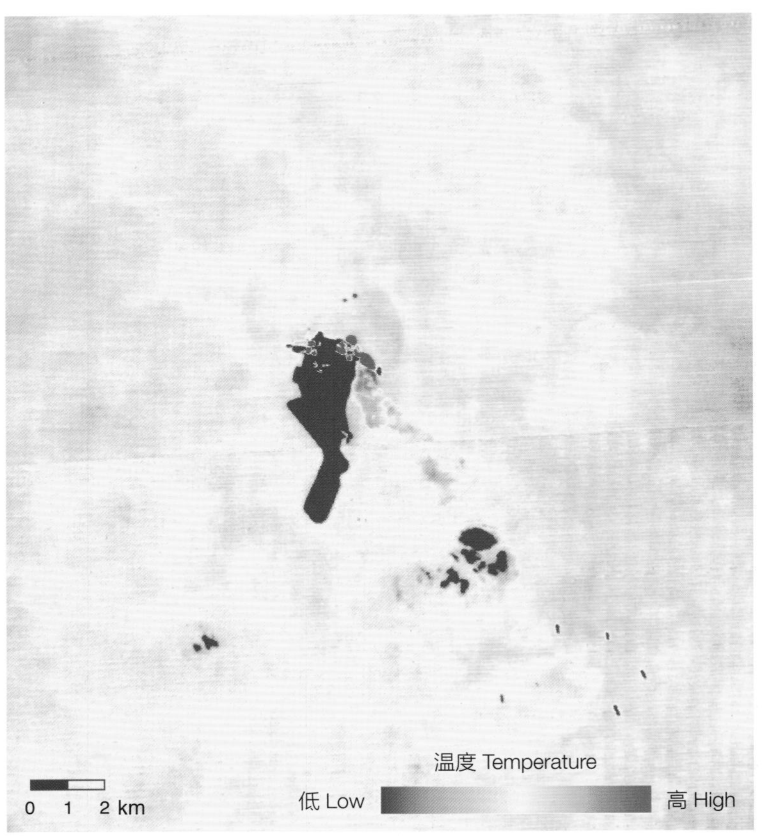

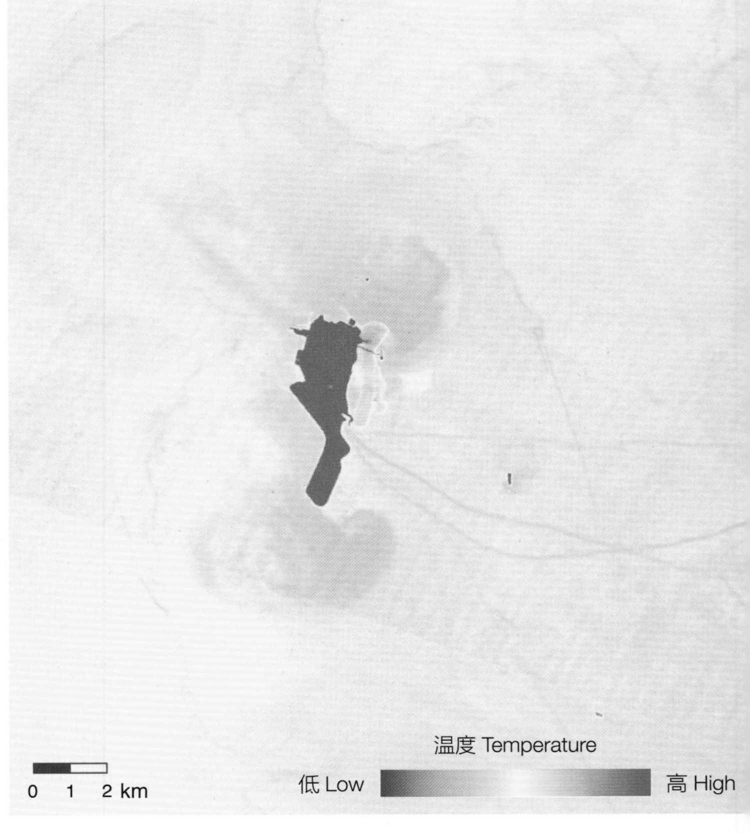

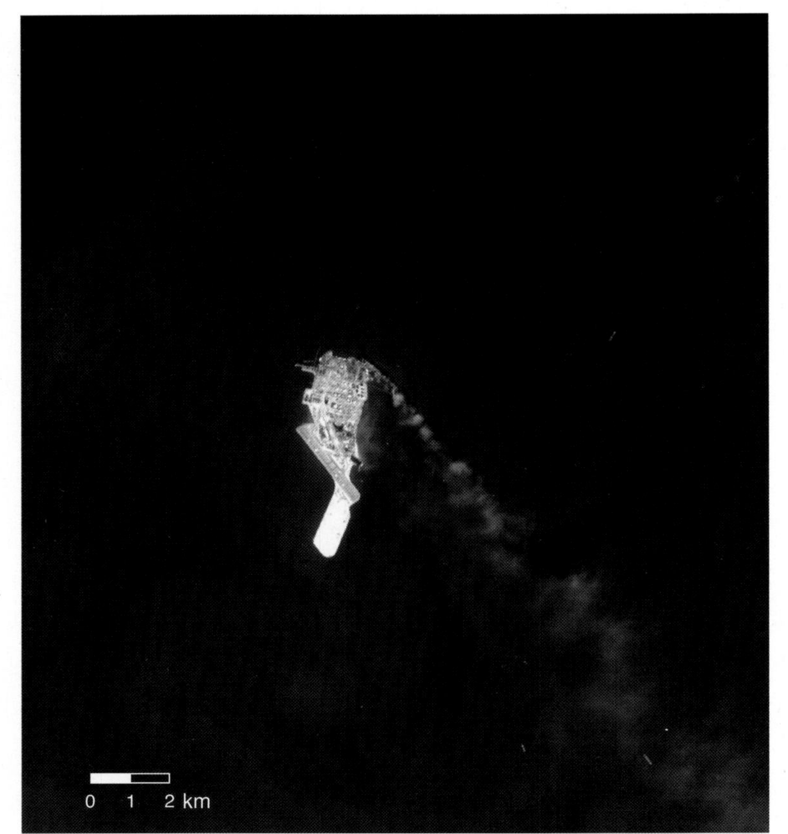

左上：单波段热红外图像（波段 2） 成像时间：2024-02-29 夜间

右上：单波段热红外图像（波段 2） 成像时间：2024-03-14 日间

左下：假彩色多光谱图像（波段组合：1-2-3） 成像时间：2024-02-27 日间

Upper left: Single band thermal infrared image (band 2)
Imaging time: 2024-02-29 Nighttime

Upper right: Single band thermal infrared image (band 2)
Imaging time: 2024-03-14 Daytime

Lower left: Pseudo color multispectral image (band combination: 1-2-3)
Imaging time: 2024-02-27 Daytime

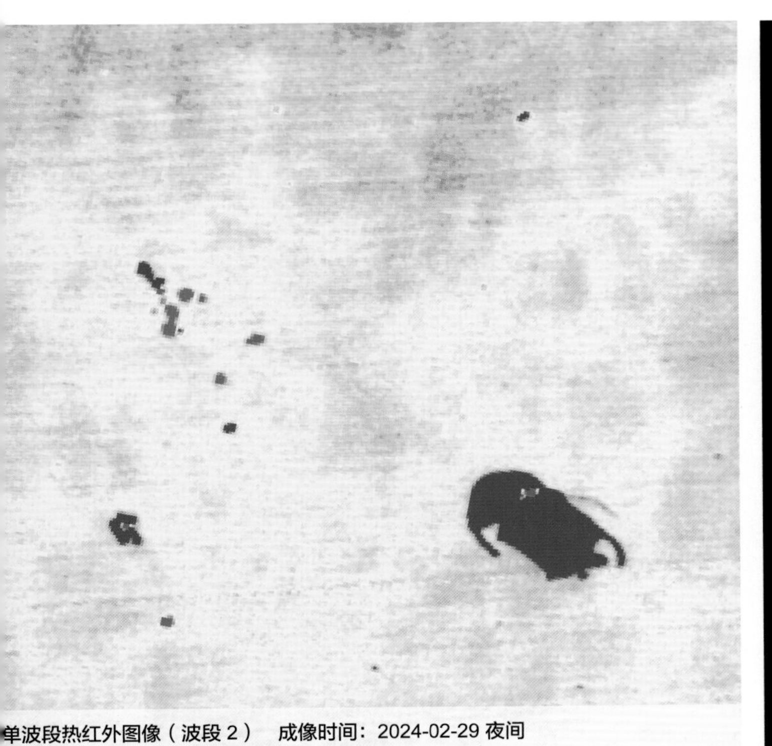

单波段热红外图像（波段 2） 成像时间：2024-02-29 夜间
Single band thermal infrared image (band 2) Imaging time: 2024-02-29 Nighttime

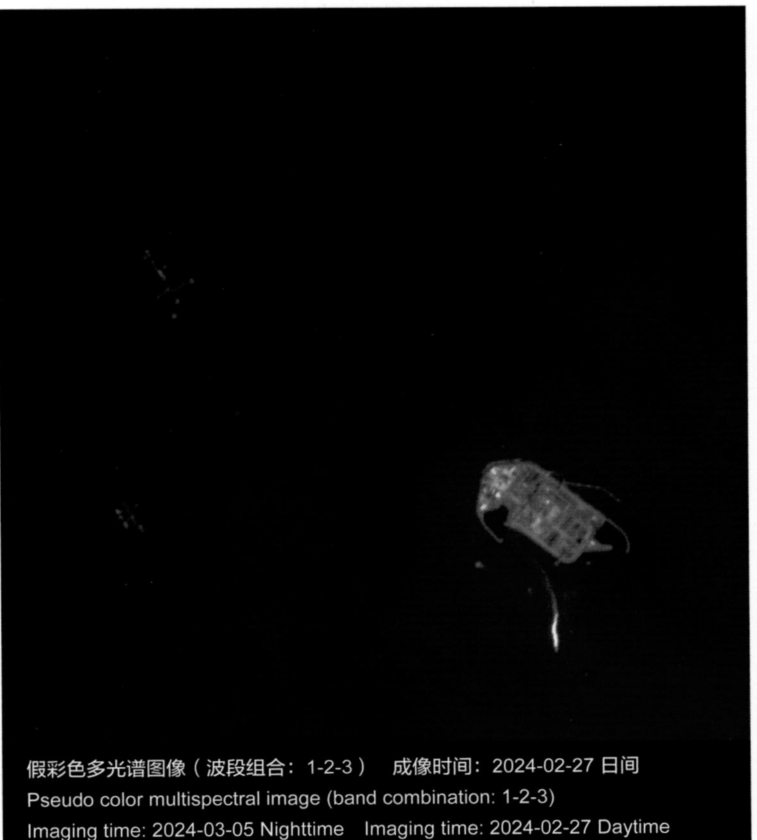

假彩色多光谱图像（波段组合：1-2-3） 成像时间：2024-02-27 日间
Pseudo color multispectral image (band combination: 1-2-3)
Imaging time: 2024-03-05 Nighttime Imaging time: 2024-02-27 Daytime

波斯湾炼油厂
Oil refinery in Persian Gulf

波斯湾海域的工业以石油和天然气开采、加工和出口为主。波斯湾地区是全球最大的天然气与石油产地之一。达斯岛是波斯湾南部小岛，属阿拉伯联合酋长国，有两个海底油田工作站和油港。在春季日、夜间热辐射图像中，白天，水体覆盖区域表征为暖色调（土黄色），建筑区域表征为红色；夜间，水体覆盖区域趋于冷色调（蓝色），建筑暖色调区域面积减少且刻画更精细（红色），能够清晰看出生产的热辐射聚集区域（红色），生产造成海水增温现象也较明显（红色渐变为土黄色）。

Industry in the Persian Gulf waters is dominated by oil and gas extraction, processing and export. The Persian Gulf region is one of the largest oil and gas producer in the world. Das Island is a small island in the southern part of the Persian Gulf, belonging to the United Arab Emirates, with two submarine oil field workstations and oil ports. Distribution of daytime and nighttime thermal radiation images in spring, During the day, the water-covered areas are characterized by warm tones (earthy yellow) and the building areas are characterized by red; at night, the water-covered areas tend to be cooler (blue), the warmer areas of the buildings are reduced in size and more finely delineated (red), and the areas of concentration of the thermal radiation from the production (red) and the warming of the sea water due to the production is more obvious (red fading to earthy yellow).

拉斯拉凡炼油厂
Oil refinery in Ras Laffan

卡塔尔是世界上最大的液化天然气（LNG）出口国之一，其经济极大程度上依赖于石油和天然气的开采、加工和出口。图中的炼油厂位于卡塔尔东北部，靠近波斯湾的海岸线，是世界上最大的炼油厂之一。在春季夜间热红外影像中，山体等植被覆盖区域表征为冷色调（蓝黄色），水体表征为暖色调（土黄色）。其中，化工企业热排放区域表征出极热现象（鲜红色），生产造成海水增温现象也较明显（红色到土黄色渐变）。

Qatar is one of the world's largest exporters of liquefied natural gas, and its economy is greatly dependent on the extraction, processing, and export of oil and gas. The refinery in the imagery is located in northeastern Qatar, near the coastline of the Persian Gulf, and is one of the largest refineries in the world. In the spring nighttime thermal infrared imagery, vegetation-covered areas, such as mountains, are characterized by cool tones (blue-yellow) and water bodies are characterized by warm tones (earthy yellow). In particular, the area of thermal discharge from the chemical company is characterized by extreme heat (bright red), and the warming of seawater due to production is also evident (red to earthy yellow gradient).

假彩色多光谱图像（波段组合：1-2-3）　成像时间：2024-01-08 日间

Pseudo color multispectral image (band combination: 1-2-3)　Imaging time: 2024-01-08 Daytime

0　2.5　5 km

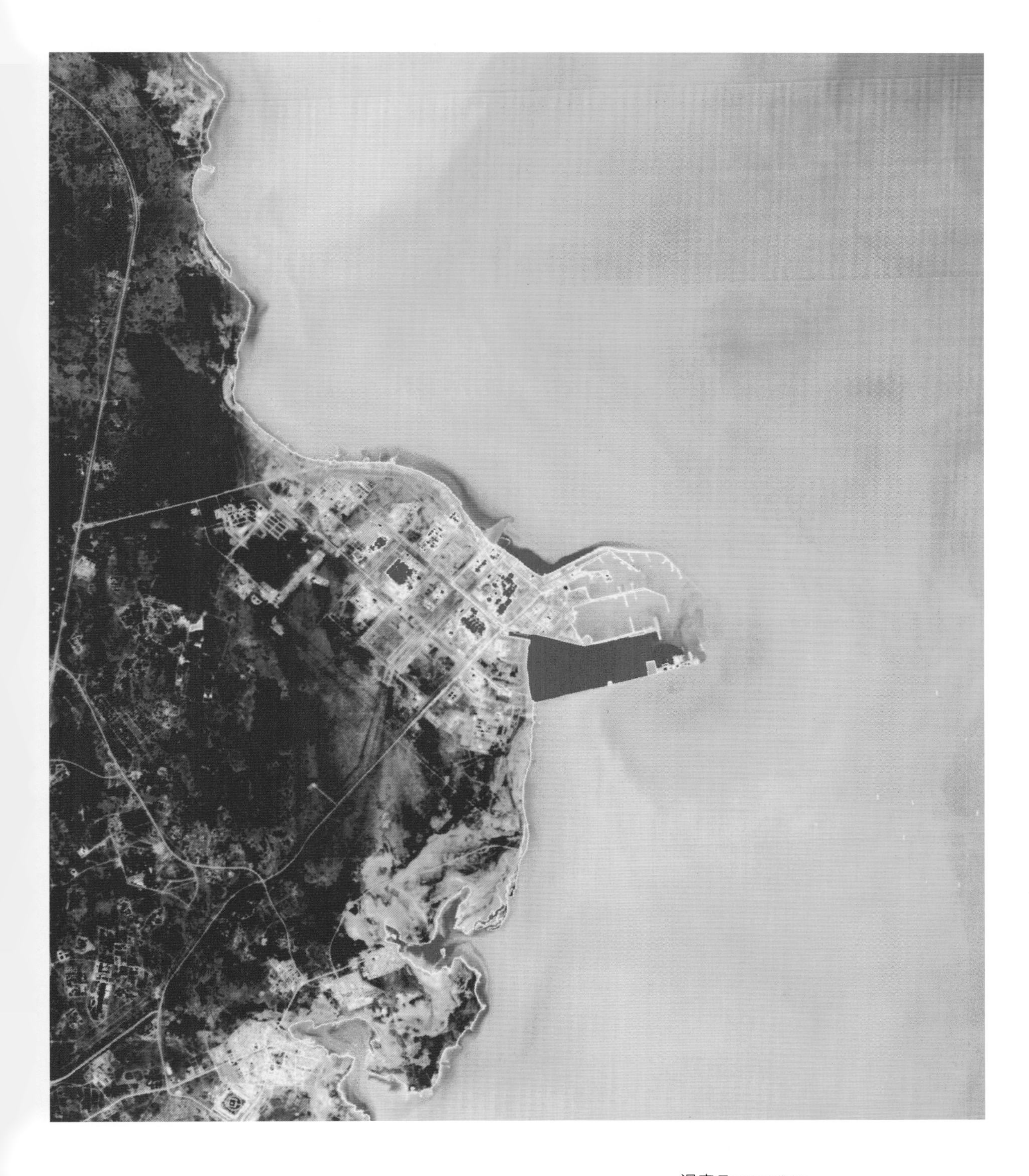

图像（波段 2）　成像时间：2024-03-05 夜间
ermal infrared image (band 2)　Imaging time: 2024-03-05 Nighttime

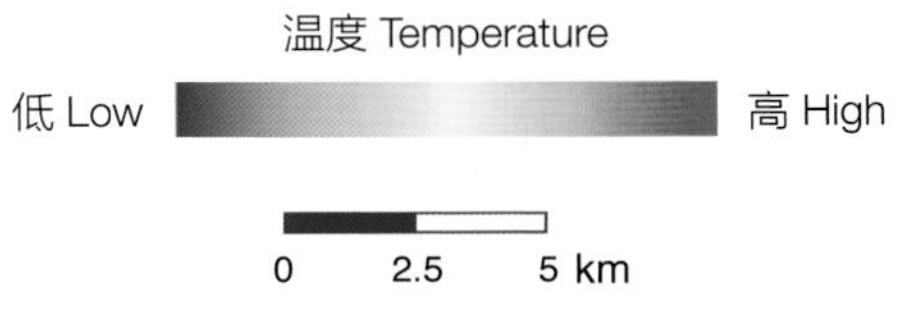

假彩色多光谱图像（波段组合：7-5-4） 成像时间：2023-06-19 日间

TPseudo color multispectral image (band combination: 7-5-4) Imaging time: 2023-06-19 Daytime

波洛夸内炼油厂
Oil refinery in Polokwane

该炼油厂位于南非的波洛夸内自治市，波洛夸内是南非林波波省的首府，也是北部地区的最大城市。在夏季夜间热红外影像中，山脉表征为冷色调（蓝色），植被覆盖区域表征为冷色调（浅绿色），水体表征为暖色调（橙色）。其中，矿井热排放区域表征出极热现象（红色），其他人造地表表征为暖色（橙色）。

The refinery is located in Polokwane of South Africa, which is the capital of the Limpopo Province of South Africa and the largest city in the Northern Region. In the summer nighttime thermal infrared imagery, mountain ranges are characterized in cool tones (blue), areas of vegetation cover are characterized in cool tones (light green), and water bodies are characterized in warm tones (orange). In particular, areas of thermal discharge from mines are characterized as extremely hot (red) and other man-made surfaces are characterized as warm (orange).

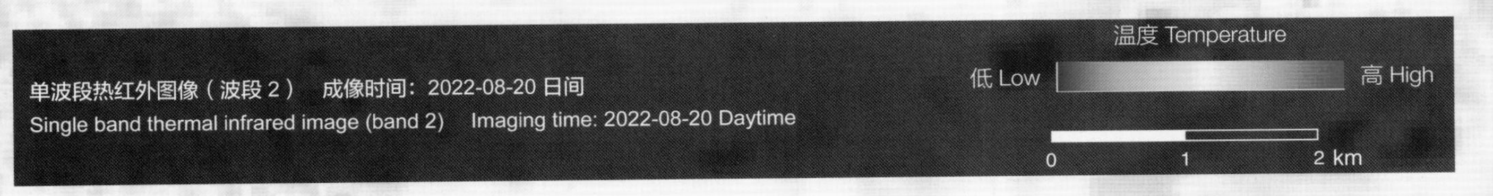

单波段热红外图像（波段 2） 成像时间：2022-08-20 日间
Single band thermal infrared image (band 2) Imaging time: 2022-08-20 Daytime

假彩色多光谱图像（波段组合：7-5-4）　成像时间：2023-06-19 日间
TPseudo color multispectral image (band combination: 7-5-4)　Imaging time: 2023-06-19 Daytime

约翰内斯堡矿井
Mine in Johannesburg

该矿井位于南非的约翰内斯堡。该地矿物丰富，金、铂、锑、金刚石、石棉的产量和铀、锰、铬、萤石的储量均居世界前列，还有煤、铁、铜、铝、锌等。在夏季夜间热红外影像中，植被覆盖区域表征为冷色调（浅绿色），水体表征为暖色调（橙色）。其中，矿井热排放区域表征出极热现象（红色），其他人造地表表征为暖色（橙色）。

The mine is located in Johannesburg, South Africa. The site is rich in minerals, with world-leading production of gold, platinum, antimony, diamonds, asbestos, and reserves of uranium, manganese, chromium and fluorite, as well as coal, iron, copper, aluminum and zinc. In the summer nighttime thermal infrared image, the vegetation-covered area is characterized by a cool tone (light green), and the water body is characterized by a warm tone (orange). In particular, areas of thermal discharge from mines are characterized as extremely hot (red), and other man-made surfaces are characterized as warm (orange).

单波段热红外图像（波段 2） 成像时间：2022-08-20 日间
Single band thermal infrared image (band 2) Imaging time: 2022-08-20 Daytime

温度 Temperature
低 Low 高 High
0 1 km

0
0.7
1.4 km
0
0.5
1 km
0
0.7
1.4 km
0
0.5
1 km

京津冀地区工业热源
Industrial heat sources in Beijing-Tianjin-Hebei region

京津冀地区工业热源以钢铁厂、油气开发为主，集中于天津市渤海湾入海口处。在秋季夜间热红外影像中，山体等植被覆盖区域表征为冷色调（蓝黄色），水体表征为暖色调（土黄色），而工业热源企业呈现出高温现象（红色）。其中，化工企业热排放区域虽呈现暖黄色，但其生产造成海水增温现象也较明显（红色到土黄色渐变）。

Industrial heat sources in Beijing-Tianjin-Hebei region are mainly iron and steel mills and oil and gas refineries, concentrated at the Bohai Bay estuary in Tianjin. In the fall nighttime thermal infrared images, vegetation-covered areas such as mountains are characterized by cool tones (blue-yellow), water bodies are characterized by warm tones (earthy-yellow), and industrial heat source enterprises show high temperature phenomena (red). Among them, the thermal discharge area of chemical enterprises is characterized by warm yellow color, but the warming of seawater caused by their production is also more obvious (red to earthy yellow gradient).

单波段热红外图像（波段 2） 成像时间：2023-10-20 夜间
Single band thermal infrared image (band 2) Imaging time: 2023-10-20 Nighttime

惠州市工业热源
Industrial heat sources in Huizhou city

惠州大湾石化化工企业群，位于惠州市淡澳河入海口处。在春季夜间热红外影像中，山体等植被覆盖区域表征为冷色调（蓝黄色），水体表征为暖色调（土黄色）。其中，化工企业热排放区域表征出极热现象（鲜红色），生产造成海水增温现象也较明显（红色到土黄色渐变）。

A cluster of petrochemical and chemical enterprises are located at the mouth of the Danao River in Huizhou City. In the spring nighttime thermal infrared imagery, vegetation-covered areas such as mountains are characterized by cool tones (blue-yellow), and water bodies are characterized by warm tones (earthy yellow). Among them, the area of thermal discharge from chemical enterprises is characterized by extreme heat (bright red), and the warming of seawater caused by production is also more obvious (red to earthy yellow gradient).

假彩色多光谱图像（波段组合：7-5-4） 成像时间：2022-12-21 日间

Pseudo color multispectral image (band combination: 7-5-4) Imaging time: 2022-12-21 Daytime

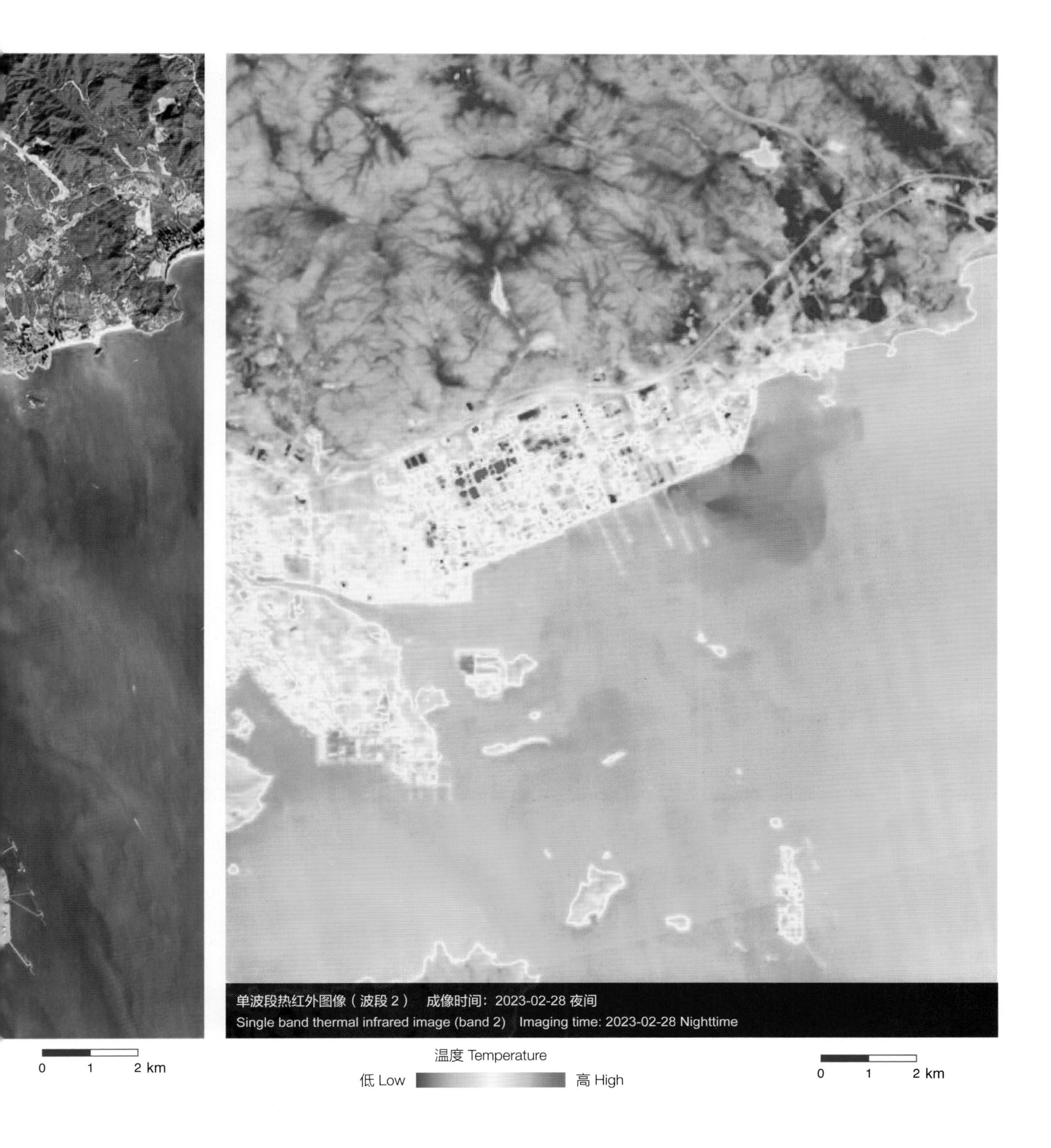

单波段热红外图像（波段 2） 成像时间：2023-02-28 夜间
Single band thermal infrared image (band 2) Imaging time: 2023-02-28 Nighttime

假彩色多光谱图像（波段组合：7-5-4） 成像时间：2022-12-25 日间
Pseudo color multispectral image (band combination: 7-5-4) Imaging time: 2022-12-25 Daytime

单波段热红外图像（波段 2）
成像时间：2022-10-22 夜间
Single band thermal infrared image (band 2)
Imaging time: 2022-10-22 Nighttime

防城港市工业热源
Industrial heat sources in Fangchenggang region

广西北海湾防城港区域工业以炼油厂、石化化工为主。该图像展示了石化化工企业群在秋冬季节日、夜间热辐射分布情况。白天，水体区域表征为冷色调（蓝色或淡绿色），植被、建筑区域表征为暖色调（红色）；夜间，水体趋于暖色调（土黄色），植被覆盖区域趋于冷色调（蓝色），建筑暖色调区域面积减少且刻画更精细（红色），生产造成海水增温现象也较明显（红色渐变为土黄色）。

Guangxi Beihai Bay Fangchenggang area's industry is dominated by oil refineries and petrochemical processing. This image shows the Petrochemical and chemical enterprises cluster, and their distribution of thermal radiation images during the fall and winter festivals and nighttime. During the daytime, the water body area is characterized as cold tone (blue or light green), and the vegetation and building area is characterized as warm tone (red); during the nighttime, the water body tends to be warmer (earthy yellow), the area covered by vegetation tends to be cooler (blue), the area of the building's warm area is reduced and more finely delineated (red), and the warming of seawater caused by the production is more obvious (red fades to earthy yellow).

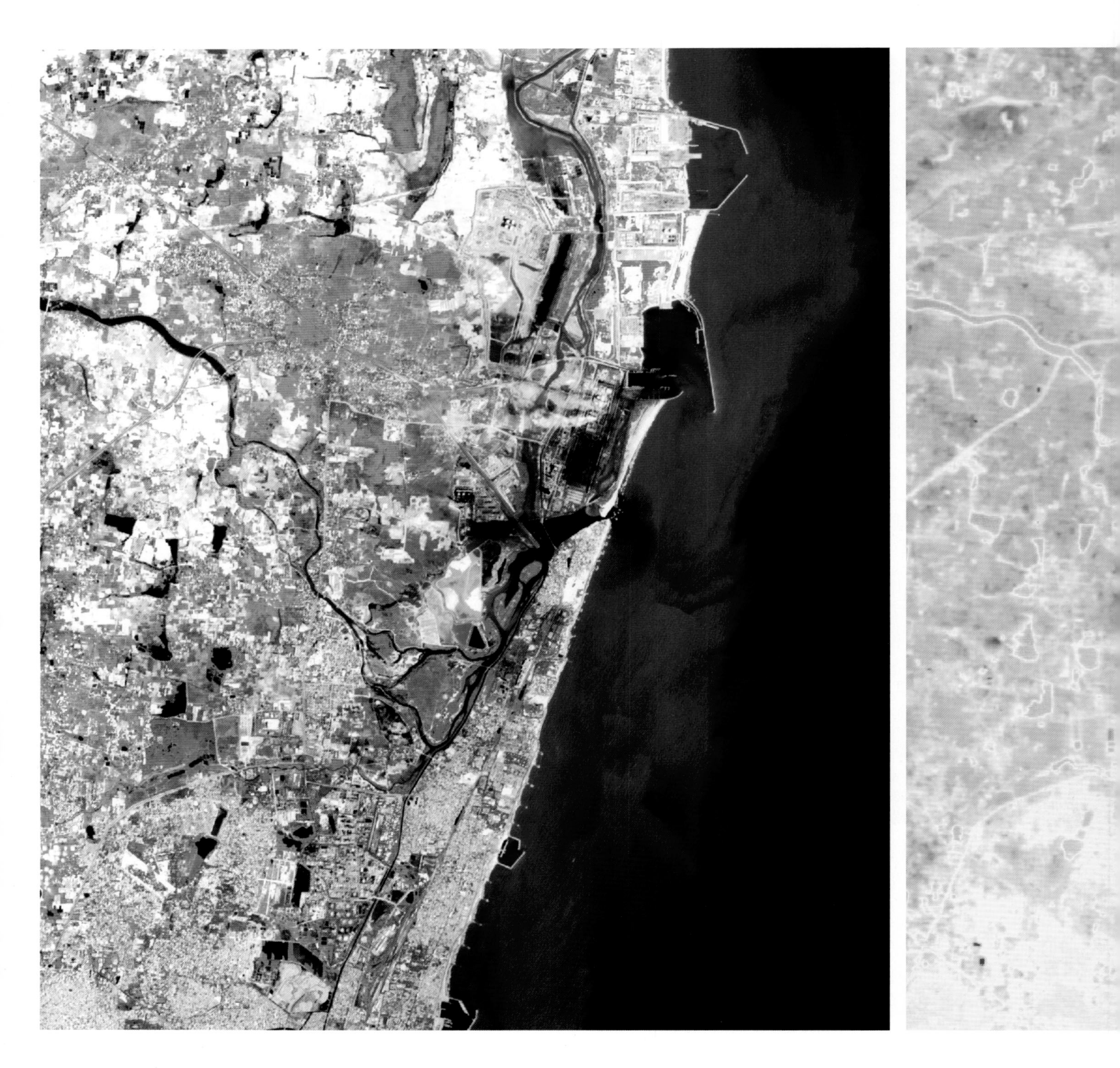

假彩色多光谱图像（波段组合：7-5-4）　成像时间：2023-03-02 日间
Pseudo color multispectral image (band combination: 7-5-4)　Imaging time: 2023-03-02 Daytime

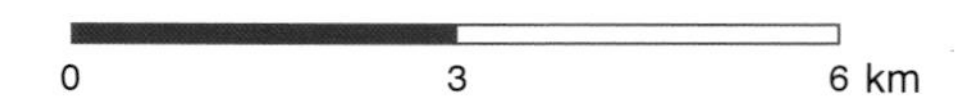

泰米尔纳德邦发电站
Power station in Tamil Nadu region

该热电站位于印度泰米尔纳德邦的钦奈市北部，在印度东海岸线上，靠近孟加拉湾，是该区域最大的热电站之一。在春季夜间热红外影像中，山体等植被覆盖区域表征为土黄色，海水表征为暖红色，而供电站和供电站南面的化工企业呈现出热辐射异常现象（红色），其生产造成海水增温现象也较明显（红色到暖红色渐变）。

This power station, located north of the city of Chennai in the Indian state of Tamil Nadu, on the eastern coastline of India, near the Bay of Bengal, is one of the largest thermal power stations in the region. In the spring nighttime thermal infrared imagery, vegetation-covered areas such as mountains are characterized as earthy yellow, seawater is characterized as warm red, and the power plant and the chemical company south of the power plant show thermal radiation anomalies (red) and warming of the seawater due to their production is more pronounced (red to warm red gradient).

单波段热红外图像（波段 2）　成像时间：2024-02-07 夜间
Single band thermal infrared image (band 2)　Imaging time: 2024-02-07 Nighttime

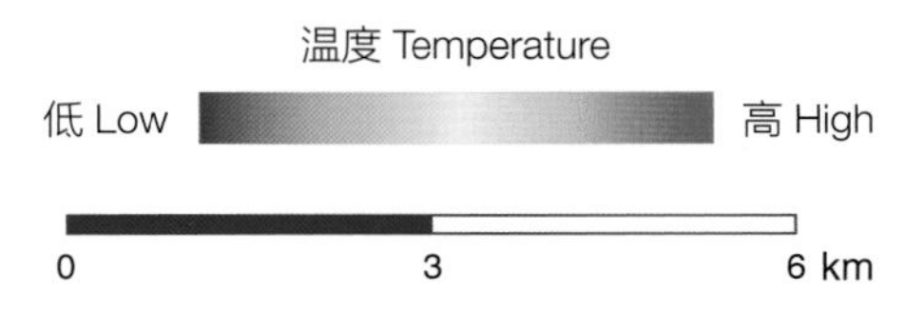

澳大利亚西部氧化铝精炼厂
Alumina refinery in western Australia

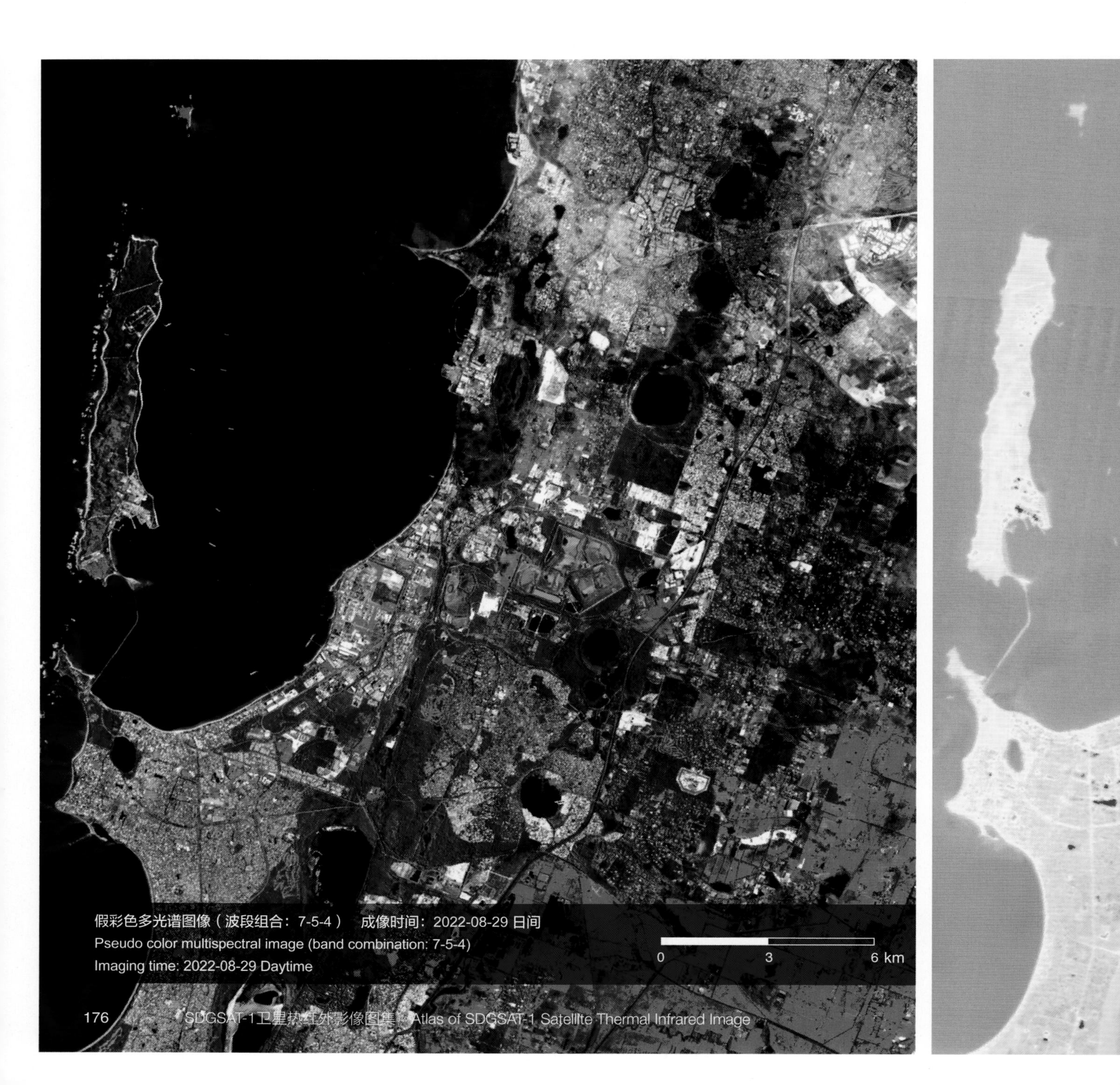

图中氧化铝精炼厂位于澳大利亚西部，珀斯州的印度洋入海口，是澳大利亚规模较大的氧化铝生产基地之一。在秋季夜间热红外影像中，山体等植被覆盖区域表征为冷色调（蓝色），水体表征为暖色调（暖红色）。其中，化工企业热排放区域表征出极热现象（鲜红色）。

The alumina refinery shown is located in western Australia, at the mouth of the Indian Ocean in the state of Perth, and is one of the largest alumina production sites in Australia. In the fall nighttime thermal infrared imagery, vegetation-covered areas such as mountains are characterized by cool tones (blue) and water bodies are characterized by warm tones (warm red). Among them, the area of thermal emissions from chemical companies is characterized by extreme heat (bright red).

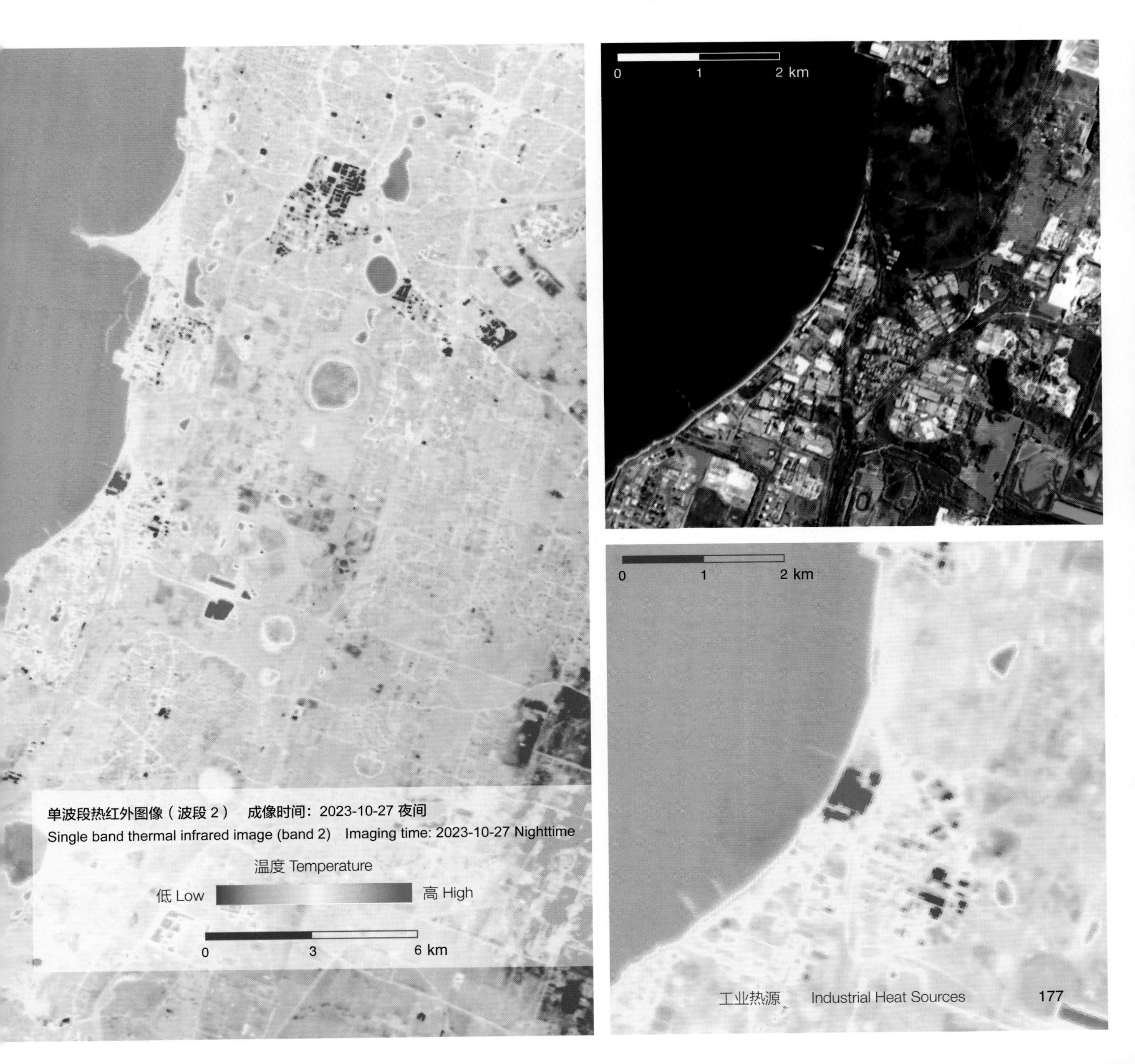

单波段热红外图像（波段 2） 成像时间：2023-10-27 夜间

Single band thermal infrared image (band 2) Imaging time: 2023-10-27 Nighttime

城市

Cities

柏林 / 200
Berlin
阿姆斯特丹 / 202
Amsterdam
芝加哥 / 191
Chicago
伦敦 / 205
London
丹佛 / 182
Denver
多伦多 / 193
Toronto
巴黎 / 207
Paris
纽约 / 195
New York
旧金山 / 184
San Francisco
丹吉尔 / 208
Tangier
华盛顿 / 196
Washington D.C.
洛杉矶 / 186
Los Angeles
墨西哥城 / 189
Mexico City
布宜诺斯艾利斯 / 199
Buenos Aires

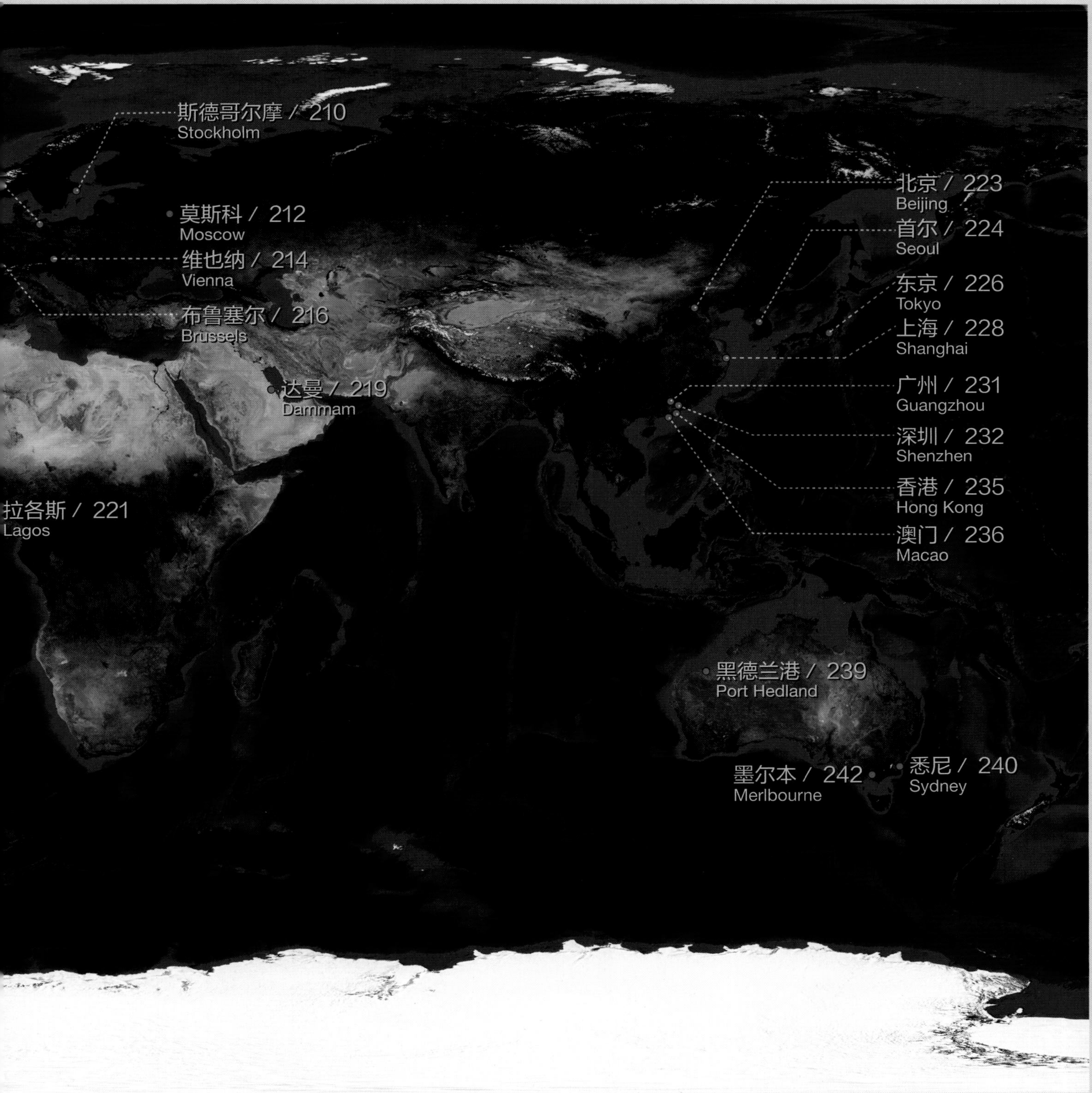

斯德哥尔摩 / 210
Stockholm

北京 / 223
Beijing

莫斯科 / 212
Moscow

首尔 / 224
Seoul

维也纳 / 214
Vienna

东京 / 226
Tokyo

布鲁塞尔 / 216
Brussels

上海 / 228
Shanghai

广州 / 231
Guangzhou

达曼 / 219
Dammam

深圳 / 232
Shenzhen

香港 / 235
Hong Kong

拉各斯 / 221
Lagos

澳门 / 236
Macao

黑德兰港 / 239
Port Hedland

悉尼 / 240
Sydney

墨尔本 / 242
Merlbourne

丹佛
Denver

丹佛是美国科罗拉多州的首府和最大城市，位于一片紧邻落基山脉的高原上，市中心位于南佩雷特河东岸，接近南佩雷特河与樱桃溪的交接口，形成丹佛 - 奥罗拉大都会区的核心。丹佛拥有航空、航天、生物技术、能源、金融服务和信息技术软件六个支柱产业，有全美占地面积最大的机场丹佛机场，经济发达，环境优美，文化体育活动。

Denver is the capital and largest city of the U.S. state of Colorado, located on a plateau adjacent to the Rocky Mountains. The city center is situated on the east bank of the South Peret River, near the intersection of the South Peret River and Cherry Creek, forming the core of the Denver Aurora Metropolitan Area. Denver is home to six pillar industries: aviation, aerospace, biotechnology, energy, financial services and information technology software. The city has a developed economy, a beautiful environment and a vibrant cultural and sports scene. Denver International Airport is the largest airport in the United States.

三波段热红外图像（波段组合：1-2-3）　成像时间：2022-09-13 夜间
Three band thermal infrared image (band combination: 1-2-3)　Imaging time: 2022-09-13 Nightt

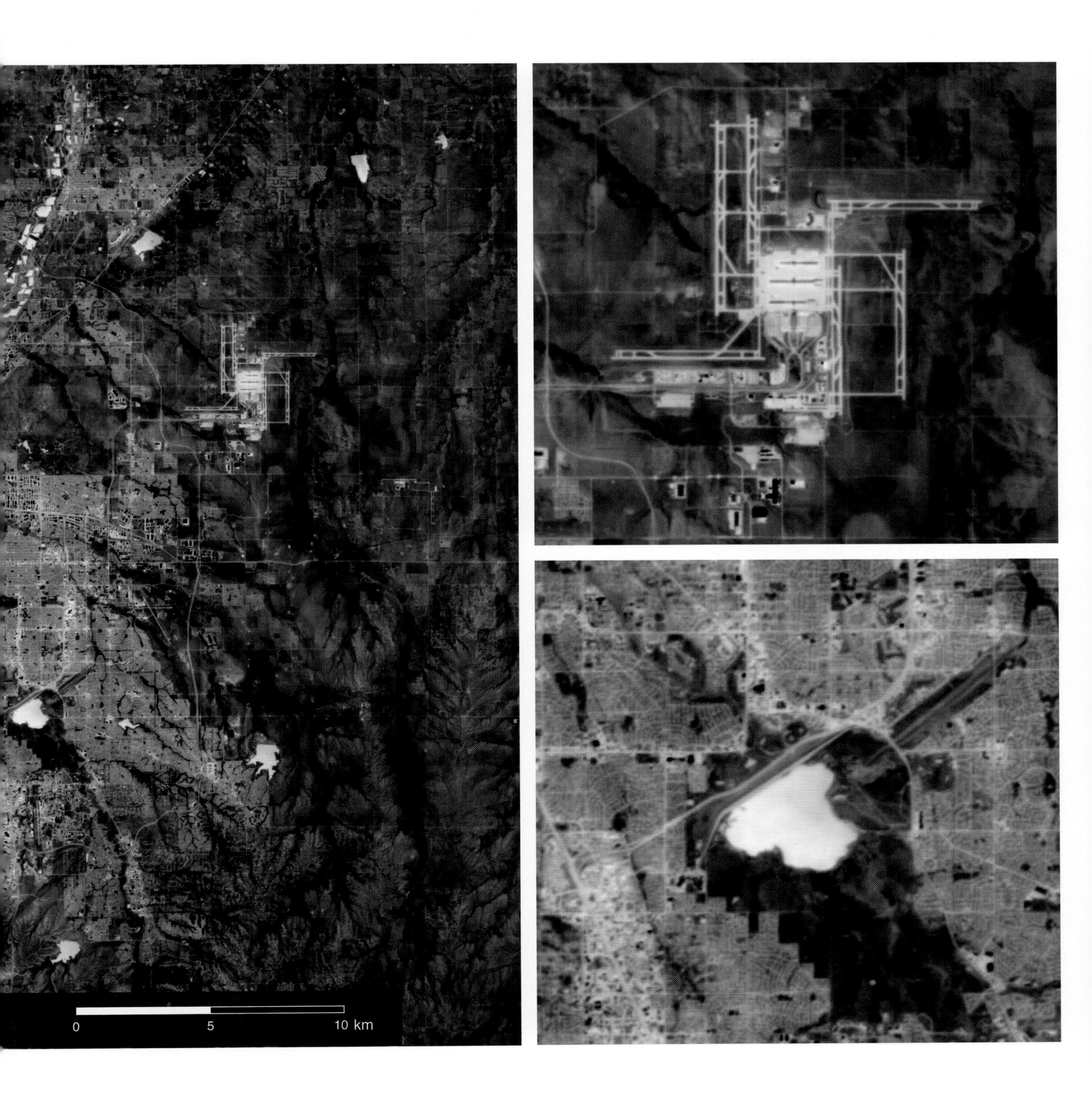

0
5
10 km

旧金山
San Francisco

旧金山是美国北加利福尼亚州和湾区重要的经济和文化中心，太平洋沿岸港口城市，世界著名旅游胜地，有著名景点金门大桥和世界上跨度最大的桥梁之一海湾大桥。旧金山邻近世界著名技术产业区硅谷和旧金山地区航空货运中心——奥克兰国际机场，有大型航空枢纽旧金山国际机场，还有唐·爱德华兹旧金山湾国家野生动物保护区。

San Francisco is an important economic and cultural center of Northern California and the Bay Area in the United States, a port city along the Pacific coast, a world-renowned tourist destination, with famous attractions such as the Golden Gate Bridge and the Bay Bridge, one of the largest bridges in the world. It is adjacent to the world-renowned technology industry area, Silicon Valley, and the Oakland International Airport, the air cargo center of San Francisco. It has a large aviation hub called San Francisco International Airport, as well as the Don Edwards San Francisco Bay National Wildlife Refuge.

三波段热红外图像（波段组合：1-2-3） 成像时间：2023-09-29 夜间
Three band thermal infrared image (band combination: 1-2-3) Imaging time: 2023-09-29 Nighttime

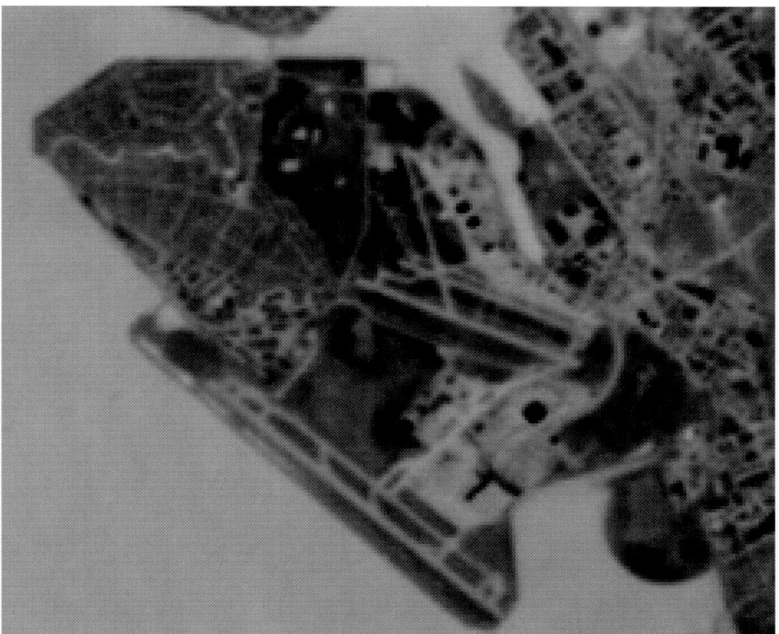

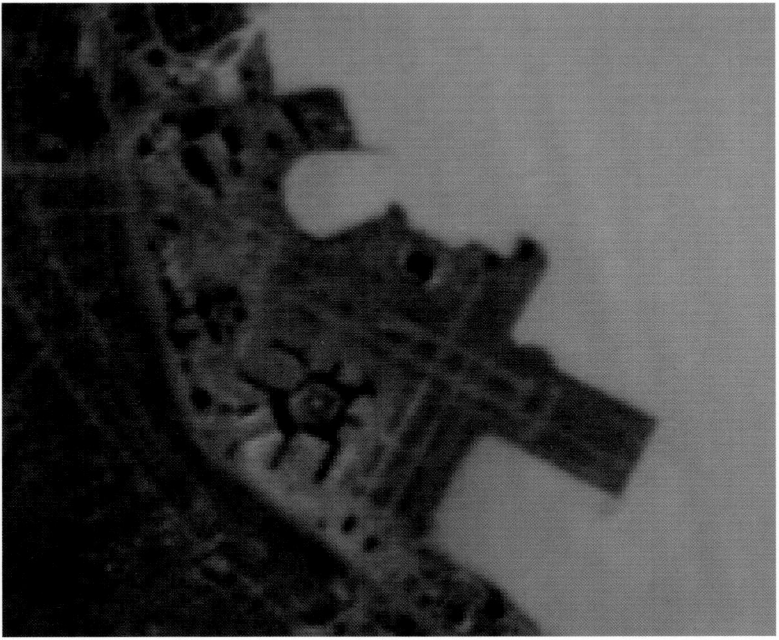

洛杉矶
Los Angeles

洛杉矶位于美国加利福尼亚州西南部，是美国第二大城市，也是美国西部最大的城市。美国重要的工商业、国际贸易、科教、娱乐和体育中心之一，有美国门户型国际航空枢纽洛杉矶机场与世界上最繁忙的通用航空机场之一长滩机场。洛杉矶是世界电影中心好莱坞的所在地，也是美国石油化工、海洋、航天和电子工业的主要基地之一，还有世界知名高校。

Los Angeles, located in the southwestern part of California, the United States, which is the second largest city in the United States and the largest city in the western United States. It is also one of the important centers of industry, commerce, international trade, science and education, entertainment, and sports in the United States. Los Angeles International Airport serves as a gateway international aviation hub, and Long Beach Airport is one of the busiest general aviation airports in the world. Los Angeles is home to the world's film center, Hollywood, and is one of the main bases for the U.S. petrochemical, ocean, aerospace, and electronics industries, with world-renowned universities.

三波段热红外图像（波段组合：1-2-3） 成像时间：2023-02-10 夜间
Three band thermal infrared image (band combination: 1-2-3) Imaging time: 2023-02-10 Nighttime

0
5
10 km

段热红外图像（波段组合：1-2-3）　成像时间：2023-02-27 日间
e band thermal infrared image (band combination: 1-2-3)　Imaging time: 2023-02-27 Daytime

墨西哥城
Mexico City

墨西哥城是墨西哥首都，政治、经济、文化和交通中心，著名国际化大都市，有全国乃至拉丁美洲最繁忙的机场墨西哥城国际机场，城市海拔较高，位于墨西哥中南部高原的山谷中。

Mexico City is the capital of Mexico, and also its political, economic, cultural and transportation center. It is a renowned international metropolis with the busiest airport in the country and even in Latin America, Mexico City International Airport. The high-altitude city is located in a valley on the central southern plateau of Mexico.

三波段热红外图像（波段组合：1-2-3） 成像时间：2023-09-04 夜间
Three band thermal infrared image (band combination: 1-2-3) Imaging time: 2023-09-04 Nighttime

芝加哥
Chicago

芝加哥位于美国密歇根湖南岸，美国第三大城市、世界金融中心之一，也是美国最重要的文化科教中心之一，拥有世界顶级学府和享誉世界的芝加哥学派。芝加哥地处北美大陆中心地带，有美国最繁忙的国际航空枢纽奥黑尔国际机场、主要服务于国内航班的中途国际机场与主要服务于芝加哥大都市区的盖瑞机场，还有美国内陆航线的重要港口伯恩斯港。

Chicago is on the south bank of Lake Michigan in the United States, which is the third largest city in the U.S. and one of the world's financial centers. It is also one of the most important cultural, scientific and educational centers in the U.S. The city is home to world-class universities and is known for its influential Chicago School of economics and sociology. Located in the heart of the North American continent, Chicago has O'Hare International Airport, the busiest international aviation hub in the United States, Midway International Airport, primarily serving domestic flights, Gary Airport, primarily serving the metropolitan area of Chicago, and the important Burns port for inland routes in the United States.

三波段热红外图像（波段组合：1-2-3） 成像时间：2023-07-04 日间

Three band thermal infrared image (band combination: 1-2-3) Imaging time: 2023-07-04 Daytime

多伦多
Toronto

多伦多坐落在安大略湖西北岸，是加拿大最大城市、安大略省省会，也是加拿大的政治、经济、文化和交通中心之一。城中建有比利·毕晓普·多伦多市机场和多伦多皮尔逊国际机场两座当地主要的国际机场。

Located on the northwestern shore of Lake Ontario, Toronto is the largest city in Canada, the capital of Canadian province of Ontario, and one of political, economic, cultural and transportation centers of Canada. The city has two major local international airports, Billy Bishop Toronto City Airport and Toronto Pearson International Airport.

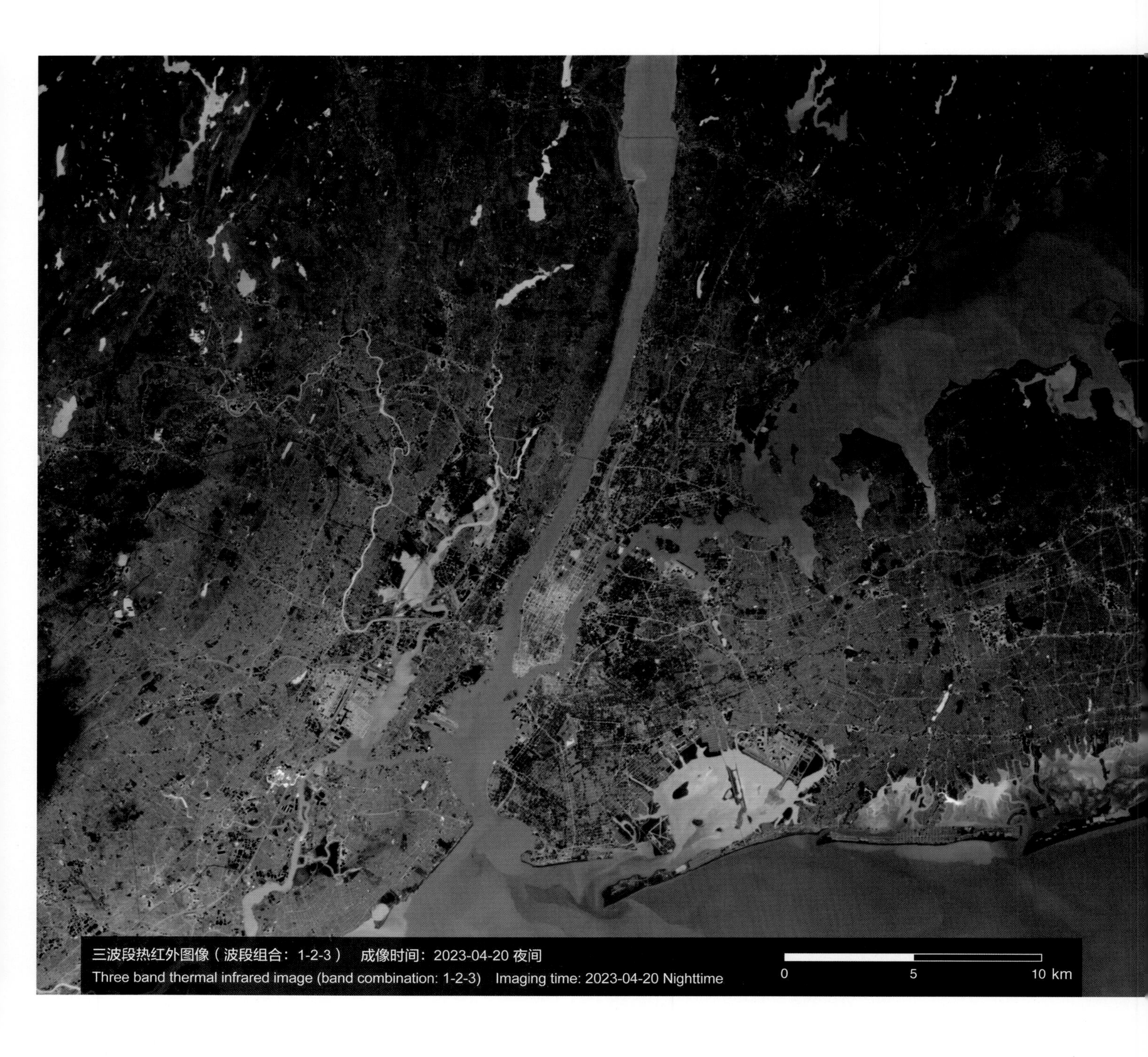

三波段热红外图像（波段组合：1-2-3）　成像时间：2023-04-20 夜间
Three band thermal infrared image (band combination: 1-2-3)　Imaging time: 2023-04-20 Nighttime

纽约
New York

纽约是美国第一大城市，美国乃至全球的经济、金融、商业、贸易、文化和传媒中心。纽约位于美国东北部沿海哈德逊河口，濒临大西洋，有北美洲最繁忙的港口，交通便利，有国际航空枢纽肯尼迪国际机场、纽瓦克国际机场与拉瓜迪亚机场。纽约拥有许多国际知名高等教育机构，还有时代广场、自由女神像、大都会博物馆等著名景点。

New York is the largest city in the United States, which is the economic, financial, commercial, trade, cultural and media center of the United States and even the world. New York City is located at the mouth of the Hudson River on the northeastern coast of the United States. It borders the Atlantic Ocean and has the busiest port in North America. The city is also a major transportation hub, with three international airports: Kennedy International Airport, Newark Liberty International Airport and LaGuardia Airport. New York City is home to many internationally renowned higher education institutions, as well as famous attractions such as Times Square, Statue of Liberty and Metropolitan Museum.

华盛顿
Washington D.C.

华盛顿是美国首都，位于美国东北部、中大西洋地区，是大多数美国政府机关与各国驻美大使馆所在地，也是许多国际组织总部的所在地，博物馆与文化史迹众多。城中有罗纳德·里根华盛顿国家机场和华盛顿联合车站等重要交通基础设施。

Washington D.C. is the capital of the United States, located in the northeast and central Atlantic region. The city is home to most U.S. government agencies and embassies, as well as the headquarters of many international organizations. Washington D.C. is also a major cultural center, with numerous museums and historical landmarks. The city hosts critical transportation infrastructure, including Ronald Reagan Washington National Airport and Washington Union Station.

三波段热红外图像（波段组合：1-2-3） 成像时间：2023-09-01 夜间

Three band thermal infrared image (band combination: 1-2-3) Imaging time: 2023-09-01 Night

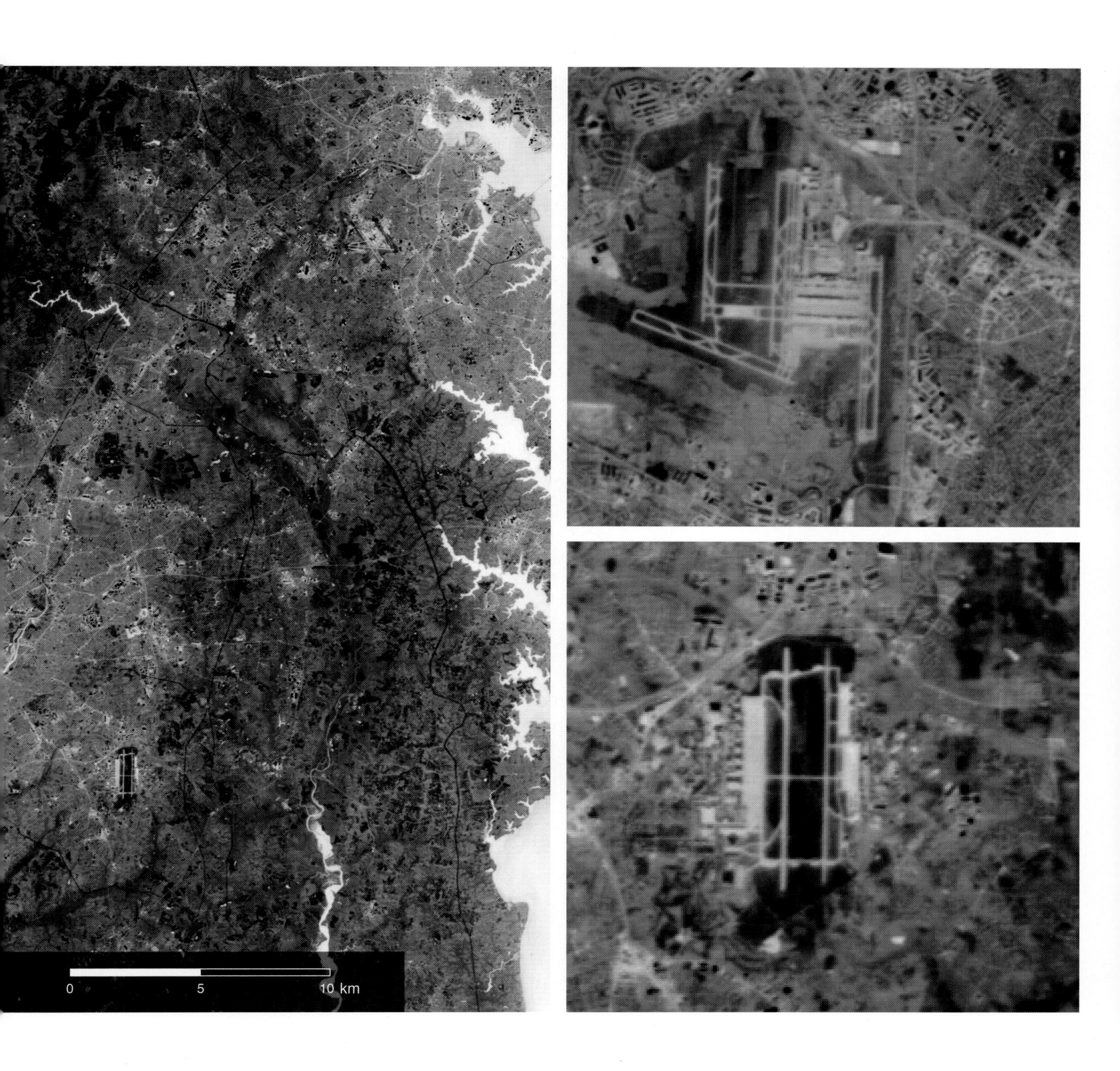
0
5
10 km

三波段热红外图像（波段组合：1-2-3） 成像时间：2022-10-10 夜间
Three band thermal infrared image (band combination: 1-2-3) Imaging time: 2022-10-10 Nighttime

布宜诺斯艾利斯
Buenos Aires

布宜诺斯艾利斯是阿根廷首都和最大城市，政治、经济、科技、文化和交通中心，人口众多，拥有多家工业企业，还有阿根廷最大的国际机场之一埃塞萨国际机场与乔治·纽伯里机场，在国民经济中具有举足轻重的地位，享有“南美洲巴黎”的盛名。布宜诺斯艾利斯位于南美洲东南部、拉普拉塔河南岸，是南美洲第二大都会区，对岸为乌拉圭。

Buenos Aires is the capital and largest city of Argentina, which is a political, economic, technological, cultural and transportation center. The city has a large population and is home to multiple industrial enterprises. It is also home to one of Argentina's largest international airports, Ezeiza International Airport, and another major airport, Jorge Newbery Airport. Buenos Aires plays a crucial role in the national economy and is renowned as the "Paris of South America". The city is located in the southeast of South America, on the south bank of the La Plata River. It is part of the second largest metropolitan area in South America, with Uruguay on the opposite bank of the river.

柏林
Berlin

柏林位于德国东北部平原，是德国首都及最大城市，兼具政治、文化、交通和经济中心的角色，新开放的勃兰登堡机场即位于此。这座城市历史悠久，经济影响力显著，尤其在机械制造业和汽车行业领域发展高度发达。四面被勃兰登堡州环绕，施普雷河和哈弗尔河流经，湖泊众多，有由停止运营的滕珀尔霍夫机场改造的城市公园。

Berlin, situated in the northeastern plains of Germany, is the capital and the largest city of the country. It serves as the political, cultural, transportation, and economic center of Germany, home to the newly opened Berlin Brandenburg Airport. With a rich history, Berlin has a significant economic presence, highlighted by its highly developed machinery manufacturing and automotive industries. It is encircled by the state of Brandenburg, with the Spree and Havel rivers flowing through it, along with numerous lakes. The city also features an urban park at the site of the decommissioned Tempelhof Airport.

三波段热红外图像（波段组合：1-2-3） 成像时间：2022-03-13 日间
Three band thermal infrared image (band combination: 1-2-3) Imaging time: 2022-03-13 Day

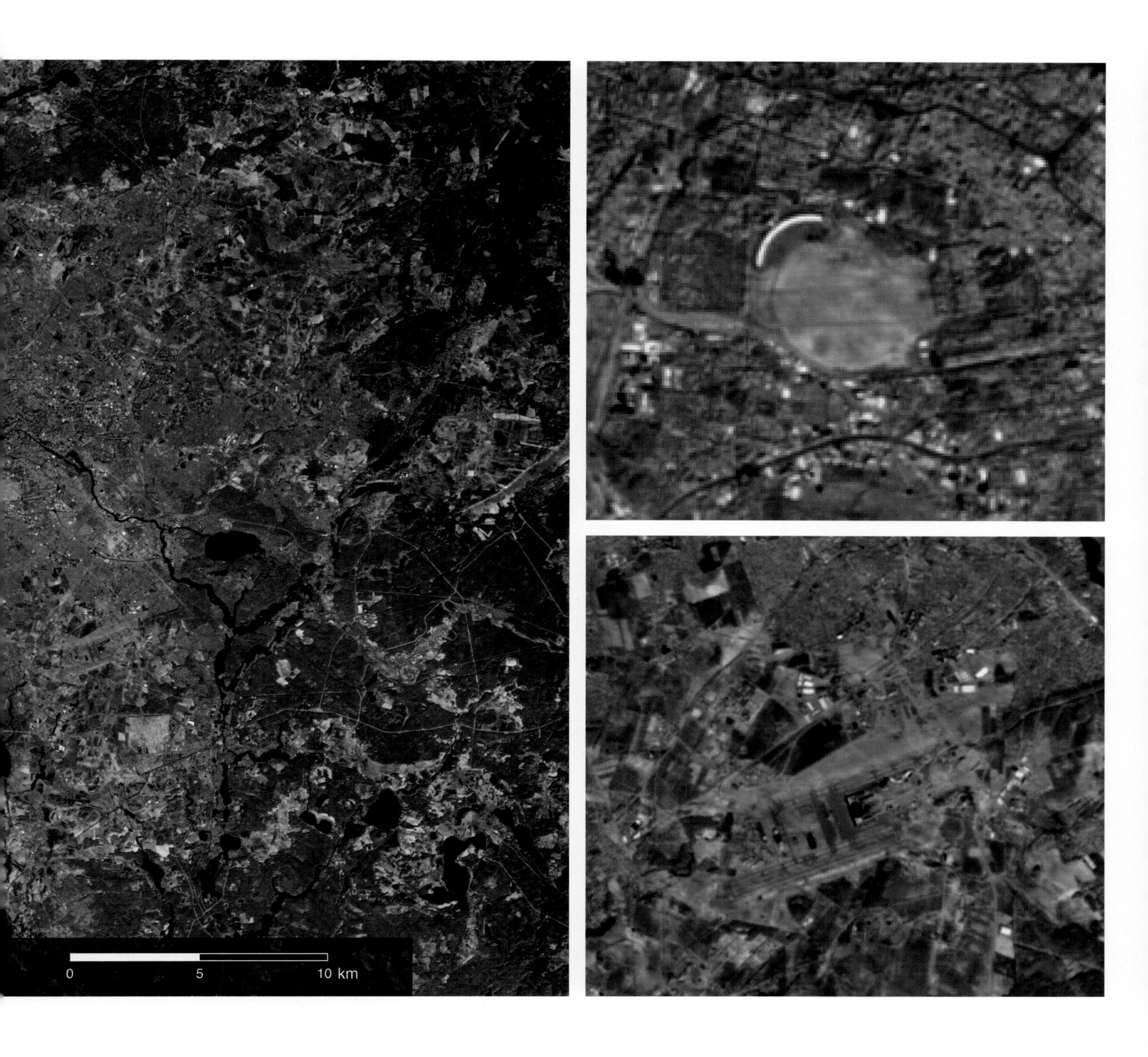
0
5
10 km

阿姆斯特丹
Amsterdam

阿姆斯特丹是荷兰首都及最大城市，位于荷兰西部的北荷兰省。它是世界著名旅游城市和国际大都市，拥有众多旅游景点，其斯希普霍尔机场为欧洲第三大航空港，临近中央政府、议会和最高法院的所在地海牙与全国第二大城市鹿特丹，鹿特丹有欧洲内陆航线的重要港口荷兰角港。

Amsterdam is the capital and largest city of the Netherlands, located in the province of North Holland in the western part of the country. It is a world-famous tourist destination and an international metropolis, boasting numerous attractions and Schiphol Airport, which ranks as the third largest airport in Europe. Amsterdam is situated near The Hague, which hosts the central government, parliament, and Supreme Court, and Rotterdam, the country's second-largest city. Rotterdam has an important port on the European inland route, Hook of Holland.

三波段热红外图像（波段组合：1-2-3） 成像时间：2022-05-19 夜间
Three band thermal infrared image (band combination: 1-2-3) Imaging time: 2022-05-19 Nigh

0
5
10 km

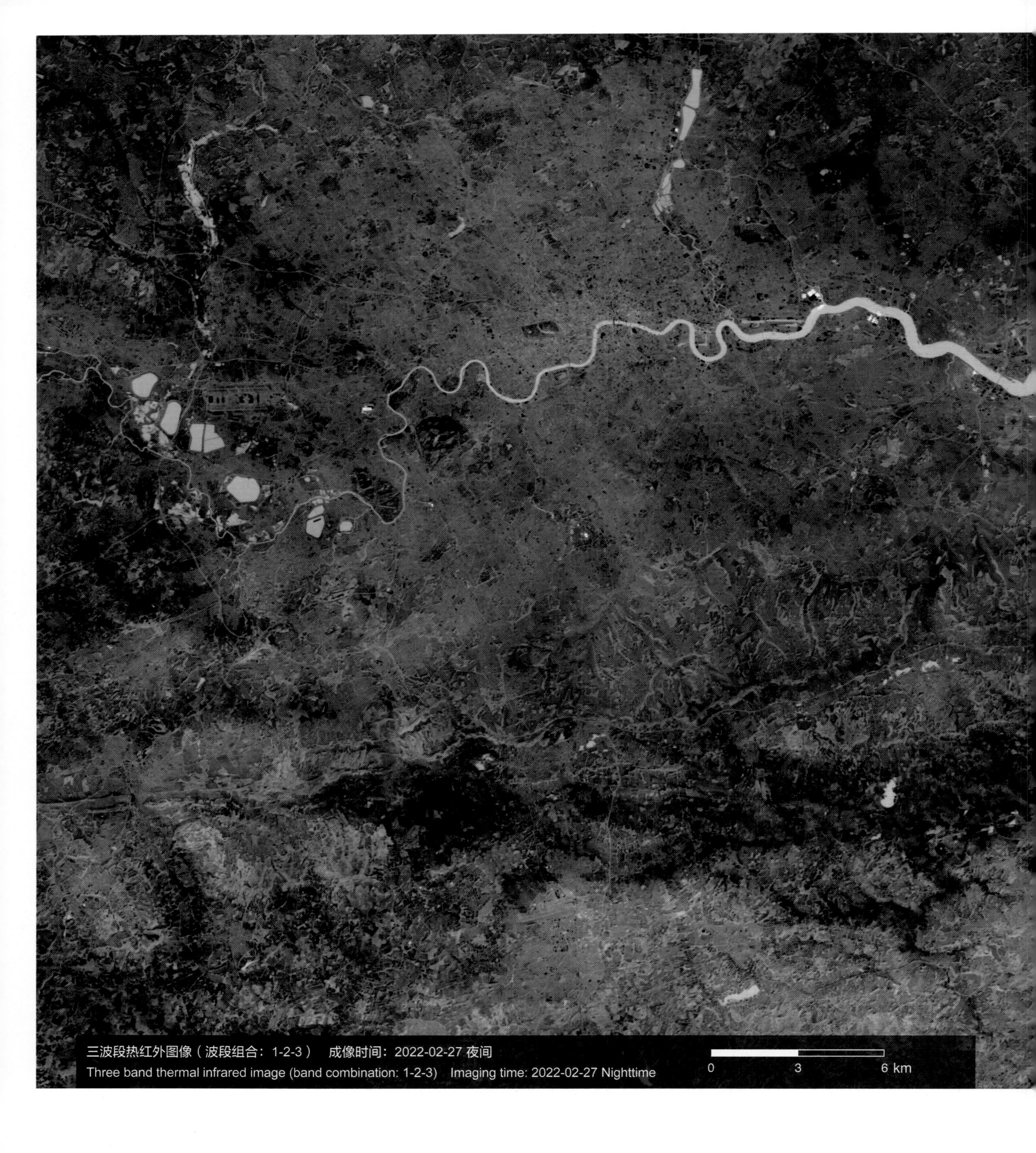

三波段热红外图像（波段组合：1-2-3） 成像时间：2022-02-27 夜间
Three band thermal infrared image (band combination: 1-2-3) Imaging time: 2022-02-27 Nighttime

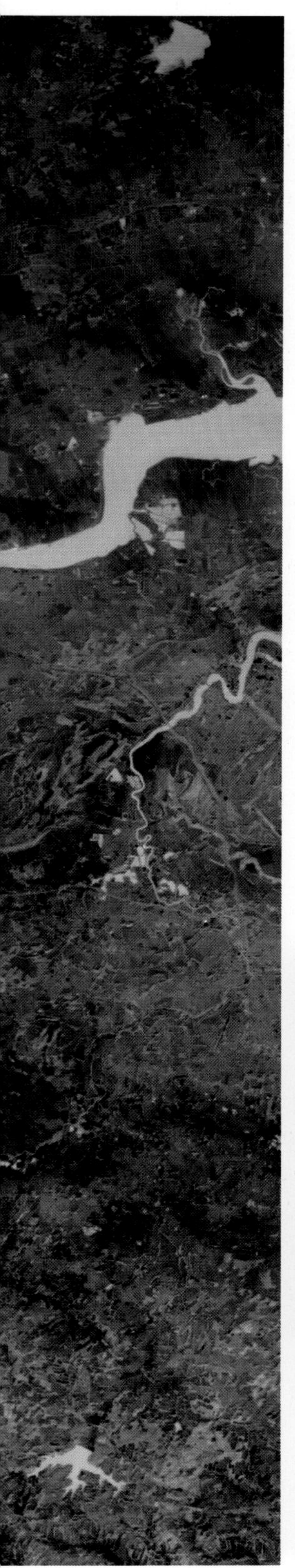

伦敦
London

伦敦是英国首都、政治中心，欧洲第一大城和最大经济中心，全世界博物馆、图书馆、电影院、戏剧院、体育场馆和五星级酒店数量最多的城市，有众多知名大学。世界金融中心之一，在世界上保持着巨大的影响力。伦敦位于英格兰东南部平原，泰晤士河贯穿其中，人口众多，有欧洲最繁忙的机场希思罗机场。

London is the capital and political center of the United Kingdom, the largest city and economic center in Europe. It boasts the highest number of museums, libraries, cinemas, theaters, sports venues and five-star hotels in the world, alongside numerous renowned universities. As one of the world's financial centers, London maintains a significant influence globally. Situated on the southeastern plains of England, with the Thames River flowing through it, London has a large population and is home to Heathrow Airport, the busiest airport in Europe.

0
5
10 km

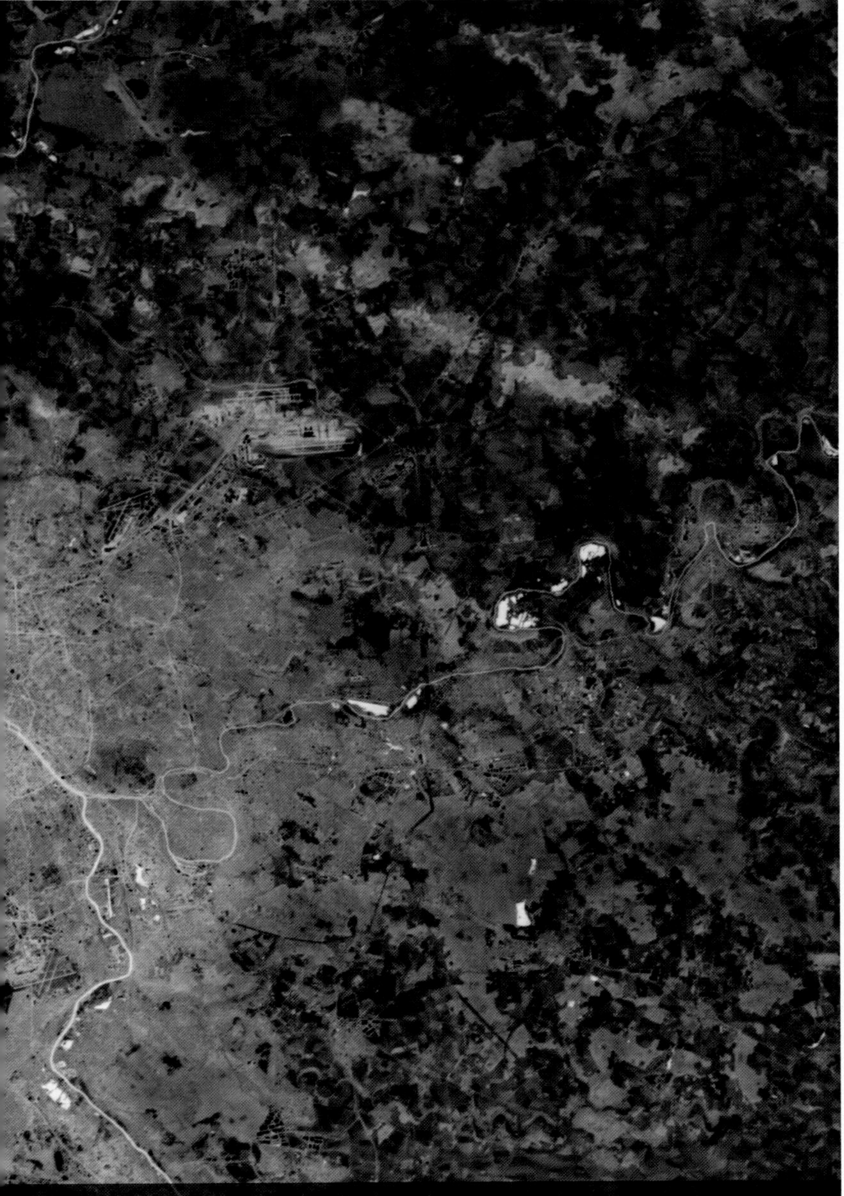

段热红外图像（波段组合：1-2-3） 成像时间：2023-09-30 夜间

e band thermal infrared image (band combination: 1-2-3) Imaging time: 2023-09-30 Nighttime

巴黎
Paris

巴黎是法国首都和最大城市，政治、经济、文化和商业中心，世界五个国际大都市之一，拥有悠久的历史和丰富的文化遗产，人口繁密，有欧洲最主要的航空枢纽之一戴高乐机场与全国第二大机场奥利机场。巴黎位于法国北部巴黎盆地的中央，横跨塞纳河两岸，被认为是现代奥林匹克运动的起点，也是 2024 年奥运会主办城市。

Paris is the capital and largest city of France, as well as its political, economic, cultural and commercial center. It is one of the five international metropolises in the world, with a long history and a rich cultural heritage. With a dense population, Paris has one of the most important aviation hubs in Europe, Charles de Gaulle Airport and the second largest airport in the country, Orly Airport. Located in the center of the Paris Basin in northern France, spanning both banks of the Seine River, Paris is the host city of the 2024 Olympic Games and is considered the starting point of modern Olympic sports.

丹吉尔
Tangier

丹吉尔是摩洛哥北部古城、海港，丹吉尔省省会，全国最大旅游中心。它位于直布罗陀海峡丹吉尔湾口，坐落在世界交通的十字路口，东进地中海和西出大西洋的船只均经此或停泊，大西洋东岸南来北往的船只也在此调整航向，战略地位十分重要。丹吉尔有摩洛哥六个机场之一——丹吉尔机场，还有多家国际车企入驻的非洲第一大汽车制造厂和科技城。

Tangier is an ancient city and seaport in northern Morocco, the capital of Tangier Province, and the largest tourist center in the country. The city is strategically located at the mouth of Tangier Bay in the Strait of Gibraltar. This makes it a crossroad of world transportation, with ships entering the Mediterranean to the east and exiting the Atlantic to the west passing or docking here. Ships traveling from south to north on the east coast of the Atlantic also adjust their course here, which is of great strategic importance. It is home to Tangier Airport, one of the six airports in Morocco. The city also boasts Africa's largest automobile manufacturing plant and technology city, with multiple international car companies having settled there.

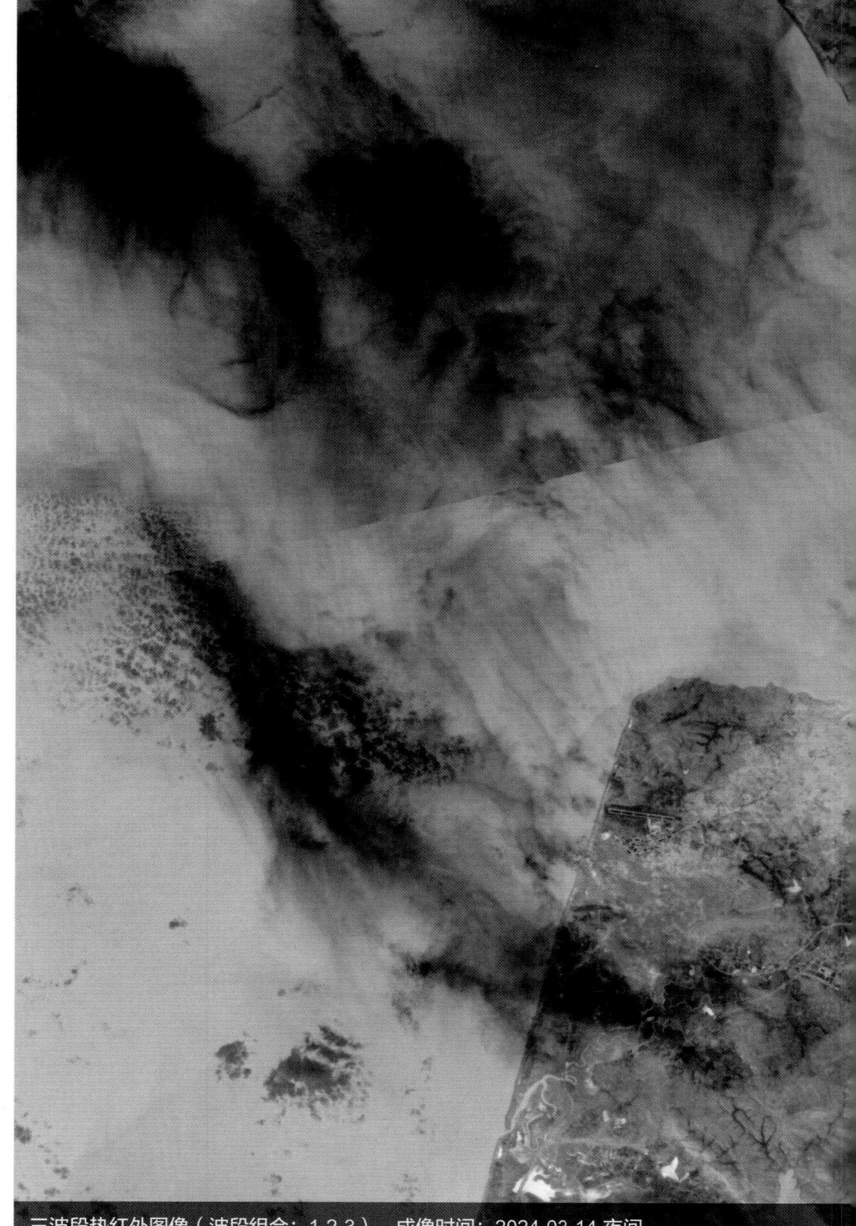

三波段热红外图像（波段组合：1-2-3） 成像时间：2024-03-14 夜间
Three band thermal infrared image (band combination: 1-2-3) Imaging time: 2024-03-14 Night

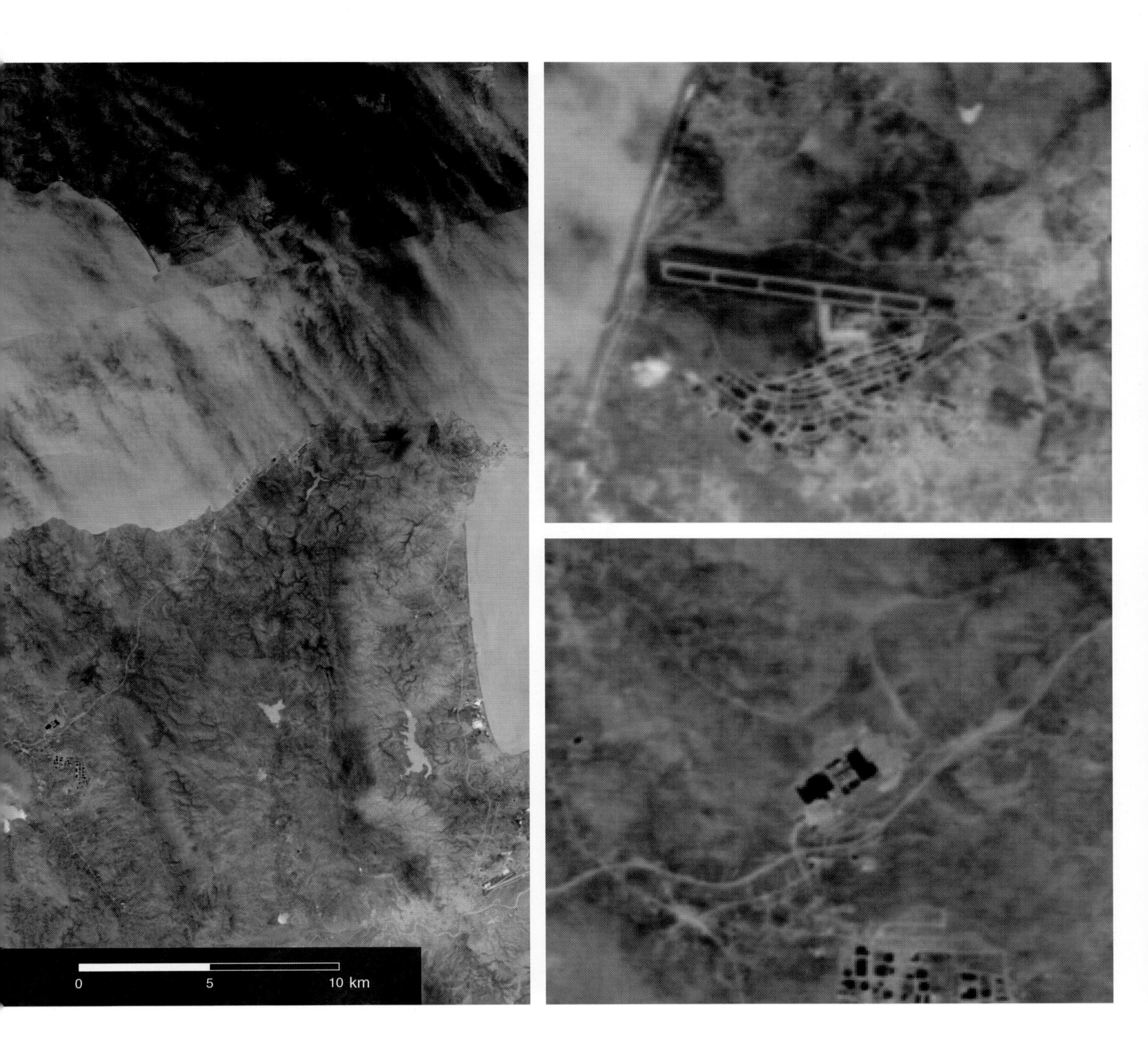
0
5
10 km

斯德哥尔摩
Stockholm

斯德哥尔摩是瑞典的首都和第一大城市，政治、经济、文化、交通中心和主要港口，也是国家政府、国会以及皇室的官方宫殿所在地，世界著名的国际大都市，有北欧航空主要的枢纽机场斯德哥尔摩阿兰达国际机场与全国第三多升降数量的机场布鲁玛机场。位于瑞典东海岸，毗邻波罗的海，梅拉伦湖入海处，风景秀丽，是著名的旅游胜地。

Stockholm, the capital and largest city of Sweden, serves as the political, economic, cultural, transportation hub, and a major port of the country. It houses the official residences of the national government, parliament, and the royal family, making it a globally renowned international metropolis. The city is served by Stockholm Arlanda Airport, the main hub for international flights in Sweden, and Bromma Stockholm Airport, the country's third busiest airport in terms of flight operations. Situated on the east coast of Sweden, adjacent to the Baltic Sea and at the entrance of Lake Mälaren, Stockholm boasts beautiful scenery and is a celebrated tourist destination.

三波段热红外图像（波段组合：1-2-3） 成像时间：2024-03-11 夜间

Three band thermal infrared image (band combination: 1-2-3) Imaging time: 2024-03-11 Night

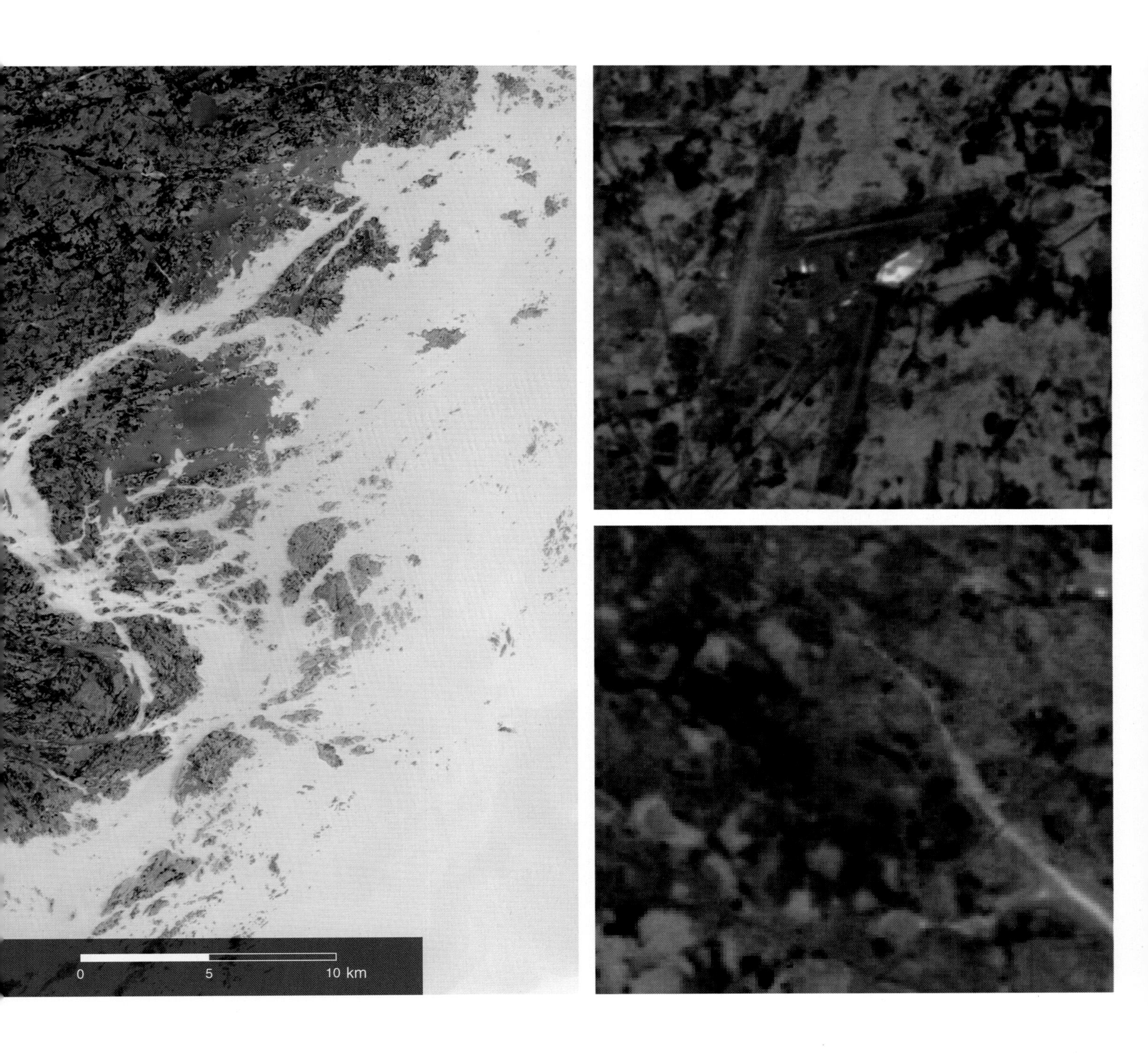
0
5
10 km

莫斯科
Moscow

莫斯科是俄罗斯的首都，也是该国的政治、经济、文化、金融和交通中心，是俄罗斯最大的综合性城市和一座国际化大都市。莫斯科地处俄罗斯欧洲部分中部、东欧平原中部，跨莫斯科河及支流亚乌扎河两岸，与伏尔加流域的上游入口和江河口处相通，是俄罗斯乃至欧亚大陆上极其重要的交通枢纽。

Moscow is the capital of Russia, serving as a pivotal political, economic, cultural, financial, and transportation center, and is the country's largest comprehensive city. As an international metropolis, it is located in the central part of European Russia and the central plains of Eastern Europe, spanning both sides of the Moscow River and its tributary, the Yauza River, connecting to the Volga Basin. It plays an extremely important role as a transportation hub in Russia and the Eurasian continent.

三波段热红外图像（波段组合：1-2-3） 成像时间：2022-04-14 日间
Three band thermal infrared image (band combination: 1-2-3) Imaging time: 2022-04-14 Daytir

0
5
10 km

维也纳
Vienna

维也纳位于多瑙河畔，是奥地利的首都和最大的城市，人口众多，有全国最繁忙和最大的机场维也纳国际机场。居民主要分布于市区东部和西部，而北部和南部主要是工业区，市中心古城区被列为世界遗产，是欧洲主要的文化中心，被誉为“世界音乐之都”，已连续多年被联合国人居署评为全球最宜居的城市之一。

Vienna is situated on the bank of the Danube River, serving as the capital and the largest city of Austria. It is renowned for its high population density and houses the busiest and largest airport in the country, Vienna International Airport. The city's residential areas are primarily located in the eastern and western parts, while the northern and southern parts are predominantly industrial zones. The ancient city center is a UNESCO World Heritage Site and acts as a major cultural hub in Europe. Known as the "Music Capital of the World," rated as one of the most livable cities in the world by the United Nations Human Settlements Programme for many consecutive years.

三波段热红外图像（波段组合：1-2-3） 成像时间：2023-09-06 夜间
Three band thermal infrared image (band combination: 1-2-3) Imaging time: 2023-09-06 Nightt

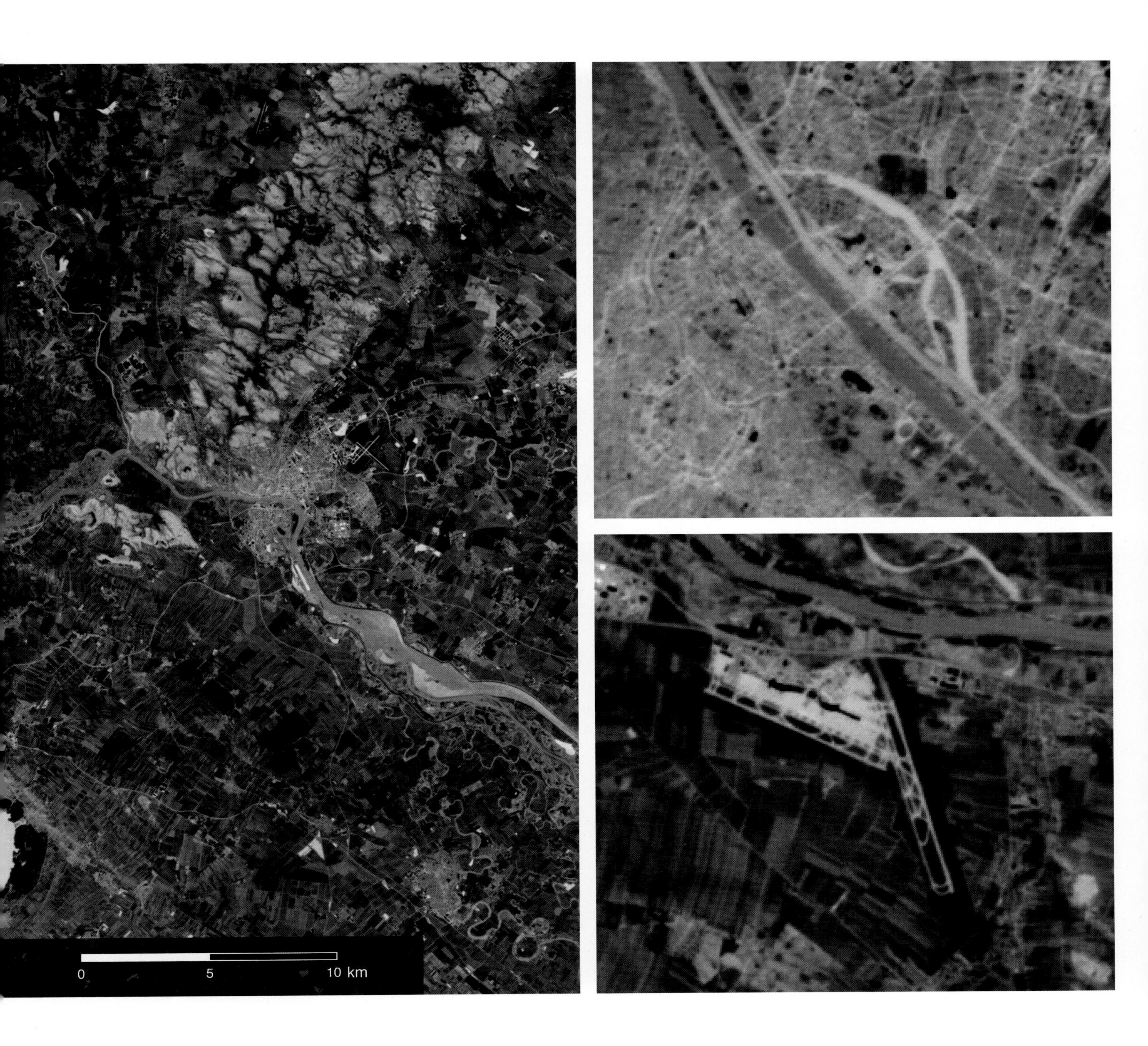
0
5
10 km

布鲁塞尔
Brussels

布鲁塞尔是比利时首都和最大的城市，有全国最主要的国际机场布鲁塞尔机场，也是欧洲联盟主要行政机构所在地，北大西洋公约组织总部驻地，有欧洲的首都之称。布鲁塞尔位于塞纳河畔，整座城市以皇宫为中心，沿“小环”而建，其中上城依坡而建，为行政区，有众多名胜；下城为繁华的商业中心，市中心屹立着许多壮观的中世纪哥特式建筑。

Brussels is the capital and largest city of Belgium, hosting the country's main international airport, Brussels Airport. It is a crucial location for international politics, serving as the main administrative center of the European Union and the headquarters of the North Atlantic Treaty Organization, earning it the title of “Capital of Europe”. Brussels is located on the bank of the Seine River, the entire city is centered around a palace and built along a “small ring”. The upper city is built on a slope and is an administrative district filled with numerous attractions, while the lower city is a bustling commercial center with many spectacular medieval Gothic buildings standing tall.

三波段热红外图像（波段组合：1-2-3） 成像时间：2023-09-14 夜间
Three band thermal infrared image (band combination: 1-2-3) Imaging time: 2023-09-14 Nightti

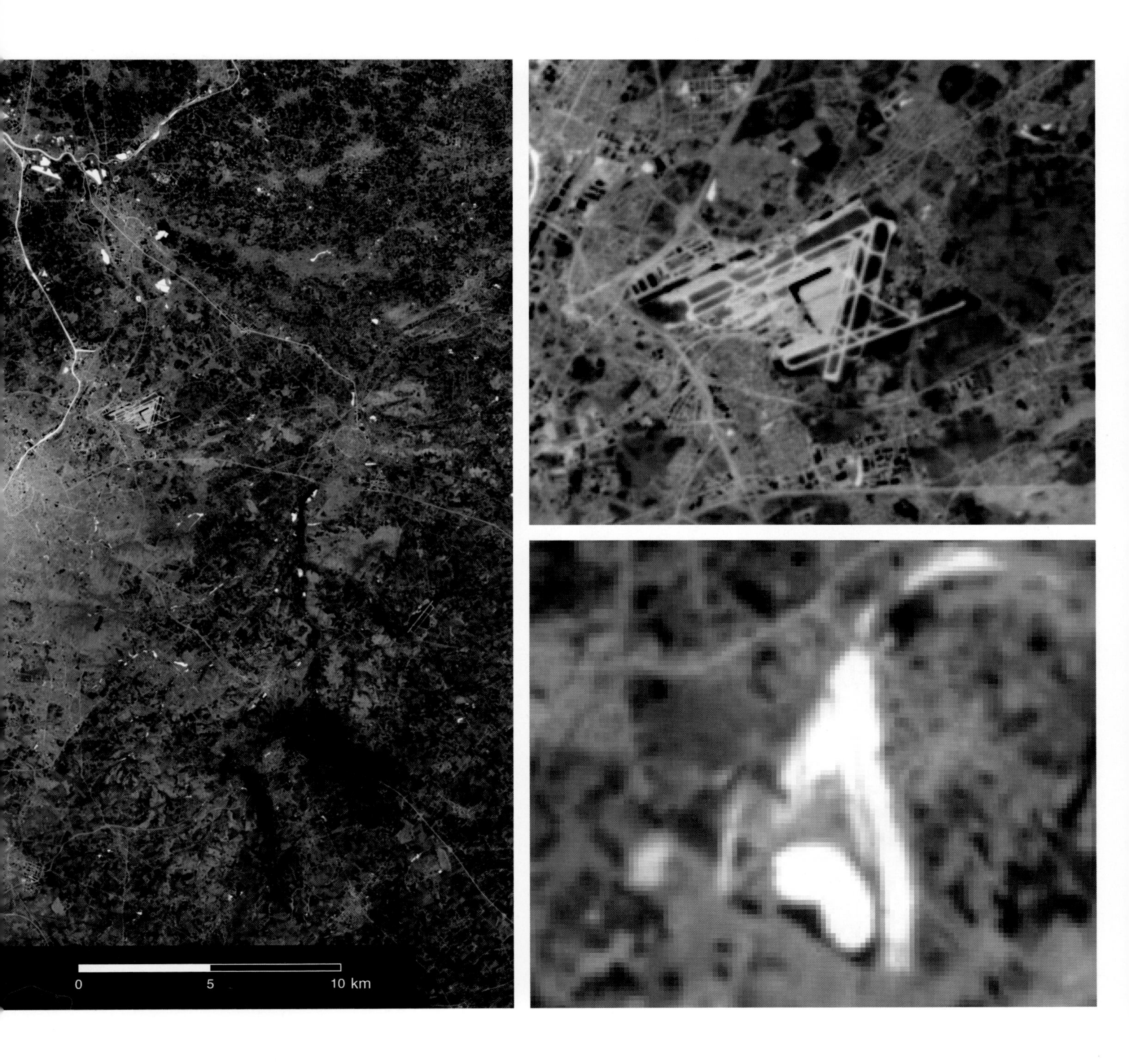
0
5
10 km

达曼

Dammam

达曼波斯湾岸边，是沙特阿拉伯东部省的省会，是东部省最大的城市，沙特石油工业的重要中心。达曼也是波斯湾最大的海港之一。城市交通便利，建有达曼火车站、法赫德国王国际机场和阿卜杜勒·阿齐兹国王港等陆海空交通枢纽。

Situated on the coast of the Persian Gulf, Dammam is the capital of the Eastern Province of Saudi Arabia. It is one of the largest city in the province and an important center of the national oil industry. It is also one of the largest seaports in the Persian Gulf. The city is well served by land, sea and air transportation hubs such as the Dammam Railway Station, King Fahd International Airport and King Abdul Aziz Port.

三波段热红外图像（波段组合：1-2-3） 成像时间：2024-02-12 夜间
Three band thermal infrared image (band combination: 1-2-3) Imaging time: 2024-02-12 Nighttime

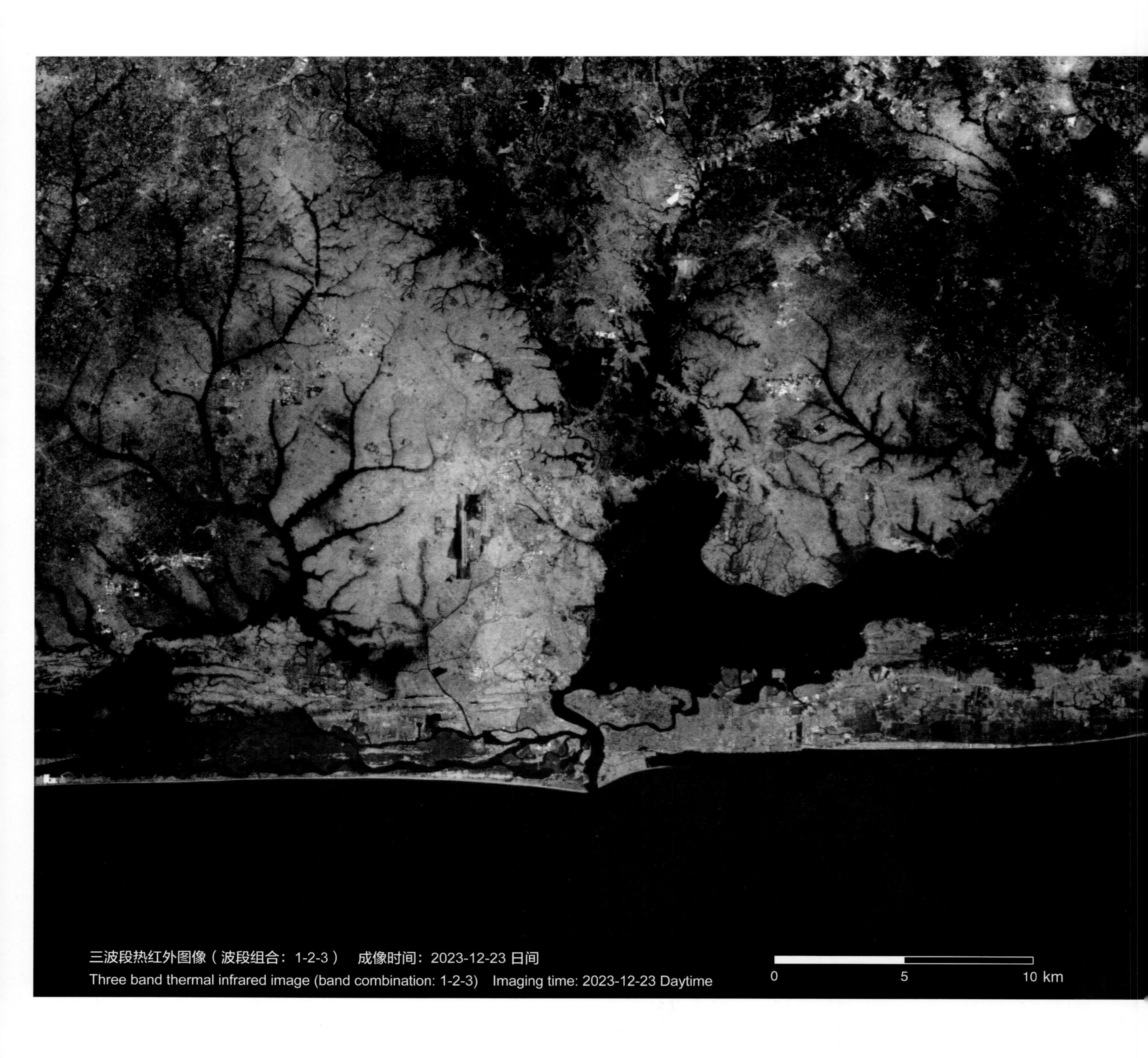

三波段热红外图像（波段组合：1-2-3） 成像时间：2023-12-23 日间
Three band thermal infrared image (band combination: 1-2-3) Imaging time: 2023-12-23 Daytime

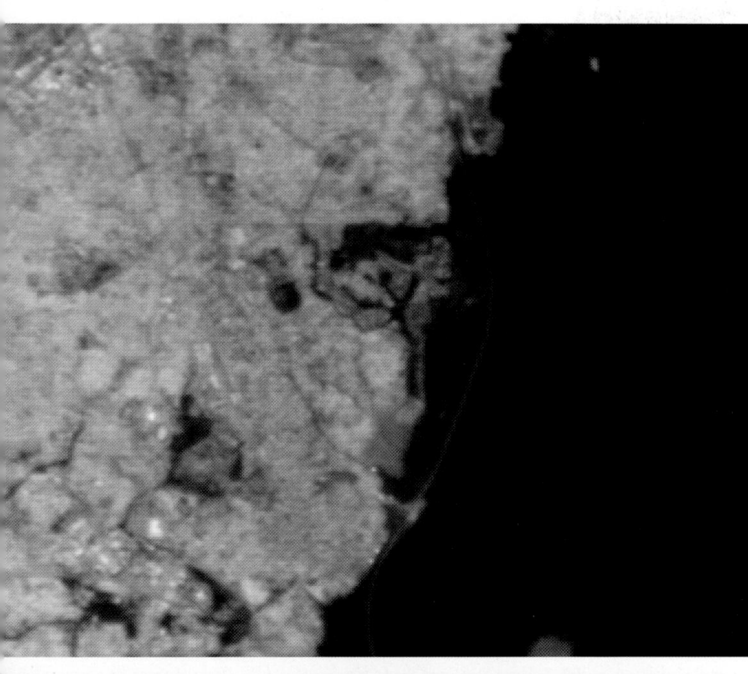

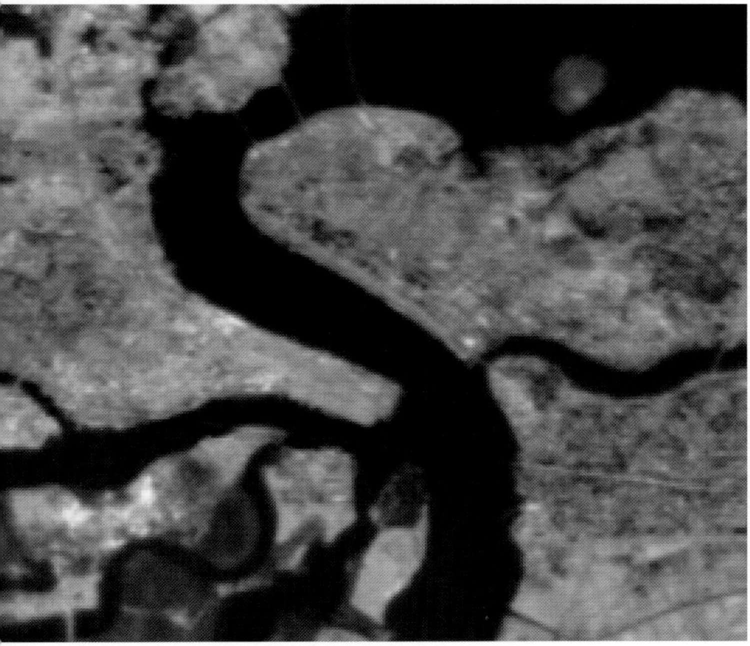

拉各斯
Lagos

拉各斯是尼日利亚旧都和最大港市，有全国最大的机场穆尔塔拉·穆罕默德国际机场。它是西非第一大城市，位于尼日利亚国境西南端，几内亚湾沿岸。拉各斯是著名的海滨疗养地、旅游中心，铁路、公路通内地扎里亚、卡诺等城市。该城市有西非最现代化海港之一，港口有设备良好的深水码头，渔业颇盛。

Lagos is the former capital and largest port city of Nigeria, boasting the country's largest airport, Murtala Muhammed International Airport. It stands as the largest city in West Africa, situated at the southwest end of Nigeria along the Gulf of Guinea coast. Renowned as a famous seaside resort and tourist center, it has railway and road connections to interior cities such as Zaria and Kano. The city is home to one of the most modern seaports in West Africa, featuring well-equipped deep-water docks and a thriving fishing industry.

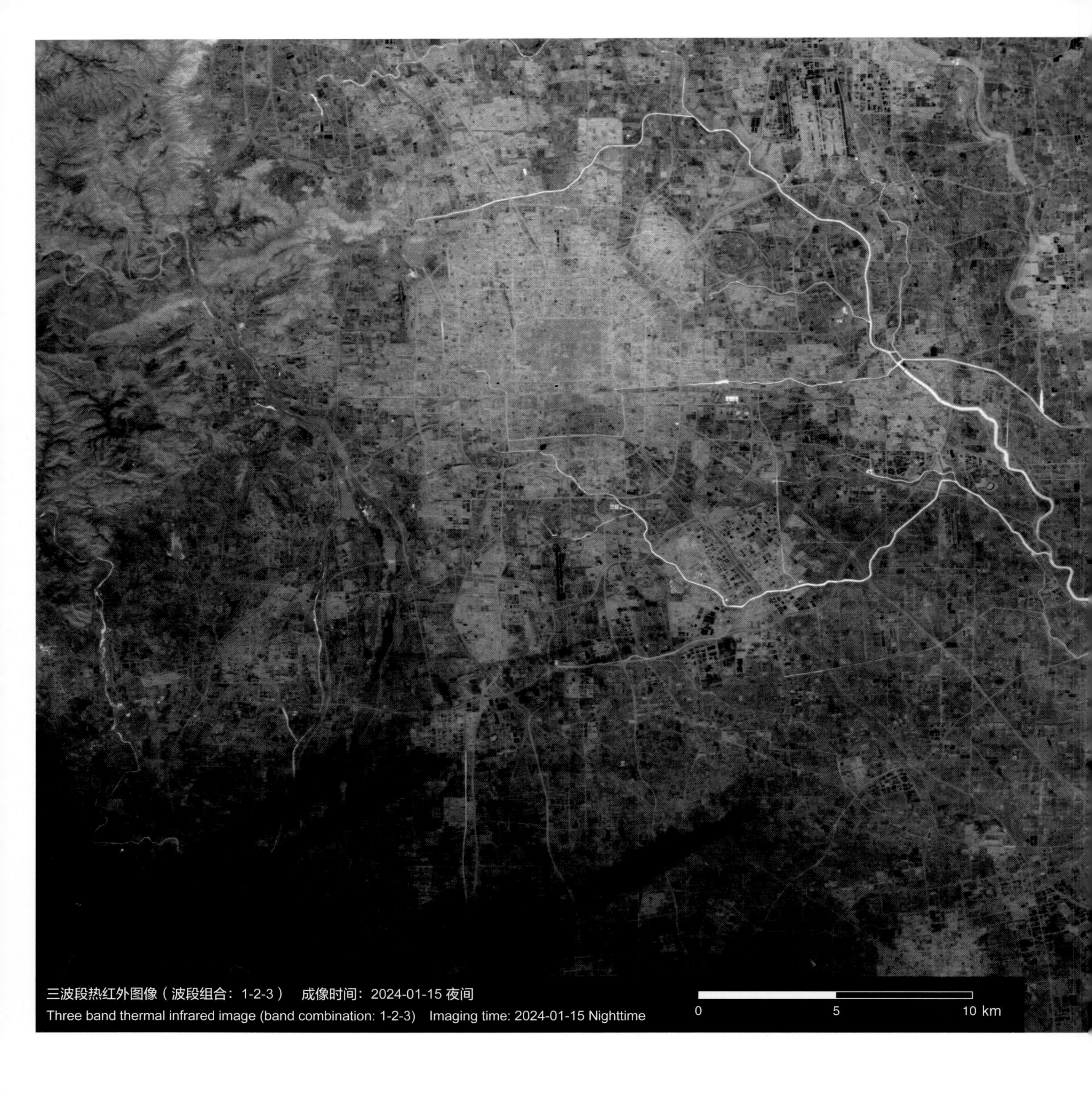

三波段热红外图像（波段组合：1-2-3）　成像时间：2024-01-15 夜间
Three band thermal infrared image (band combination: 1-2-3)　Imaging time: 2024-01-15 Nighttime

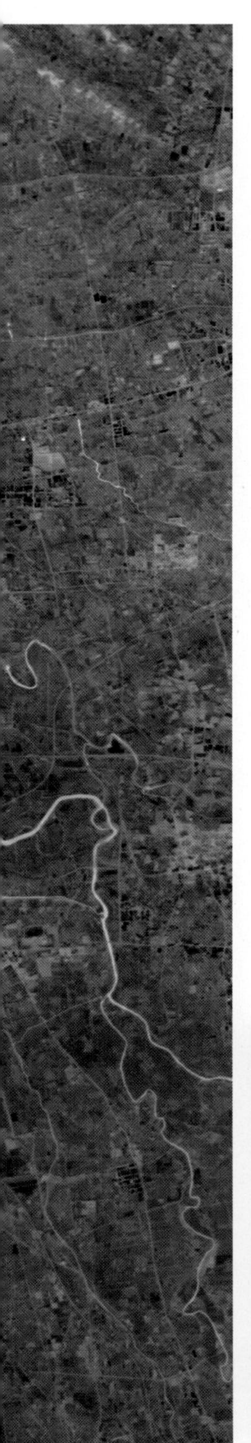

北京
Beijing

北京是中国首都、政治中心、国际交往中心，有世界超大型机场首都机场与世界级航空枢纽大兴机场，还有中国第一座军用机场南苑机场。北京地处中国北部、华北平原北部，东与天津毗连，其余均与河北相邻，境内流经的主要河流有潮白河、北运河等。成功举办夏奥会与冬奥会，是全世界第一个“双奥之城”。

Beijing, the capital of China, serves as the country's political center and an international communication hub. It is home to the world-class Capital International Airport and the world-class aviation hub, Daxing International Airport. It also hosts China's first military airport, Nanyuan Airport. Located at the northern edge of the North China Plain, Beijing is adjacent to Tianjin to the east and surrounded by Hebei on other sides. Major rivers flowing through the city include the Chaobai River, the North Canal, etc. Beijing is recognized as the world's first "Dual Olympic City", having successfully hosted both the Summer and Winter Olympic Games.

首尔
Seoul

首尔是韩国首都，也是其政治、经济、科技、教育、文化中心，亚洲主要金融城市之一，有大型国际枢纽机场仁川机场与 4E 级国际机场金浦机场，还有城南空军基地。首尔位于韩国西北部汉江流域，朝鲜半岛中部，是世界上人口密度极高城市之一，高度数字化，拥有众多著名院校，是韩国的大学之城。

Seoul is the capital of Republic of Korea, serving as its political, economic, technological, educational and cultural center. It is one of the major financial cities in Asia, home to Incheon International Airport, a major international hub, and Gimpo International Airport, classified as a 4E level airport, along with the Seongnam Air Force Base. Situated in the northwest of Republic of Korea, in the Han River basin and centrally located on the Korean Peninsula, it is one of the world's most densely populated cities. Highly digitized Seoul hosts numerous prestigious universities, earning it the reputation as South Korea's university city.

三波段热红外图像（波段组合：1-2-3） 成像时间：2023-11-20 日间
Three band thermal infrared image (band combination: 1-2-3) Imaging time: 2023-11-20 Dayti

0
5
10 km

东京
Tokyo

东京是日本的首都，世界经济中心城市、亚洲第一大都市，经济发达，人口密集，交通便利。东京国际机场（羽田）、成田国际机场是世界重要的航空枢纽。东京位于本州岛关东地区，东邻千叶县，西邻日本著名工业城市之一的川崎市。

Tokyo is the capital of Japan and one of the world's economic centers. It is also among the largest city in Asia, with a developed economy, dense population, and convenient transportation. Tokyo International Airport (Haneda) and Narita International Airport serve as major aviation hubs globally. Tokyo located in the Kanto region of Honshu Island, adjacent to Chiba Prefecture to the east. To the west, it is near Kawasaki City, one of Japan's well-known industrial cities.

三波段热红外图像（波段组合：1-2-3）　成像时间：2023-12-29 夜间
Three band thermal infrared image (band combination: 1-2-3)　Imaging time: 2023-12-29 Nighttime

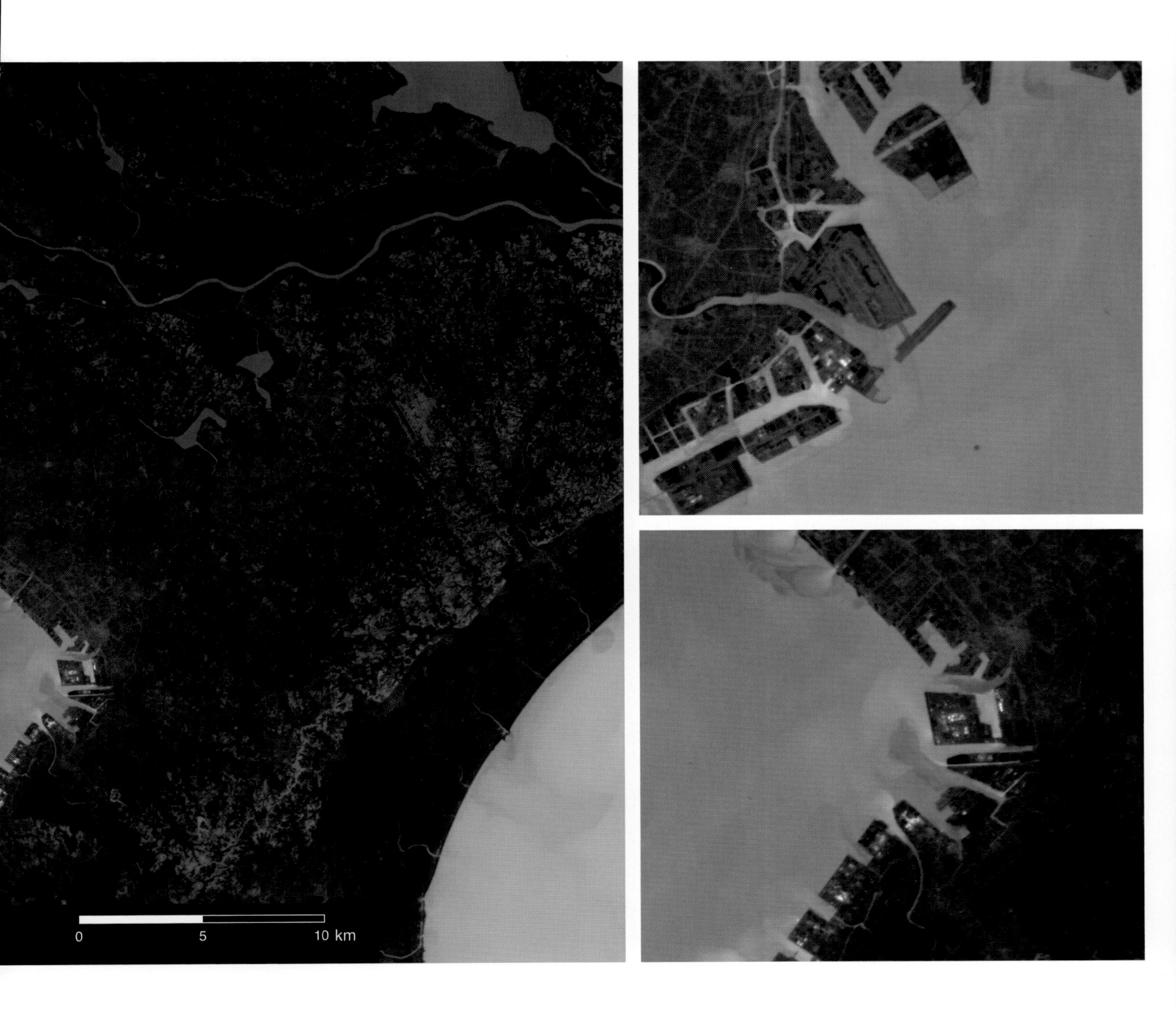
0
5
10 km

上海
Shanghai

上海是国家中心城市、超大城市，有华东区域第一大枢纽机场浦东机场与对外开放的一类航空口岸虹桥机场，还有船只进出繁忙的世界集装箱第一大港。上海位于中国华东地区，地处太平洋西岸，亚洲大陆东沿，是长江三角洲冲积平原的一部分，河网主要有流经市区的主干道黄浦江及人工河道金汇港、大治河等。

Shanghai is a national central city, a megacity, home to Pudong International Airport, the largest hub in East China, and Hongqiao International Airport, a first-class airport open to international traffic. It also boasts the world's busiest container port. Shanghai is located in the eastern of China, at the Pacific Ocean's western shore and the eastern edge of the Asian continent, it is part of the Yangtze River Delta alluvial plain. Its river network is dominated by the Huangpu River, which flows through the city, along with artificial rivers such as Jinhui Port and Dazhi River, etc.

三波段热红外图像（波段组合：1-2-3） 成像时间：2024-01-24 日间
Three band thermal infrared image (band combination: 1-2-3) Imaging time: 2024-01-24 Dayti

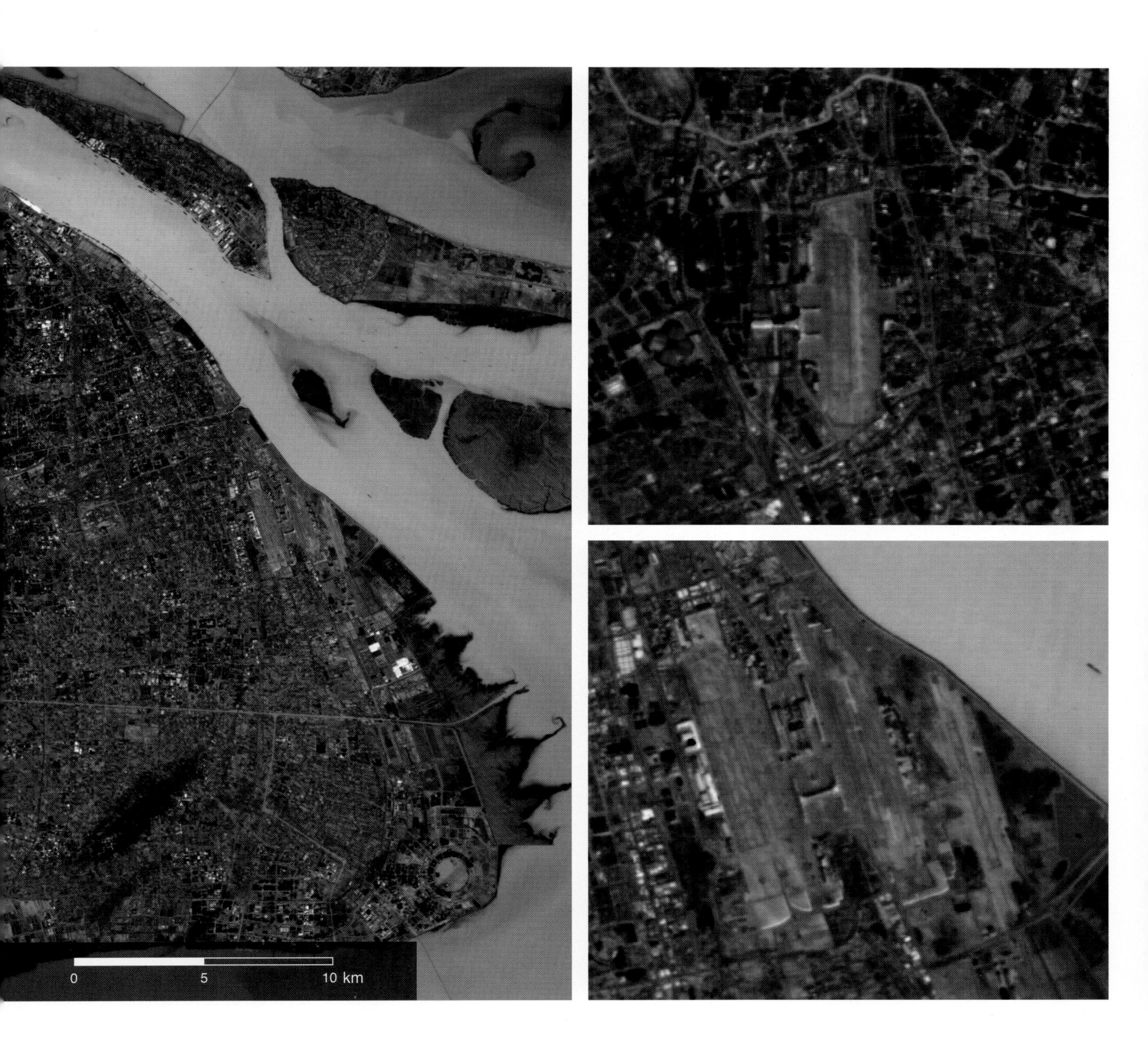
0
5
10 km

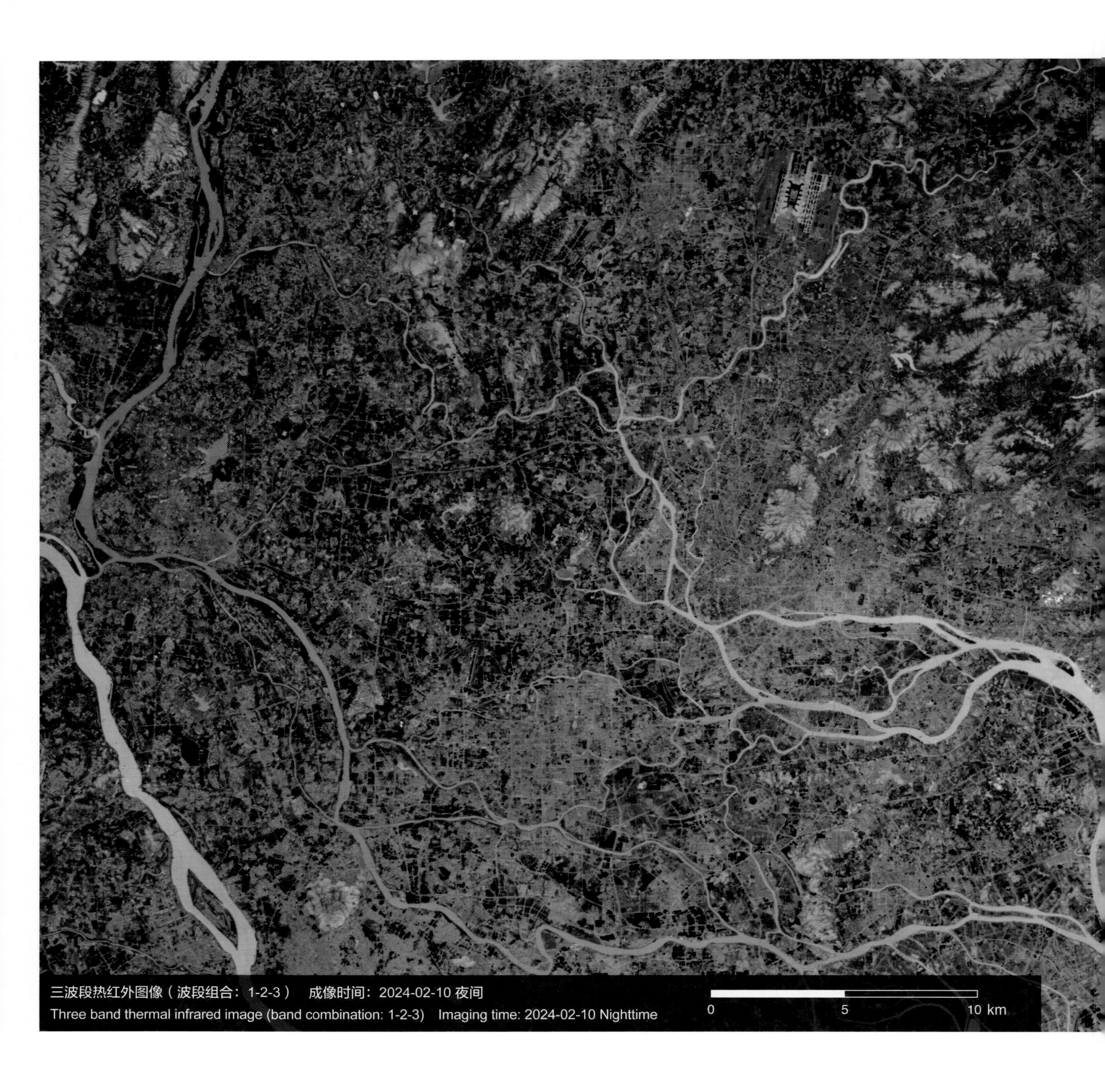

三波段热红外图像（波段组合：1-2-3） 成像时间：2024-02-10 夜间
Three band thermal infrared image (band combination: 1-2-3) Imaging time: 2024-02-10 Nighttime

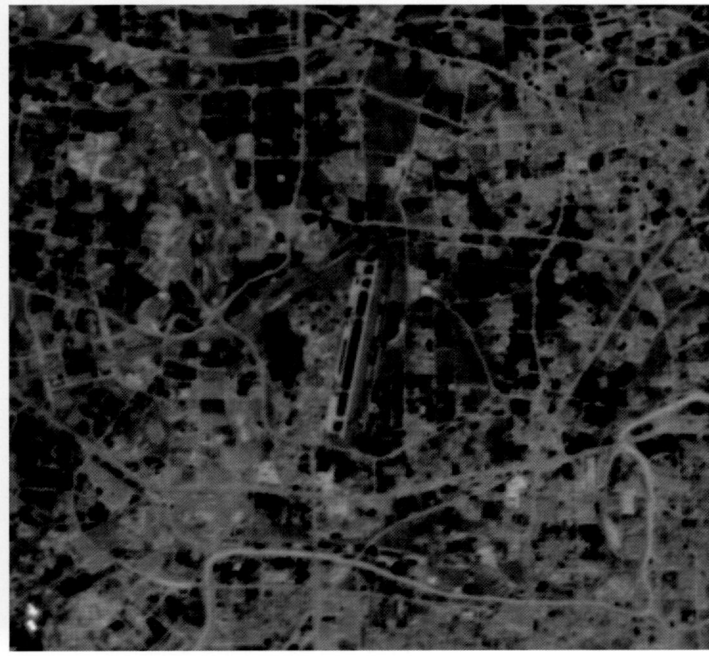

广州
Guangzhou

广州地处中国华南地区、珠江下游、濒临南海，东连惠州，西邻佛山，北靠清远与韶关，南接东莞和中山，与香港、澳门隔海相望。广州是超大城市、国家物流枢纽、国际商贸中心和综合交通枢纽，是中国南部战区司令部驻地，有世界前五十位的主要机场之一——白云机场与全球第四、华南地区最大的综合性主枢纽港和集装箱干线港口。

Guangzhou is located in the south of China, along the lower reaches of the Pearl River, and close to the South China Sea. It borders Huizhou to the east, Foshan to the west, Qingyuan and Shaoguan to the north, and Dongguan and Zhongshan to the south, situated across the Pearl River Estuary from Hong Kong and Macao. As a megacity, Guangzhou serves as a national logistics hub, an international trade center, and a comprehensive transportation hub. It is the headquarters of the Southern Theater Command of China, hosting Baiyun Airport, which is one of the world's top 50 major airports, and is home to the fourth largest comprehensive hub port and the largest container trunk port in southern China.

深圳
Shenzhen

深圳地处中国南部，东临大亚湾和大鹏湾，西濒珠江口和伶仃洋，南与香港相连，北与东莞、惠州接壤。深圳是超大城市，全国性经济中心城市和国家创新型城市，粤港澳大湾区核心引擎城市之一，被誉为“中国硅谷”。深圳是国家物流枢纽、国际性综合交通枢纽，有世界百强机场之一——宝安机场与现代化集装箱大港盐田港。

Shenzhen is located in the southern part of China, bordering Daya Bay and Dapeng Bay to the east, Pearl River Estuary and Lingdingyang Bay to the west, Hong Kong to the south, and Dongguan and Huizhou to the north. As a megacity, it is a national economic center and an innovation hub, serving as one of the core engine cities of the Guangdong-Hong Kong-Macao Greater Bay Area, and is known as the “Silicon Valley of China.” Shenzhen is a national logistics hub and an international comprehensive transportation hub, home to Bao’an Airport, ranked among the top 100 airports in the world, and Yantian Port, a leading modern container port.

三波段热红外图像（波段组合：1-2-3） 成像时间：2024-02-10 夜间
Three band thermal infrared image (band combination: 1-2-3) Imaging time: 2024-02-10 Nighttime

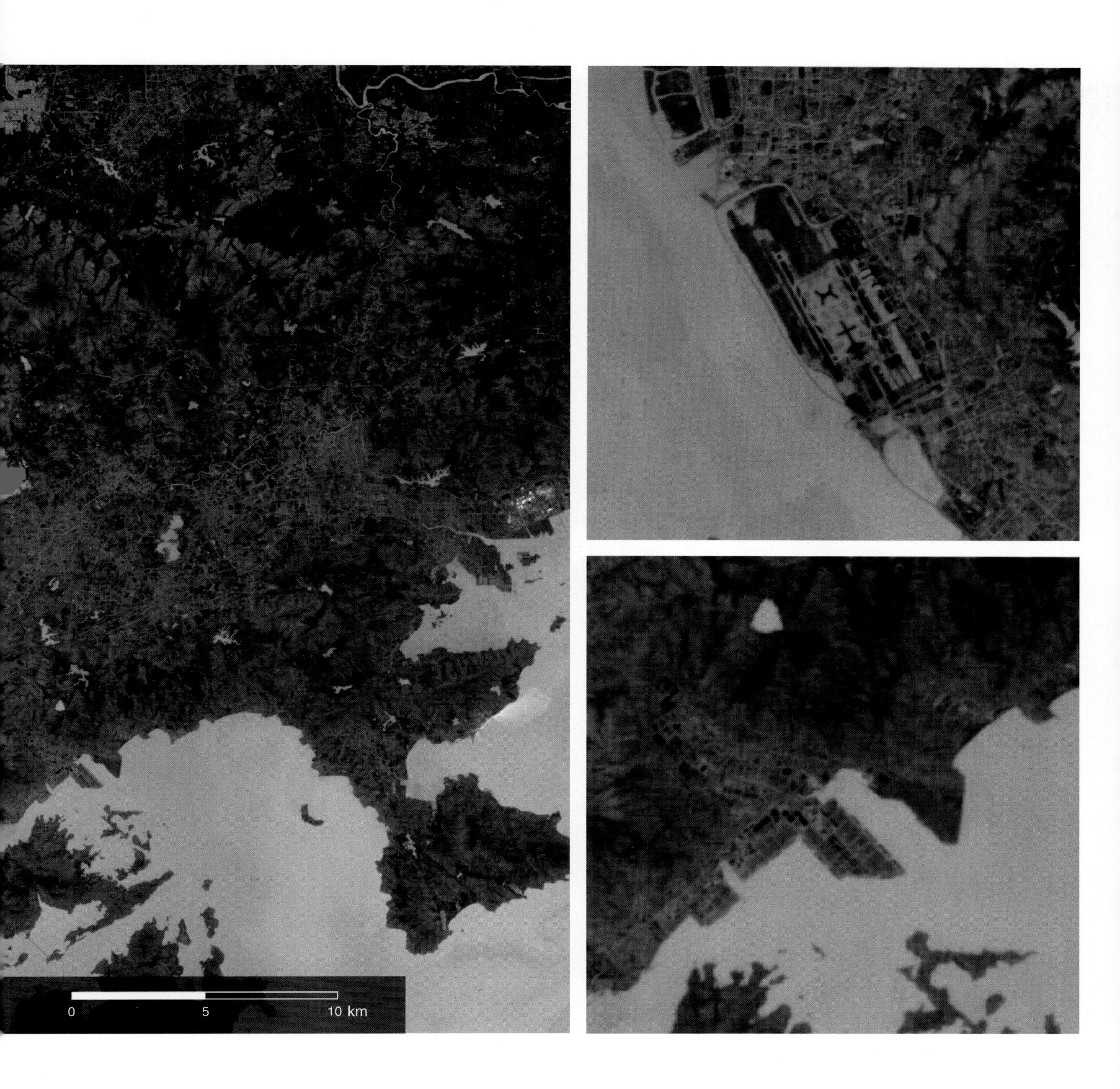
0
5
10 km

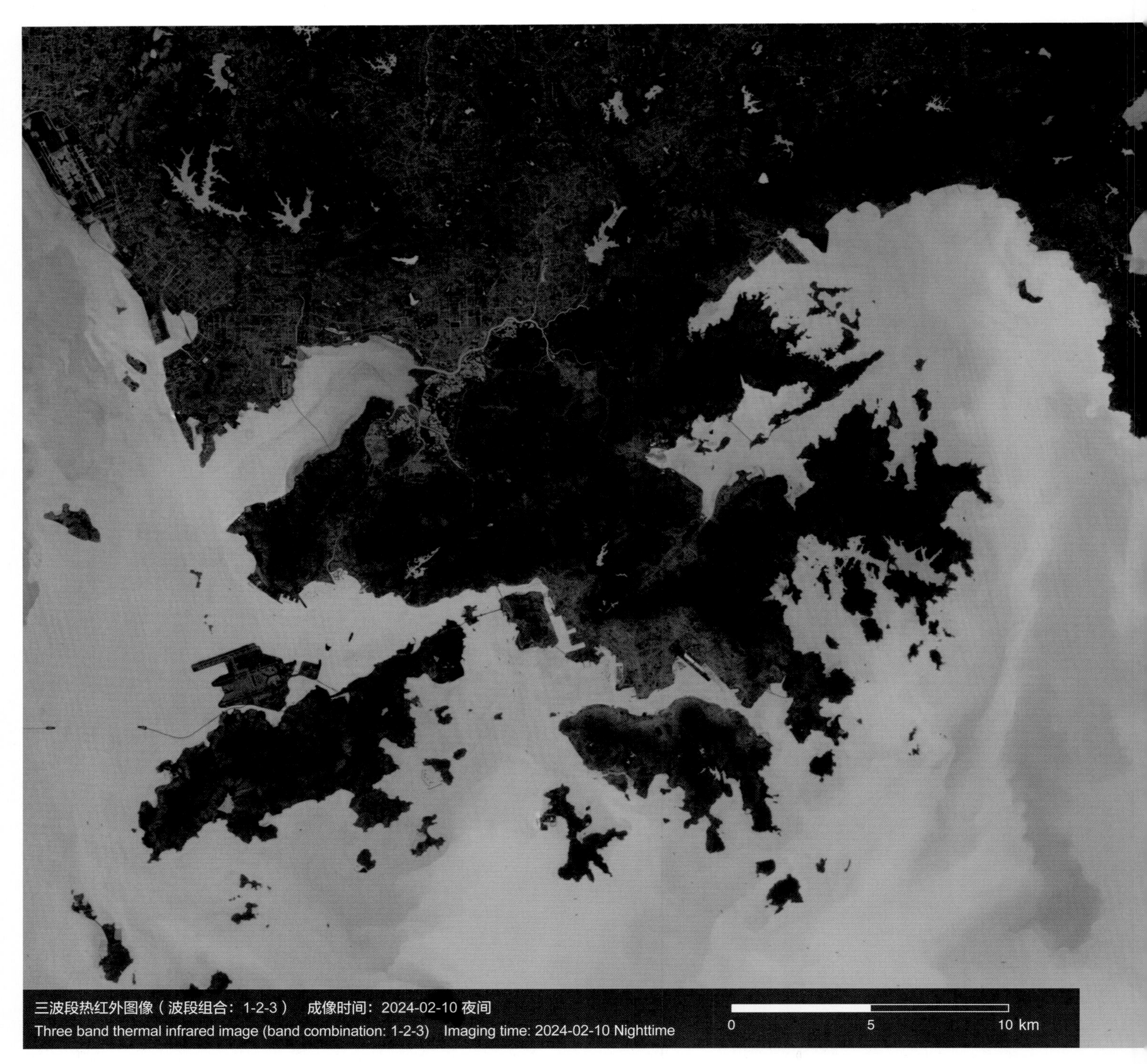

三波段热红外图像（波段组合：1-2-3） 成像时间：2024-02-10 夜间
Three band thermal infrared image (band combination: 1-2-3) Imaging time: 2024-02-10 Nighttime

香港
Hong Kong

香港位于中国南部、珠江口东岸，西与澳门隔海相望，北与深圳相邻，南临珠海万山群岛，区域范围包括香港岛、九龙、新界和周围岛屿。它是高度繁荣的自由港和国际大都市，有世界三大天然良港之一——维多利亚港与世界最繁忙的航空港之一——香港国际机场，是重要的国际贸易、航运中心和国际创新科技中心，与纽约、伦敦并称为“纽伦港”。

Hong Kong is located in the southern part of China, on the east bank of the Pearl River Estuary, offshore across from Macao to the west, adjacent to Shenzhen to the north, and close to the Zhuhai Wanshan Islands to the south. It comprises Hong Kong Island, Kowloon, the New Territories, and surrounding islands. It is a highly prosperous free port and international metropolis, featuring Victoria Harbour, one of the world's top three natural harbors, and Hong Kong International Airport, one of the busiest airports in the world. Additionally, it serves as an important international trade and shipping center, as well as an international innovation and technology hub, informally ranked alongside New York and London in the term "Nylonkong".

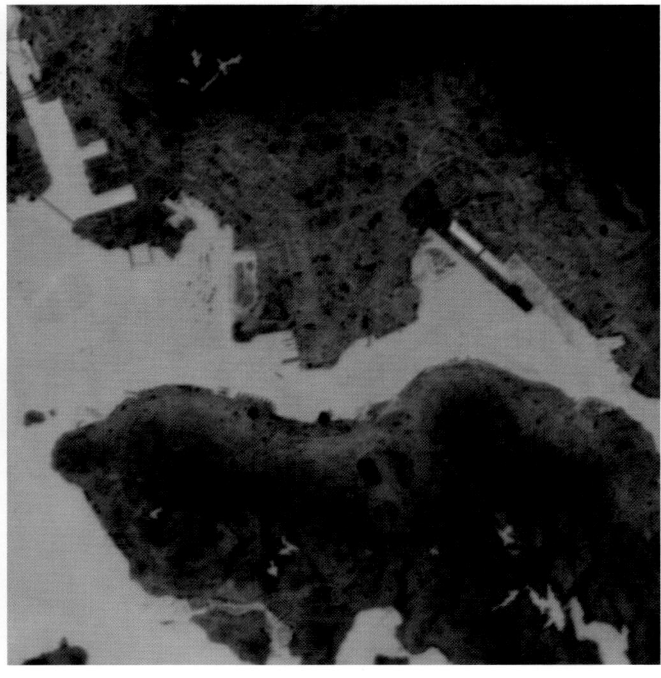

澳门
Macao

澳门位于中国南部、珠江口西侧，是中国内地与南海的水陆交汇处，毗邻广东省，东与香港隔海相望，由澳门半岛、氹仔岛、路环岛、路氹城（填海区）组成。澳门是国际自由港、世界旅游休闲中心、世界四大赌城之一，世界人口密度最高的地区之一，有全球第二个、中国第一个完全由填海造陆而建成的机场——澳门机场。

Macao is located in the south of China and west of the Pearl River Estuary, serving as a land and water junction between the mainland of China and the South China Sea. It is adjacent to Guangdong Province and faces Hong Kong across the sea to the east. Composed of the Macao Peninsula, Taipa Island, Coloane Island, and Cotai City (a reclamation area), Macao is an international free port and a world tourism and leisure center. It is one of the four major cosmopolitan gambling cities and one of the regions with the highest population density in the world. Macao is home to Macau Airport, the second airport in the world and the first in China built entirely by land reclamation.

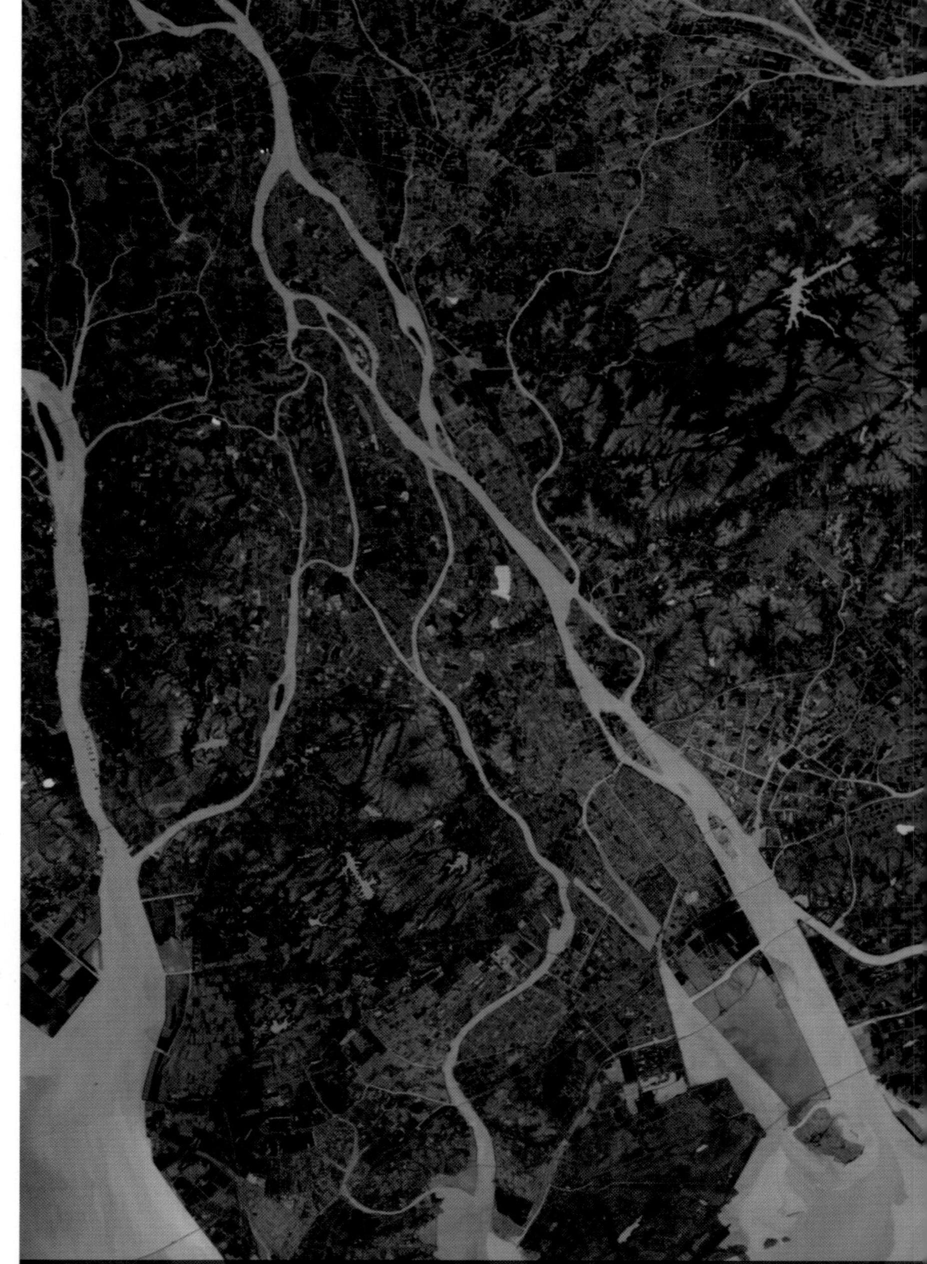

三波段热红外图像（波段组合：1-2-3）　成像时间：2024-02-10 夜间
Three band thermal infrared image (band combination: 1-2-3)　Imaging time: 2024-02-10 Nighttime

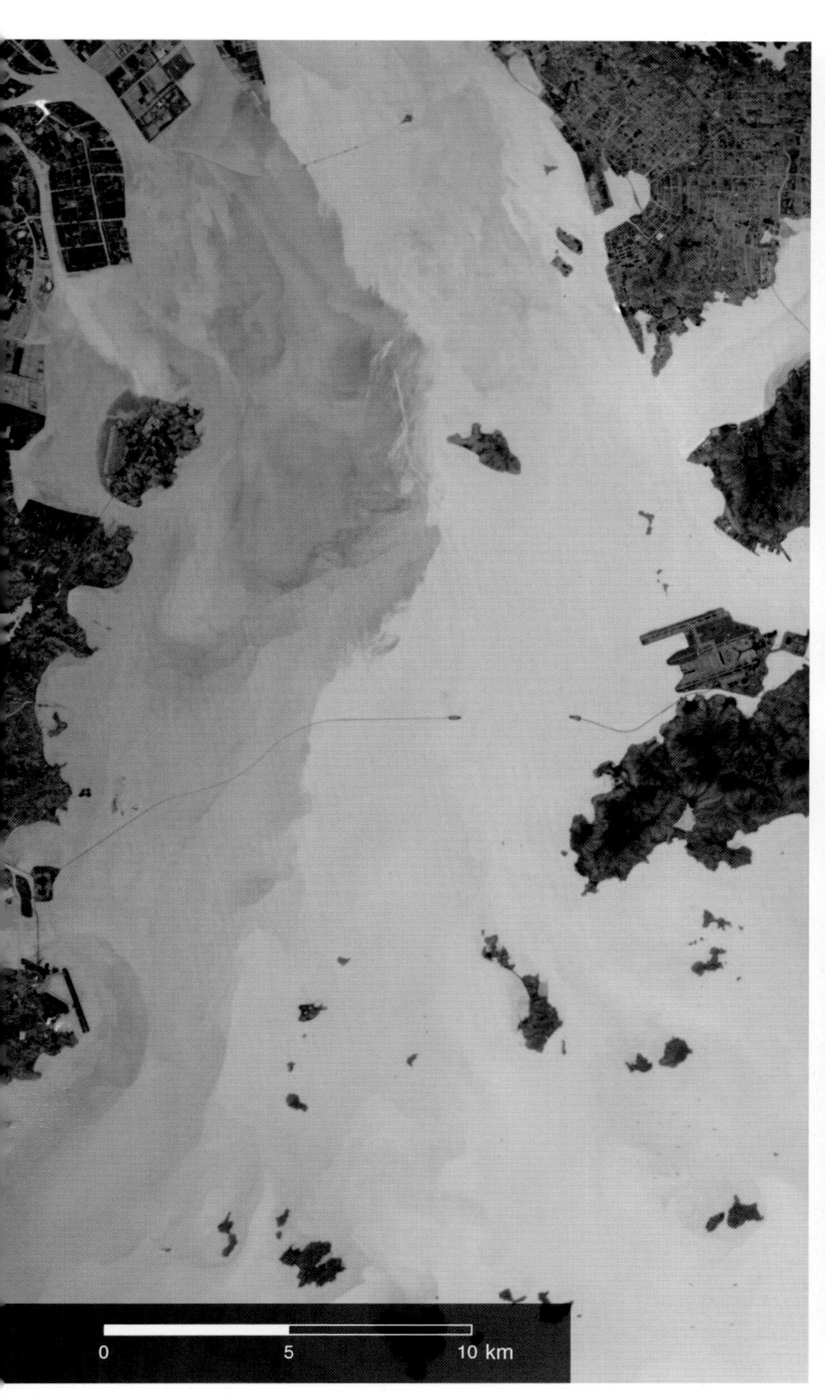
0
5
10 km

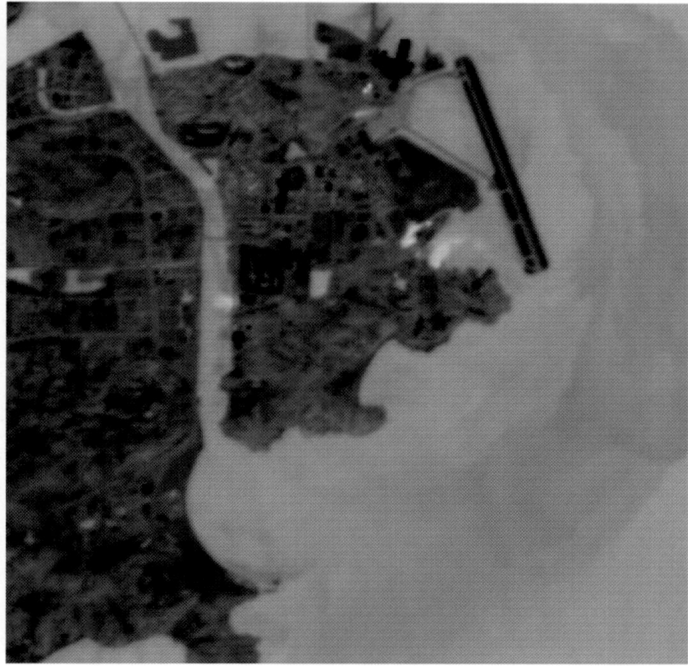

三波段热红外图像（波段组合：1-2-3） 成像时间：2023-07-12 夜间
Three band thermal infrared image (band combination: 1-2-3) Imaging time: 2023-07-12 Nighttime

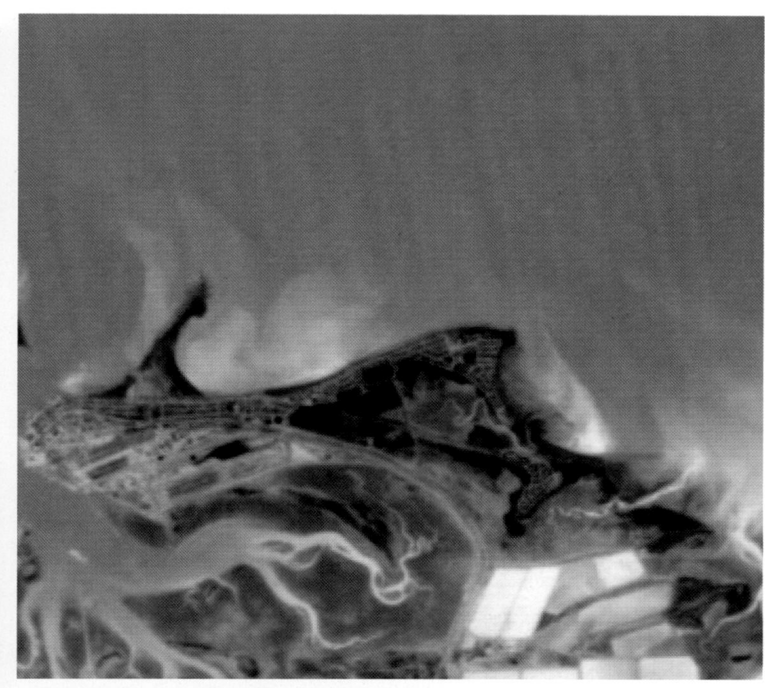

黑德兰港
Port Hedland

黑德兰港位于大洋洲、澳大利亚西北部，北临南印度洋，是天然的深水港，为该地区主要的天然气和集装箱集散地，澳大利亚主要矿石出口港之一，皮尔巴拉地区第二大市镇和商业中心。黑德兰港有黑德兰港国际机场，其是西澳大利亚州西北部地区和皮尔巴拉地区的对外门户，主营国内、地区和国际定期的客货运输业务。

Port Hedland is located in northwest Australia, Oceania, faces the South Indian Ocean to the north. It is a natural deep-water port, serving as a major natural gas and container distribution center in the region. It is also one of Australia's principal ore export ports and the second largest municipality and commercial center in the Pilbara region. The Port of Hedland is served by the Port Hedland International Airport, which acts as the gateway to the northwest region of Western Australia and the Pilbara region, primarily offering domestic, regional and international scheduled passenger and freight transportation services.

悉尼
Sydney

悉尼位于澳大利亚东南沿岸，是澳大利亚新南威尔士州的首府，也是澳大利亚面积最大的城市，已连续多年被联合国人居署评为全球最宜居的城市之一。金融业、制造业和旅游业高度发达，曾举办过多项重要国际体育赛事。

Sydney is located on the southeast coast of Australia and is the capital of New South Wales. It stands as Australia's largest city by area and has been named one of the most livable cities in the world by the United Nations Human Settlements Programme for several consecutive years. The city's finance, manufacturing and tourism industries are highly developed, and it has hosted multiple important international sports events.

三波段热红外图像（波段组合：1-2-3） 成像时间：2023-05-17 夜间
Three band thermal infrared image (band combination: 1-2-3) Imaging time: 2023-05-17 Night

0
5
10 km

墨尔本
Merlbourne

墨尔本位于澳大利亚的南海岸，是维多利亚州的首府。它是澳大利亚第二大城市，同时是该国的主要文化、艺术和工业中心。作为南半球的文化之都，墨尔本是一个世界著名的旅游目的地和国际化大都市。该市拥有澳大利亚第二繁忙的机场—墨尔本机场，它是墨尔本都会区四个机场中唯一的国际机场。此外，墨尔本还有澳大利亚最大的现代化港口—墨尔本港，它是一个关键的国际贸易港口。

Melbourne is located on the southern coast of Australia and serves as the capital of Victoria. It is the second largest city in Australia and a major cultural, artistic and industrial center. Renowned as the cultural capital of the southern hemisphere, Melbourne is a world-famous tourist destination and an international metropolis. It is served by Melbourne Airport, the second busiest airport in Australia and the only international airport among the four airports in the Melbourne metropolitan area. The city also hosts the Port of Melbourne, the largest modern port in Australia and a crucial international trade port.

三波段热红外图像（波段组合：1-2-3） 成像时间：2024-03-07 夜间
Three band thermal infrared image (band combination: 1-2-3) Imaging time: 2024-03-07 Night

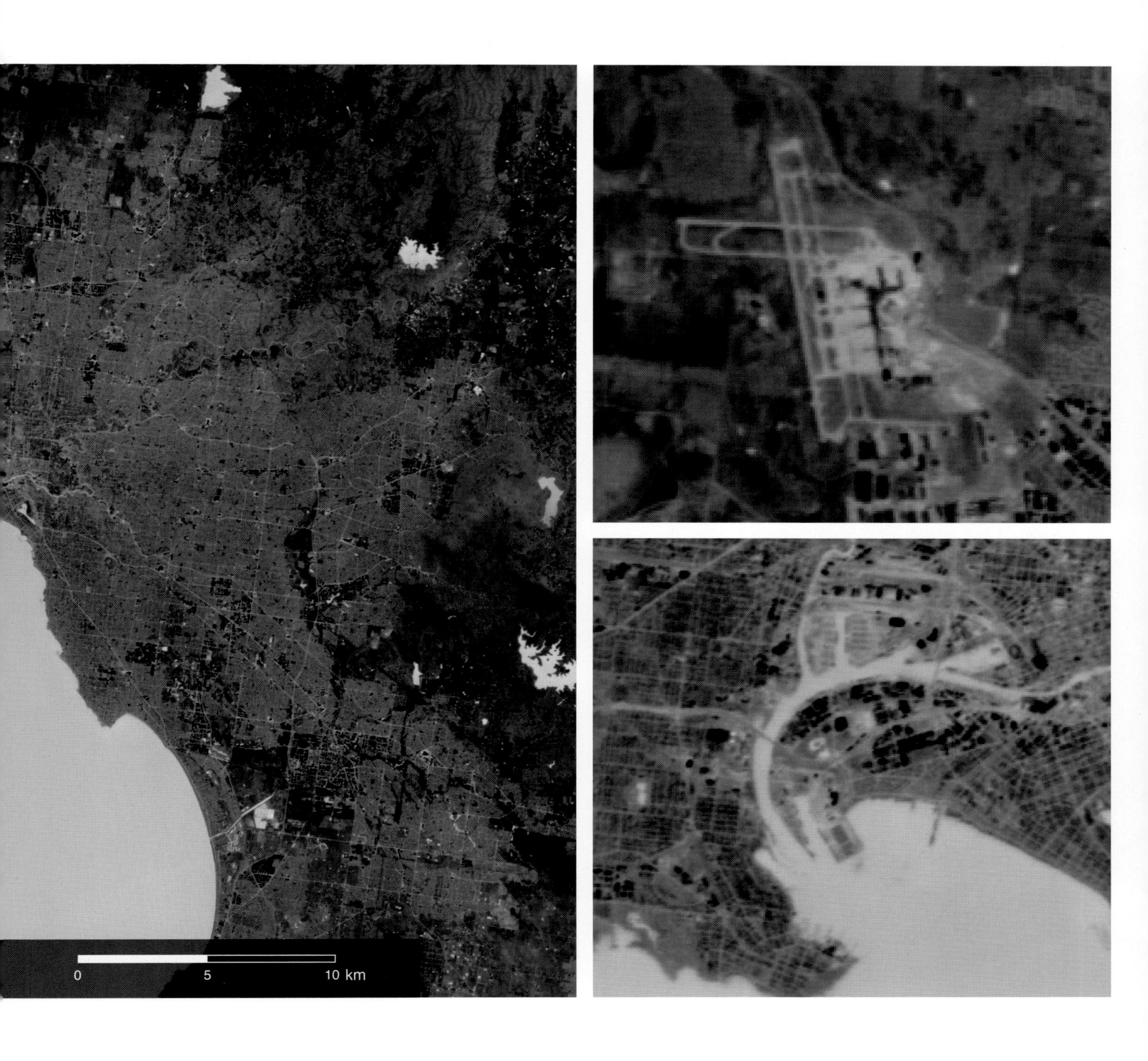
0
5
10 km

船贝

Boats and Ships

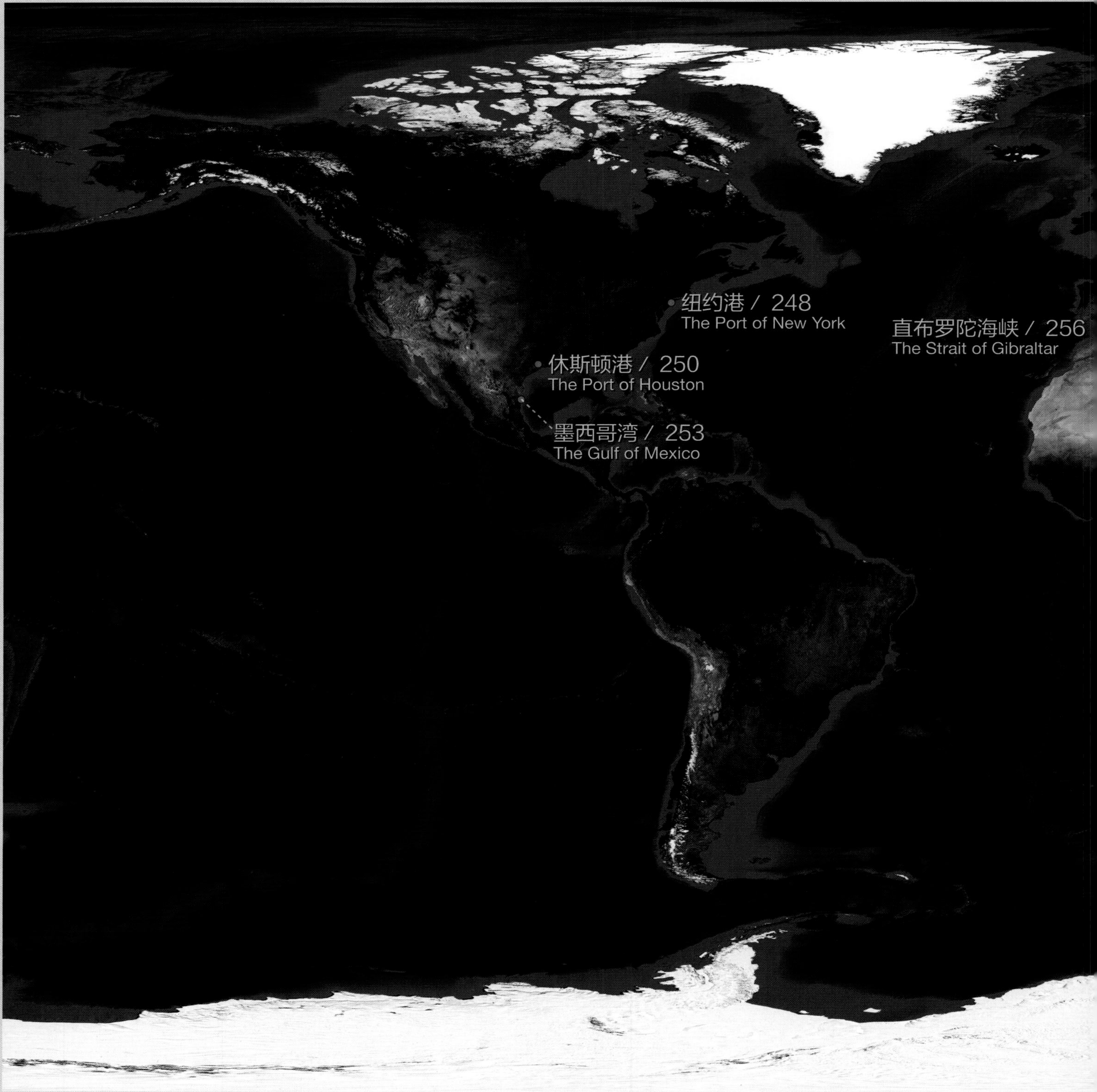

纽约港 / 248
The Port of New York

直布罗陀海峡 / 256
The Strait of Gibraltar

休斯顿港 / 250
The Port of Houston

墨西哥湾 / 253
The Gulf of Mexico

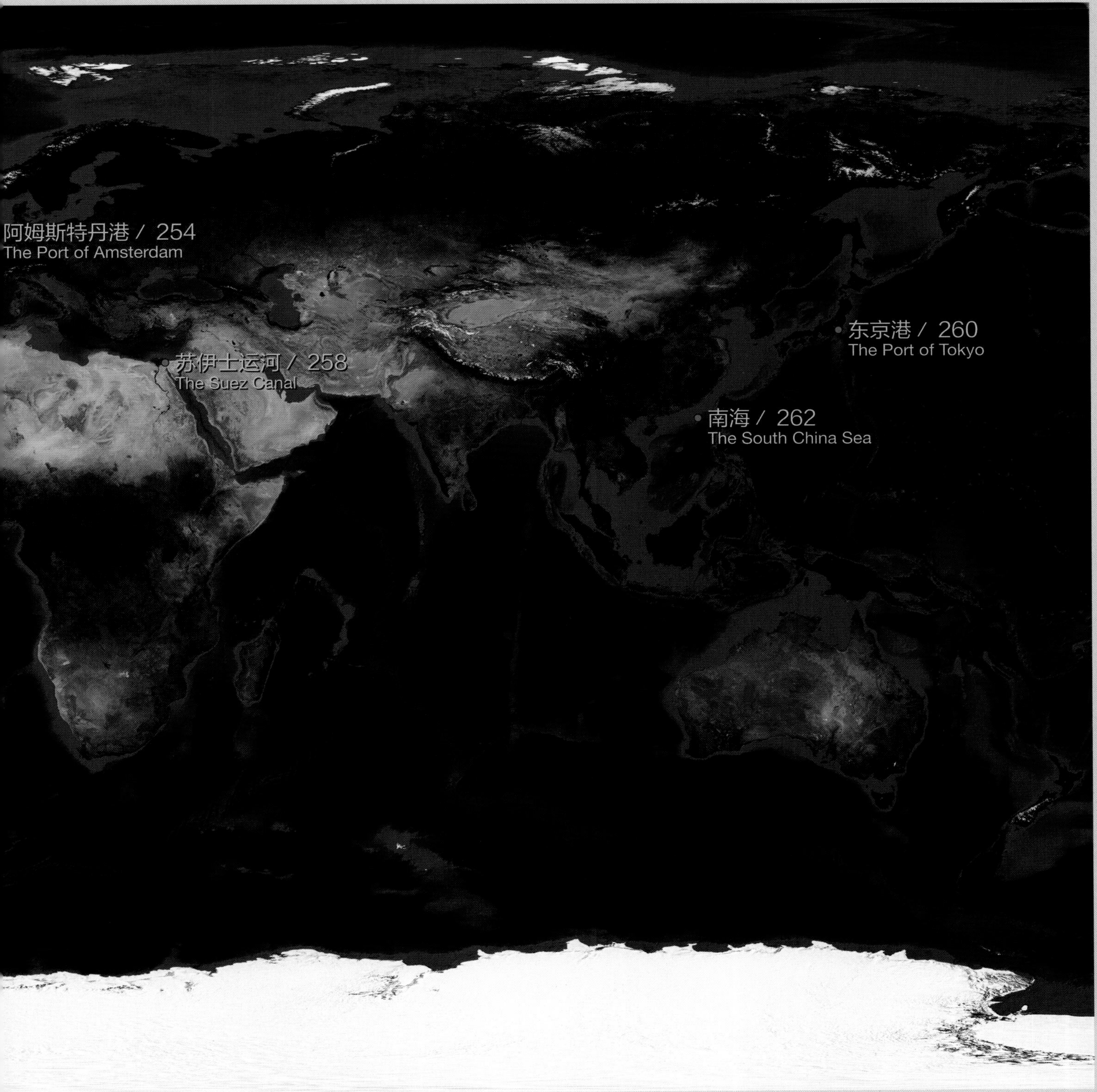

阿姆斯特丹港 / 254
The Port of Amsterdam

苏伊士运河 / 258
The Suez Canal

东京港 / 260
The Port of Tokyo

南海 / 262
The South China Sea

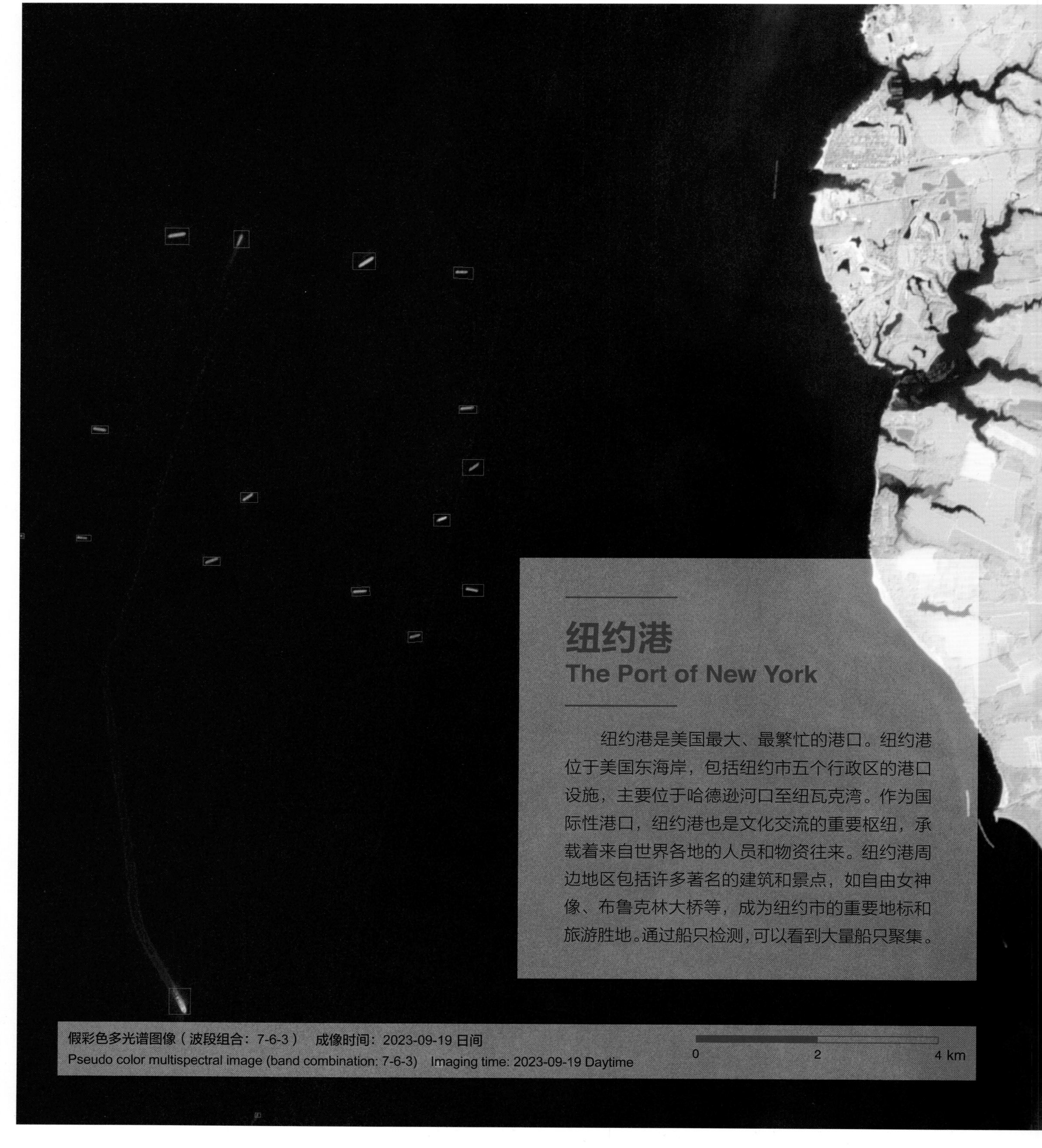

纽约港
The Port of New York

纽约港是美国最大、最繁忙的港口。纽约港位于美国东海岸，包括纽约市五个行政区的港口设施，主要位于哈德逊河口至纽瓦克湾。作为国际性港口，纽约港也是文化交流的重要枢纽，承载着来自世界各地的人员和物资往来。纽约港周边地区包括许多著名的建筑和景点，如自由女神像、布鲁克林大桥等，成为纽约市的重要地标和旅游胜地。通过船只检测，可以看到大量船只聚集。

假彩色多光谱图像（波段组合：7-6-3） 成像时间：2023-09-19 日间
Pseudo color multispectral image (band combination: 7-6-3) Imaging time: 2023-09-19 Daytime

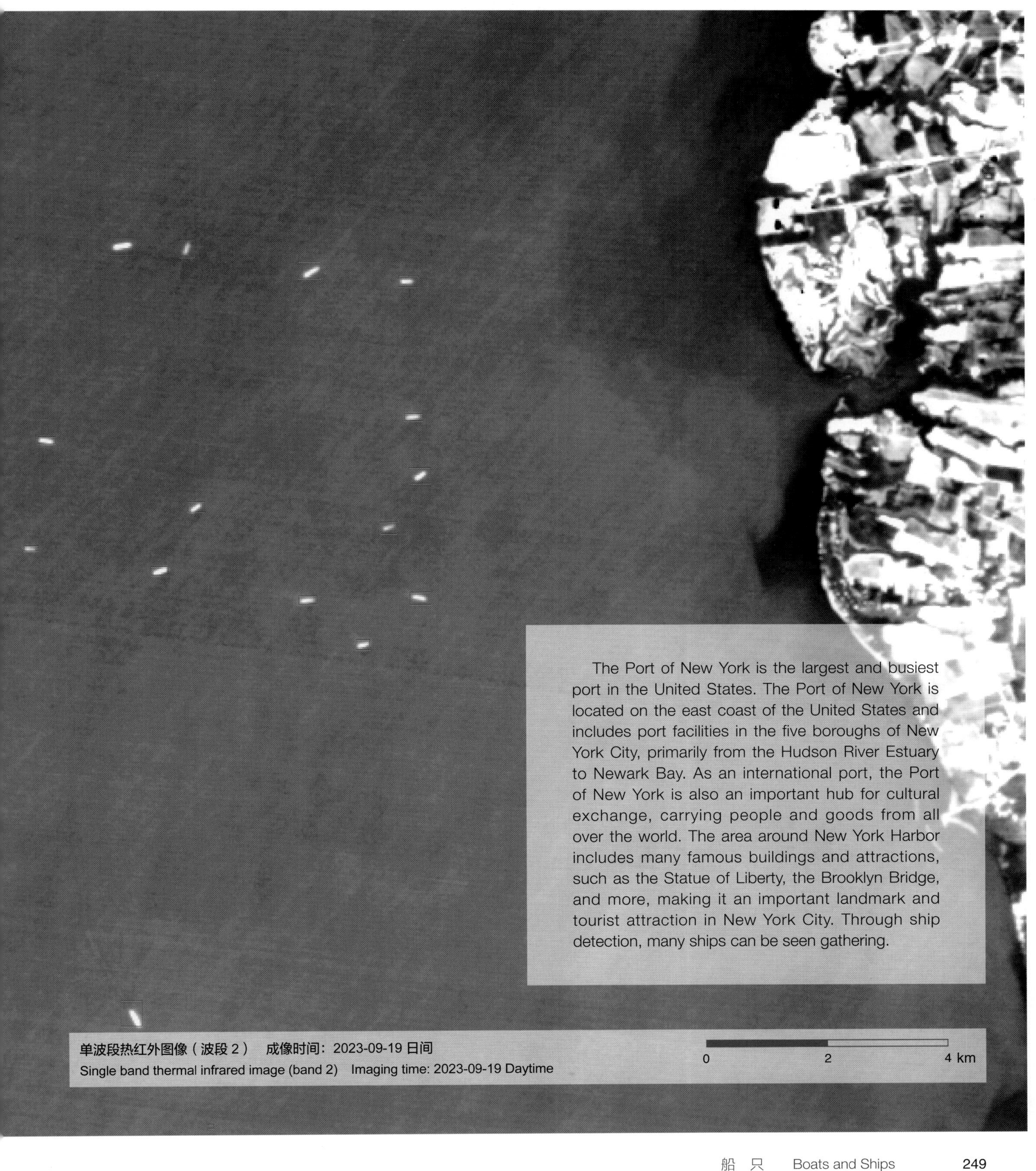

The Port of New York is the largest and busiest port in the United States. The Port of New York is located on the east coast of the United States and includes port facilities in the five boroughs of New York City, primarily from the Hudson River Estuary to Newark Bay. As an international port, the Port of New York is also an important hub for cultural exchange, carrying people and goods from all over the world. The area around New York Harbor includes many famous buildings and attractions, such as the Statue of Liberty, the Brooklyn Bridge, and more, making it an important landmark and tourist attraction in New York City. Through ship detection, many ships can be seen gathering.

单波段热红外图像（波段 2） 成像时间：2023-09-19 日间
Single band thermal infrared image (band 2) Imaging time: 2023-09-19 Daytime

休斯顿港
The Port of Houston

休斯顿港位于美国德克萨斯州东南部，沿着布拉佛河和加尔维斯顿湾。休斯顿港通过一系列航道与墨西哥湾相连，是内河和海洋航运的重要枢纽，拥有现代化的码头设施和装卸设备，能够容纳各种类型的货物和船只，是德克萨斯州最繁忙的港口之一。

The Port of Houston is located in southeastern Texas, United States, along the Braavor River and Galveston Bay. The Port of Houston is connected to the Gulf of Mexico by a series of shipping lanes and is an important hub for inland and ocean shipping. With modern terminal facilities and handling equipment capable of accommodating all types of cargo and vessels, the Port of Houston is one of the busiest ports in Texas.

假彩色多光谱图像（波段组合：7-6-3）　成像时间：2022-11-27 日间

Pseudo color multispectral image (band combination: 7-6-3)　Imaging time: 2022-11-27 Daytime

0　5　10 km

单波段热红外图像（波段 2）成像时间：2022-11-27 日间

Single band thermal infrared image (band 2)　Imaging time: 2022-11-27 Daytime

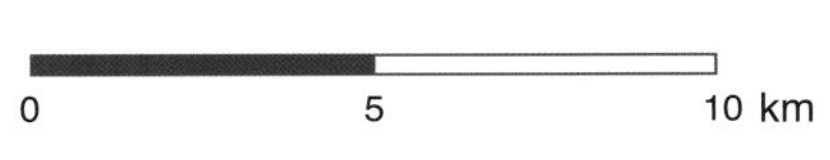

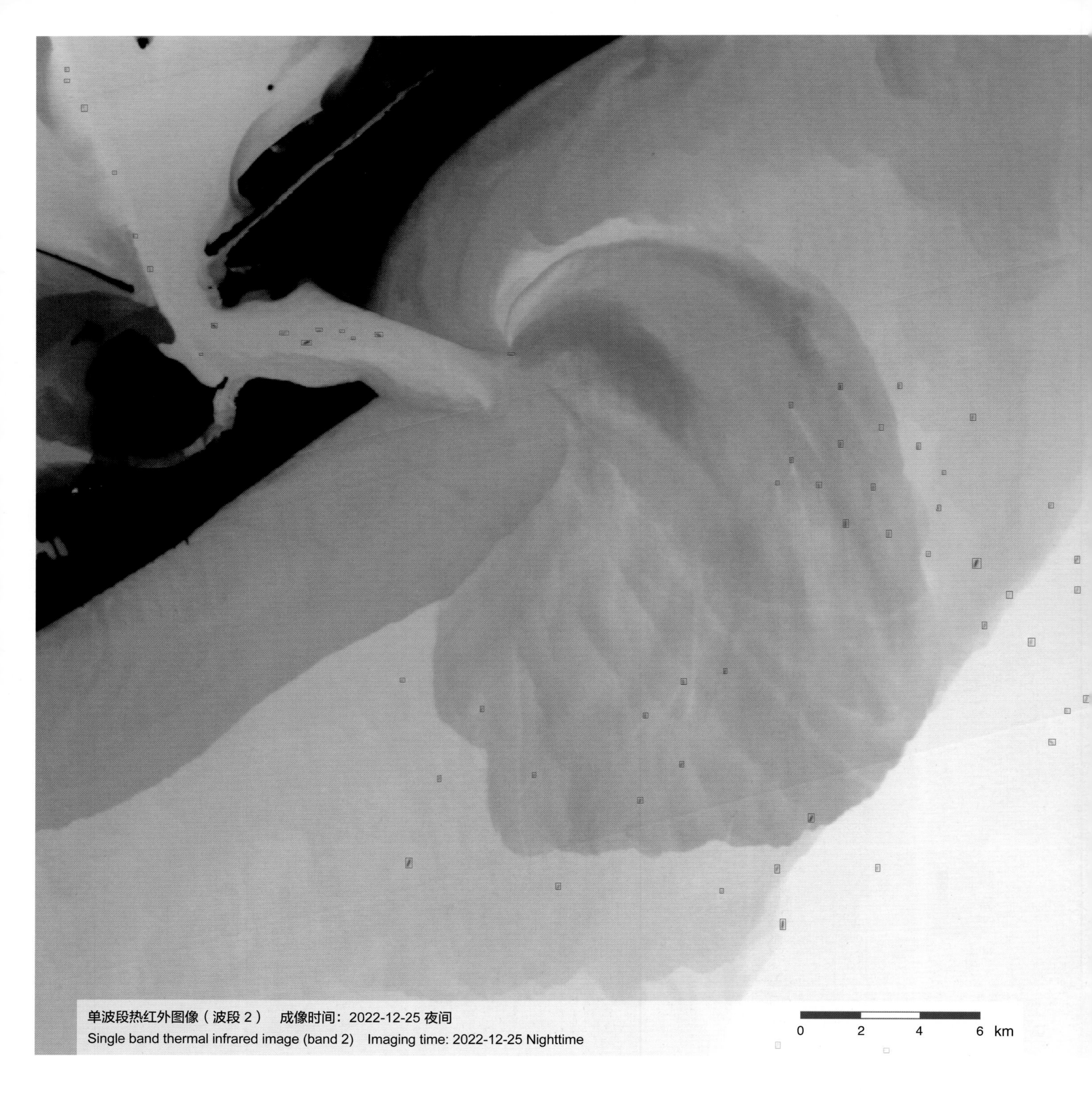

单波段热红外图像（波段 2） 成像时间：2022-12-25 夜间
Single band thermal infrared image (band 2) Imaging time: 2022-12-25 Nighttime

墨西哥湾
The Gulf of Mexico

墨西哥湾位于北美洲的墨西哥东部，向北连接着美国的德克萨斯州、路易斯安那州、密西西比州、阿拉巴马州和佛罗里达州。墨西哥湾是世界上最大的陆架边缘海之一。它拥有复杂的海岸线，包括沙滩、潟湖、沼泽地和沼泽树林。墨西哥湾是一个生态丰富的区域，拥有多样的生物多样性，包括珊瑚礁、海龟、海豚、鲨鱼等。通过热红外图像船只检测，大量船只在湾区航行。

The Gulf of Mexico is located in eastern Mexico in North America, connecting the United States to the north by states of Texas, Louisiana, Mississippi, Alabama, and Florida. The Gulf of Mexico is one of the largest shelf marginal seas in the world. It has a complex coastline that includes sandy beaches, lagoons, marshes and marshy forests. The Gulf of Mexico is an ecologically rich region with diverse biodiversity, including coral reefs, sea turtles, dolphins, sharks, and more. Using thermal infrared image for ship detection, many ships can be seen sailing in the Gulf of Mexico.

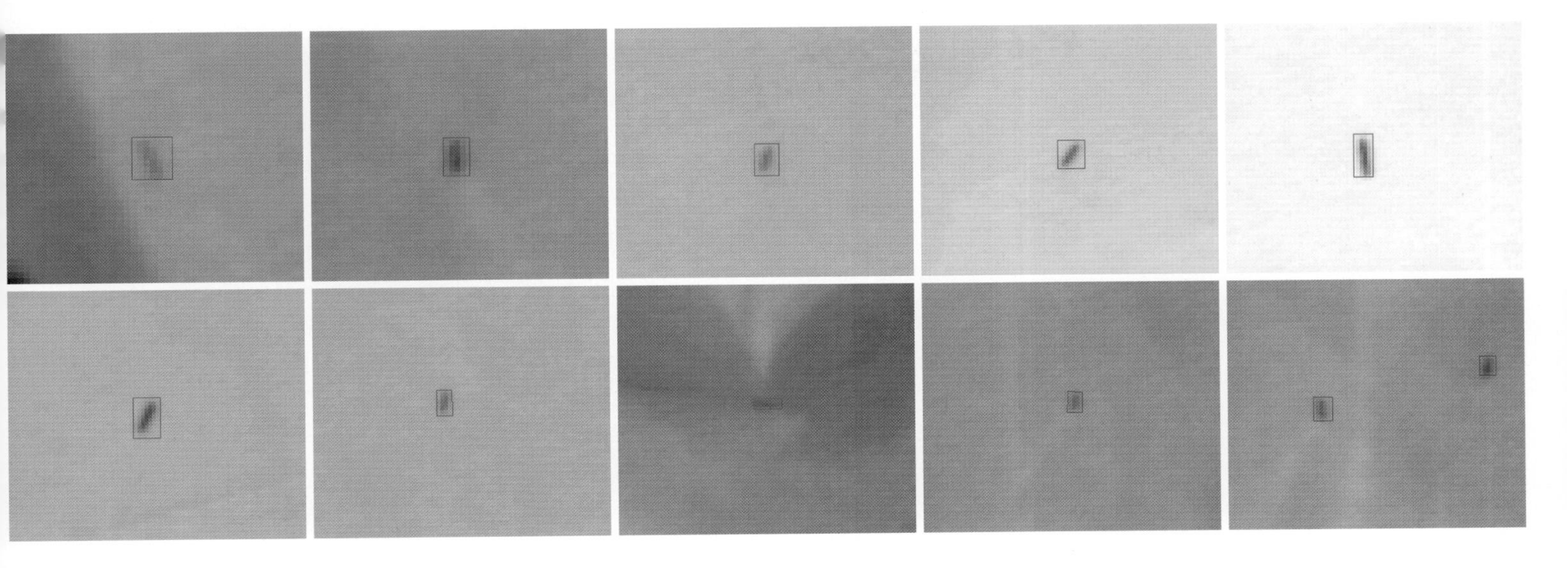

阿姆斯特丹港
The Port of Amsterdam

阿姆斯特丹港位于荷兰西部，与北海相连，是荷兰最大的港口之一。阿姆斯特丹港通过众多航道与北海和莱茵河等内陆水道相连，是重要的国际航运枢纽。阿姆斯特丹港拥有现代化的码头、装卸设备和物流设施，能够容纳大型货船和处理各类货物，港区内有大量船只通行。

The Port of Amsterdam is located in the western part of the Netherlands, connected to the North Sea, and is one of the largest ports in the Netherlands. The Port of Amsterdam is connected to inland waterways such as the North Sea and the Rhine by numerous shipping lanes, making it an important international shipping hub. The Port of Amsterdam has modern terminals, handling equipment and logistics facilities that can accommodate large cargo ships and handle all types of cargo. There are a large number of ships passing through the port area.

假彩色多光谱图像（波段组合：7-6-3） 成像时间：2022-05-08 日间
Pseudo color multispectral image (band combination: 7-6-3) Imaging time: 2022-05-08 Daytime

0 2.5 5 km

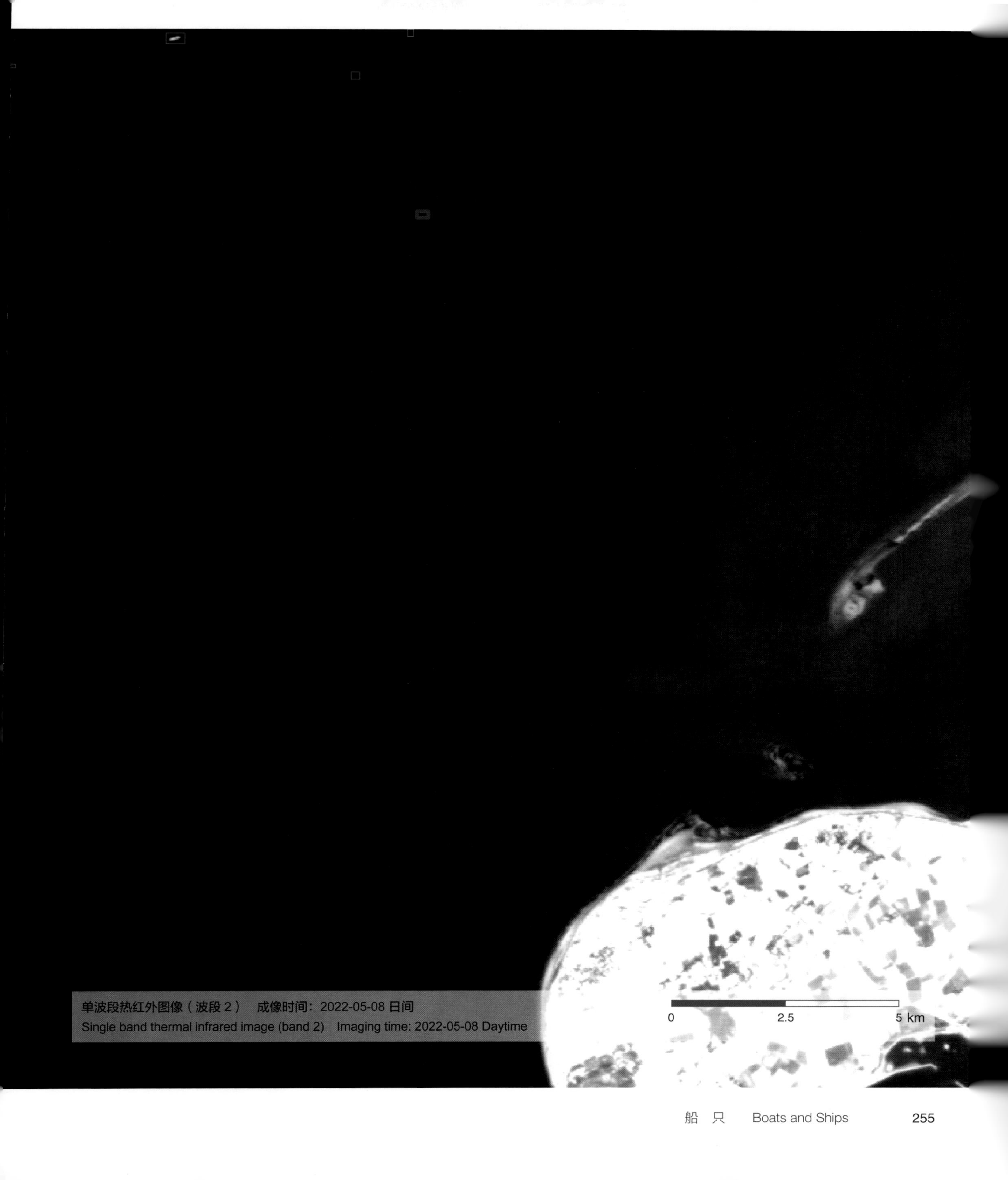
单波段热红外图像（波段 2）　成像时间：2022-05-08 日间
Single band thermal infrared image (band 2)　Imaging time: 2022-05-08 Daytime
0
2.5
5 km

直布罗陀海峡

The Strait of Gibraltar

直布罗陀海峡位于欧洲大陆的南端，连接地中海东部和大西洋西部，是地中海和大西洋之间的主要海上通道。直布罗陀海峡是世界上最狭窄的海峡之一。由于地理位置的特殊性，直布罗陀海峡成为连接地中海沿岸国家与大西洋沿岸国家的主要航道。通过船只检测，可以看到大量船只穿梭海峡的繁忙景象。

The Strait of Gibraltar is located at the southern tip of the European continent, connecting the eastern Mediterranean Sea and the western Atlantic, and is the main sea passage between the Mediterranean and the Atlantic Ocean. At its narrowest point, making it one of the narrowest straits in the world. Due to its geographical location, the Strait of Gibraltar has become the main shipping route connecting the countries bordering the Mediterranean Sea with the countries bordering the Atlantic. With the support of ship detection, a busy scene of a large number of ships shuttling through the strait can be seen.

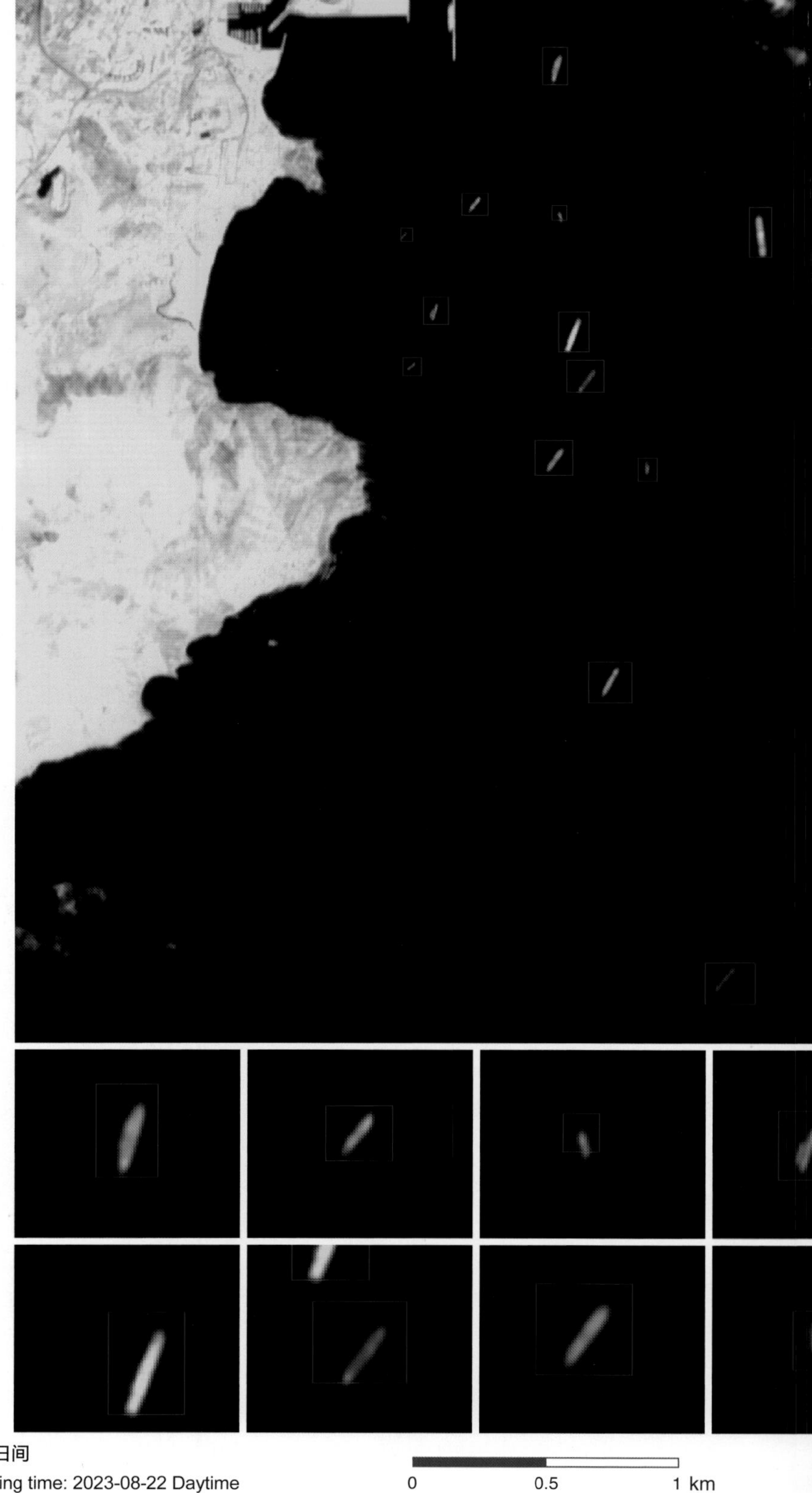

假彩色多光谱图像（波段组合：7-6-3） 成像时间：2023-08-22 日间

Pseudo color multispectral image (band combination: 7-6-3) Imaging time: 2023-08-22 Daytime

0 0.5 1 km

单波段热红外图像（波段 2） 成像时间：2023-08-22 日间
Single band thermal infrared image (band 2) Imaging time: 2023-08-22 Daytime

苏伊士运河
The Suez Canal

苏伊士运河位于埃及北部，连接地中海北部的港口城市亚历山大和红海南部的苏伊士市，总长约 193.3km。作为连接地中海和红海的主要航道，苏伊士运河大大缩短了从欧洲到印度洋的航行距离，对国际贸易具有重要意义。通过船只检测，可以看到大量船只通行其中。

The Suez Canal is located in northern Egypt, connecting the port city of Alexandria in the northern Mediterranean with the city of Suez in the southern Red Sea. The Suez Canal is about 193.3km long. As the main waterway connecting the Mediterranean Sea and the Red Sea, the Suez Canal greatly shortens the sailing distance from Europe to the Indian Ocean, and is of great importance for international trade. It can be seen that a lot of ships passing through the canal by ship detection.

假彩色多光谱图像（波段组合：7-6-3） 成像时间：2023-03-04 日间

Pseudo color multispectral image (band combination: 7-6-3) Imaging time: 2023-03-04 Daytime

0 2 4 6 km

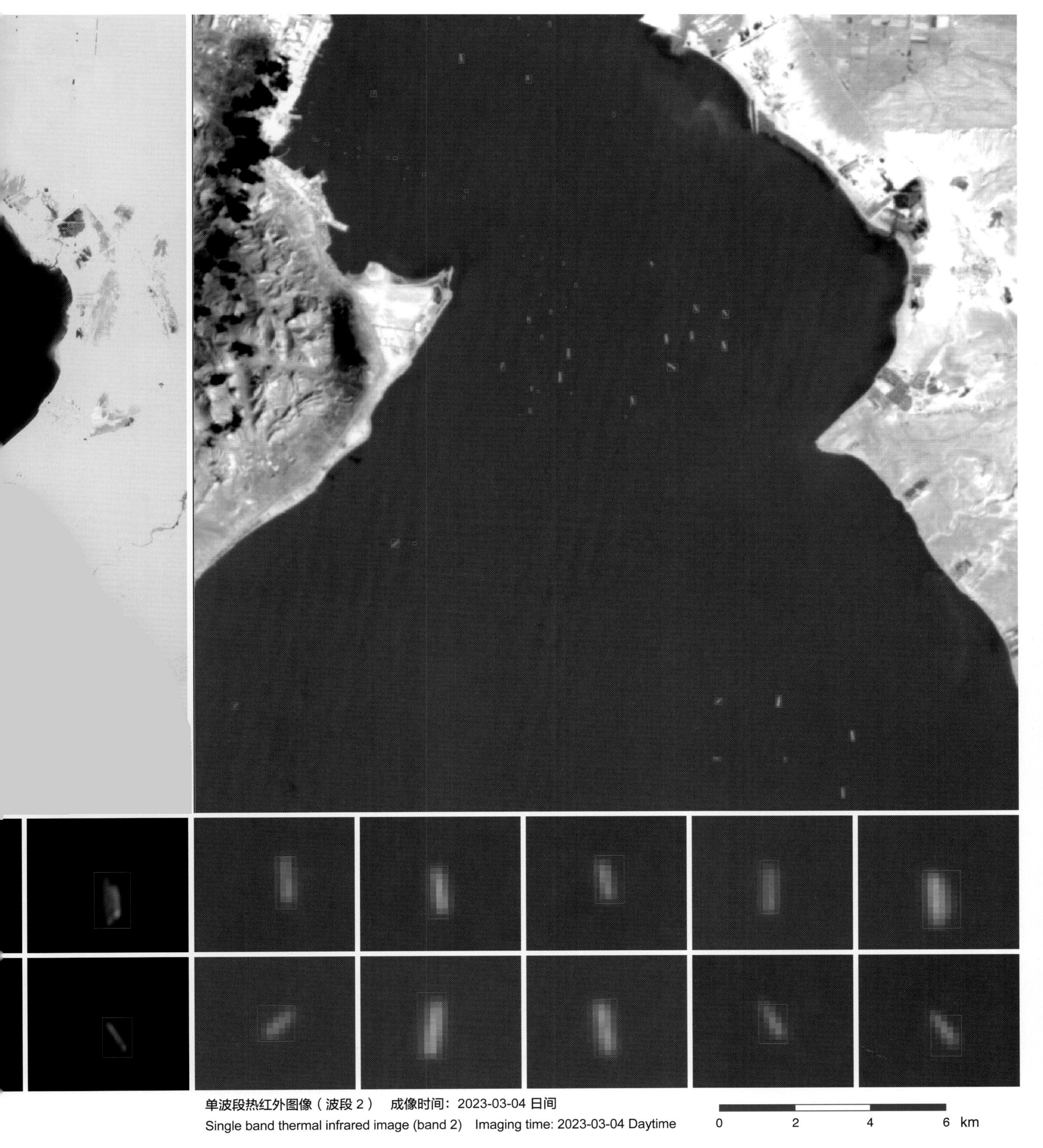

单波段热红外图像（波段 2）　成像时间：2023-03-04 日间
Single band thermal infrared image (band 2)　Imaging time: 2023-03-04 Daytime

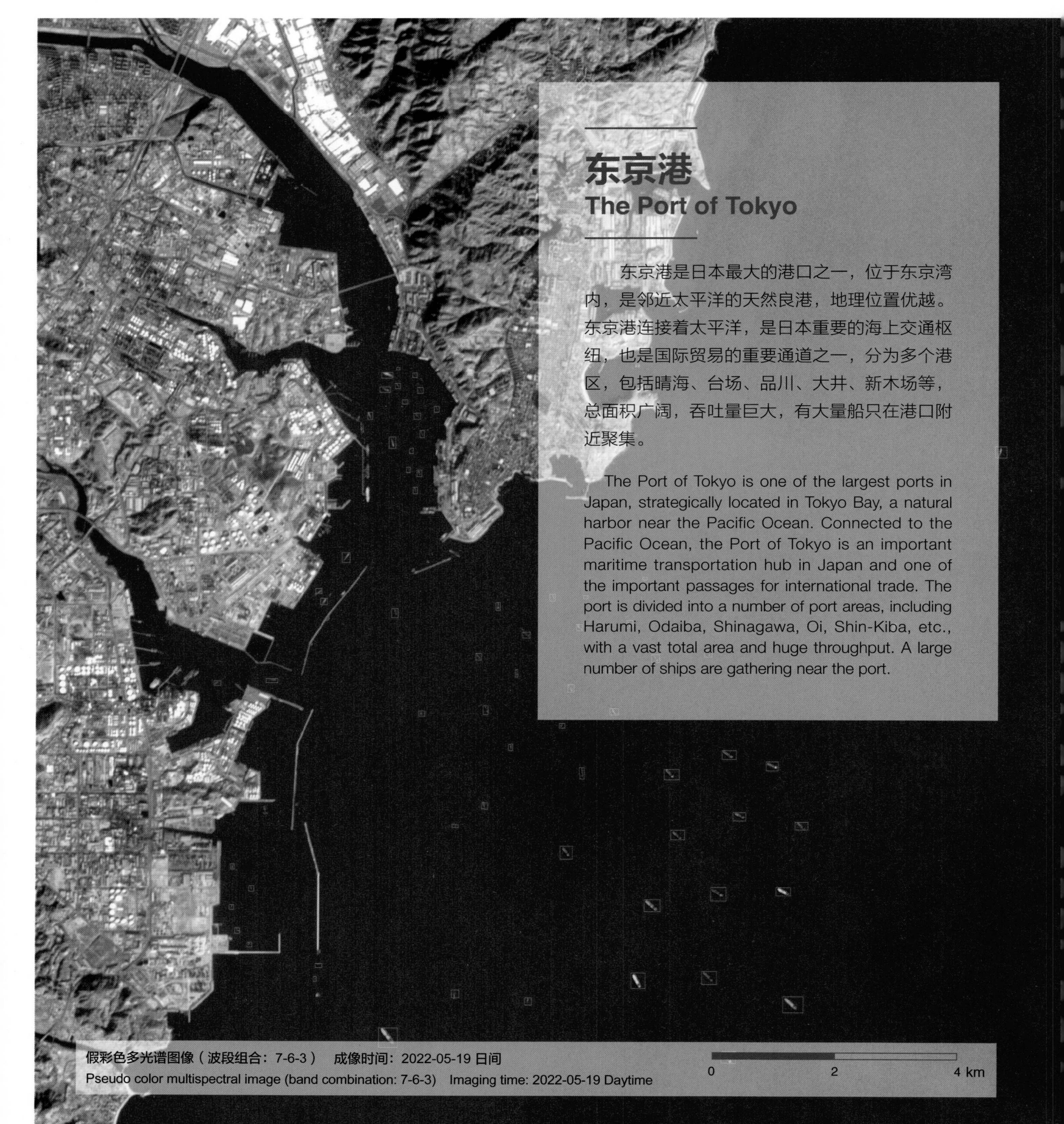

东京港
The Port of Tokyo

东京港是日本最大的港口之一，位于东京湾内，是邻近太平洋的天然良港，地理位置优越。东京港连接着太平洋，是日本重要的海上交通枢纽，也是国际贸易的重要通道之一，分为多个港区，包括晴海、台场、品川、大井、新木场等，总面积广阔，吞吐量巨大，有大量船只在港口附近聚集。

The Port of Tokyo is one of the largest ports in Japan, strategically located in Tokyo Bay, a natural harbor near the Pacific Ocean. Connected to the Pacific Ocean, the Port of Tokyo is an important maritime transportation hub in Japan and one of the important passages for international trade. The port is divided into a number of port areas, including Harumi, Odaiba, Shinagawa, Oi, Shin-Kiba, etc., with a vast total area and huge throughput. A large number of ships are gathering near the port.

假彩色多光谱图像（波段组合：7-6-3）　成像时间：2022-05-19 日间

Pseudo color multispectral image (band combination: 7-6-3)　Imaging time: 2022-05-19 Daytime

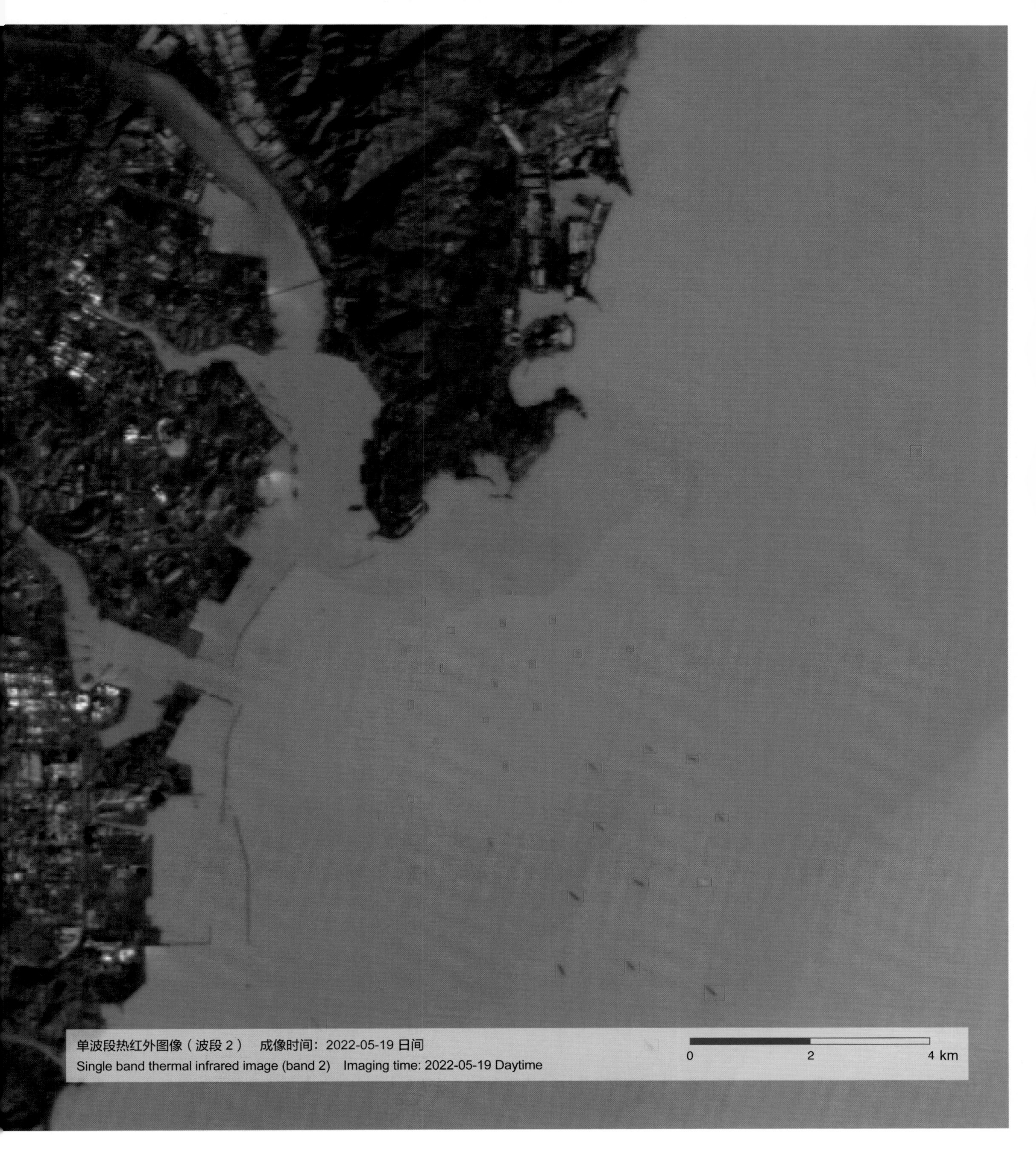
单波段热红外图像（波段 2）　成像时间：2022-05-19 日间
Single band thermal infrared image (band 2)　Imaging time: 2022-05-19 Daytime
0
2
4 km

南海
The South China Sea

南海东临菲律宾群岛、中国台湾和琉球群岛，南濒马六甲海峡，西界越南、中国、马来西亚等国的海岸线，北接华南海域。南海拥有广阔的海域，包括若开深海盆地、南海中部海盆和北部湾等地理特征，同时涵盖着数百个岛礁和海域。有许多船只在海岛附近航行。

The South China Sea is bordered by the Philippine Islands, Taiwan Province of China and the Ryukyu Islands in the east, the Strait of Malacca in the south, the coastlines of Vietnam, China, Malaysia and other countries in the west, and the South China Sea in the north. The South China Sea has a vast sea area, including the Rakhine Deep Sea Basin, the Central South China Sea Basin and the Beibu Gulf, and covers hundreds of islands, reefs and sea areas. There are many ships sailing near the island.

假彩色多光谱图像（波段组合：7-6-3） 成像时间：2022-07-22 日间
Pseudo color multispectral image (band combination: 7-6-3) Imaging time: 2022-07-22 Daytime

0 2 4 km

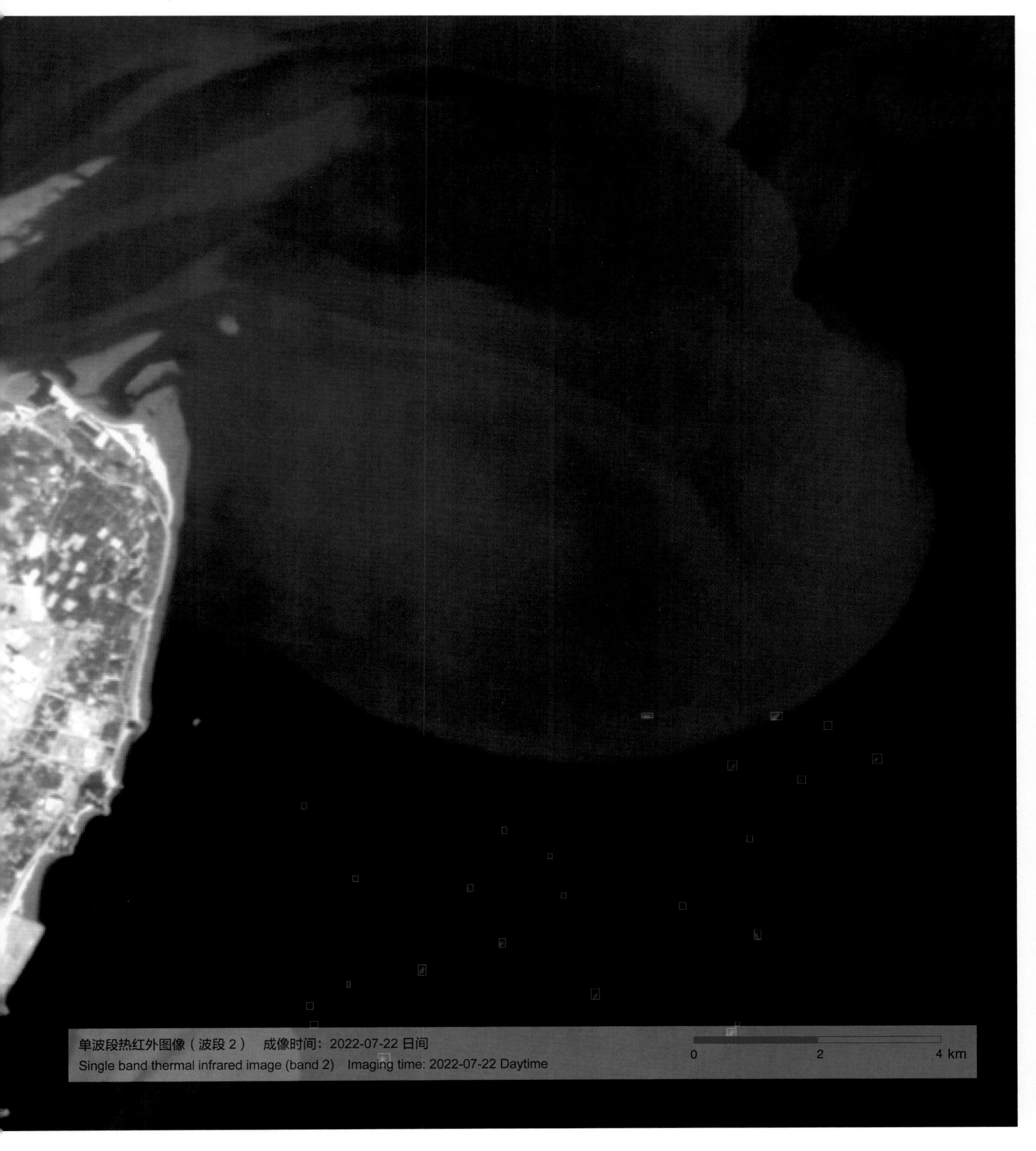

单波段热红外图像（波段 2）　成像时间：2022-07-22 日间
Single band thermal infrared image (band 2)　Imaging time: 2022-07-22 Daytime

数据产品

Data Products

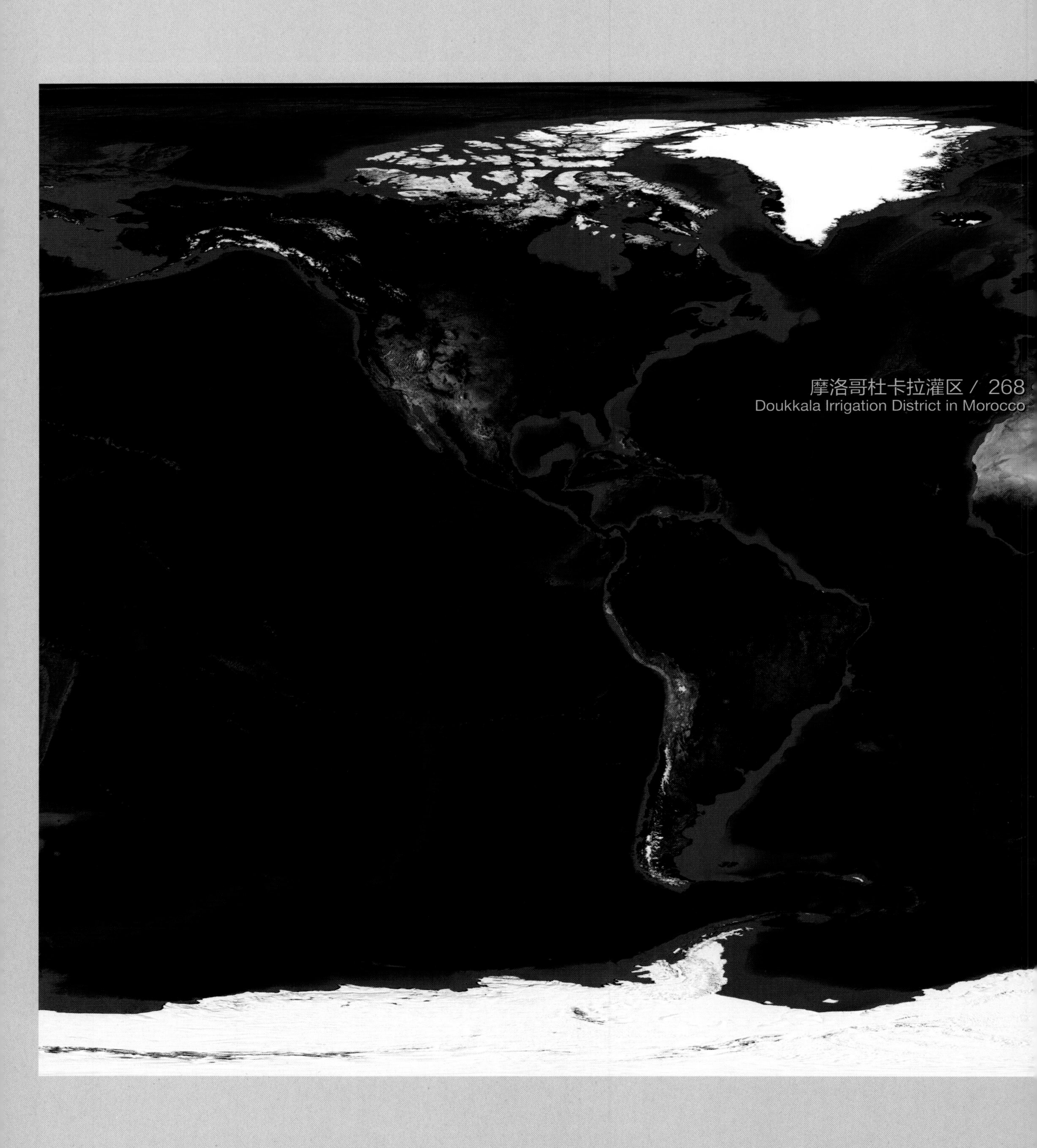
摩洛哥杜卡拉灌区 / 268
Doukkala Irrigation District in Morocco

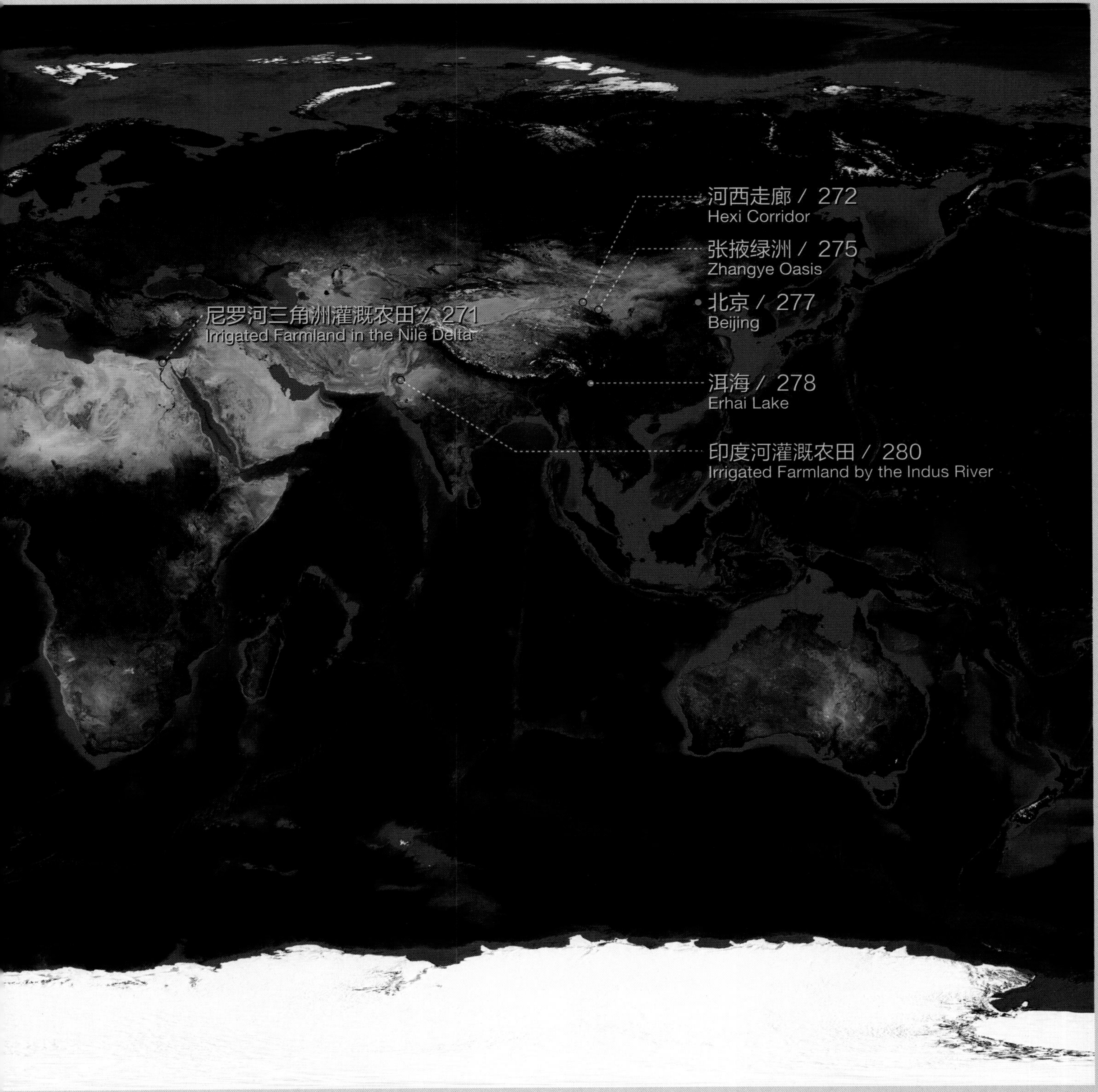

河西走廊 / 272
Hexi Corridor

张掖绿洲 / 275
Zhangye Oasis

北京 / 277
Beijing

尼罗河三角洲灌溉农田 / 271
Irrigated Farmland in the Nile Delta

洱海 / 278
Erhai Lake

印度河灌溉农田 / 280
Irrigated Farmland by the Indus River

摩洛哥杜卡拉灌区
Doukkala Irrigation District in Morocco

摩洛哥位于非洲西北端，西濒大西洋，北临地中海，是地中海出入大西洋的门户。摩洛哥水资源相对匮乏，杜卡拉灌区为摩洛哥第一大河乌姆赖比阿河流域内的主要灌区。摩洛哥杜卡拉灌区农田多依靠休耕或轮作来恢复地力，在蒸散耗水量的空间分布特征上得到较好地呈现。蒸散耗水量空间分布差异性主要由小麦、甜菜、苜蓿等当地主要作物类型的作物种植结构所决定，休耕田块的蒸散耗水量显著低于耕种田块的蒸散耗水量。

Morocco is located at the northwest end of Africa, bordering the Atlantic Ocean to the west and the Mediterranean Sea to the north, serving as the gateway for the Mediterranean to enter the Atlantic Ocean. Morocco has relatively scarce water resource, and the Doukkala Irrigation District is the main irrigation area within the Oum Er-Rbia River basin, the largest river in Morocco. The farmland in the Doukkala Irrigation District mostly relies on fallow or crop rotation to restore soil fertility, and the spatial distribution characteristics of water consumption by evapotranspiration (ET) are well presented. The spatial distribution differences in evapotranspiration are mainly determined by the planting structure of local main crop types such as wheat, sugar beet and alfalfa. The evapotranspiration of fallow fields is significantly lower than that of cultivated fields.

数据获取时间：2022-02-16

Data capture time: 2022-02-16

0 1 2 km

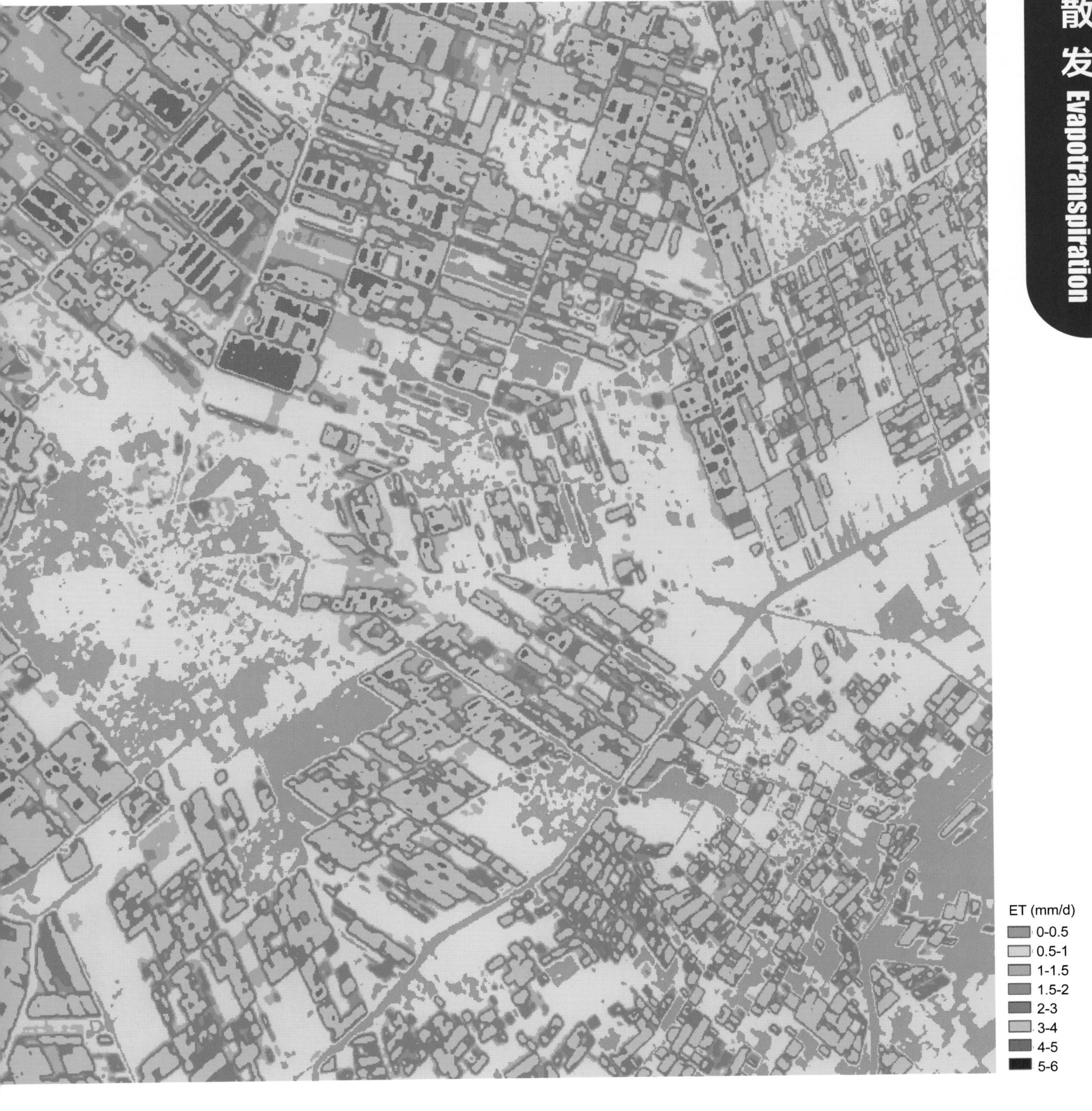
ET (mm/d)
0-0.5
0.5-1
1-1.5
1.5-2
2-3
3-4
4-5
5-6

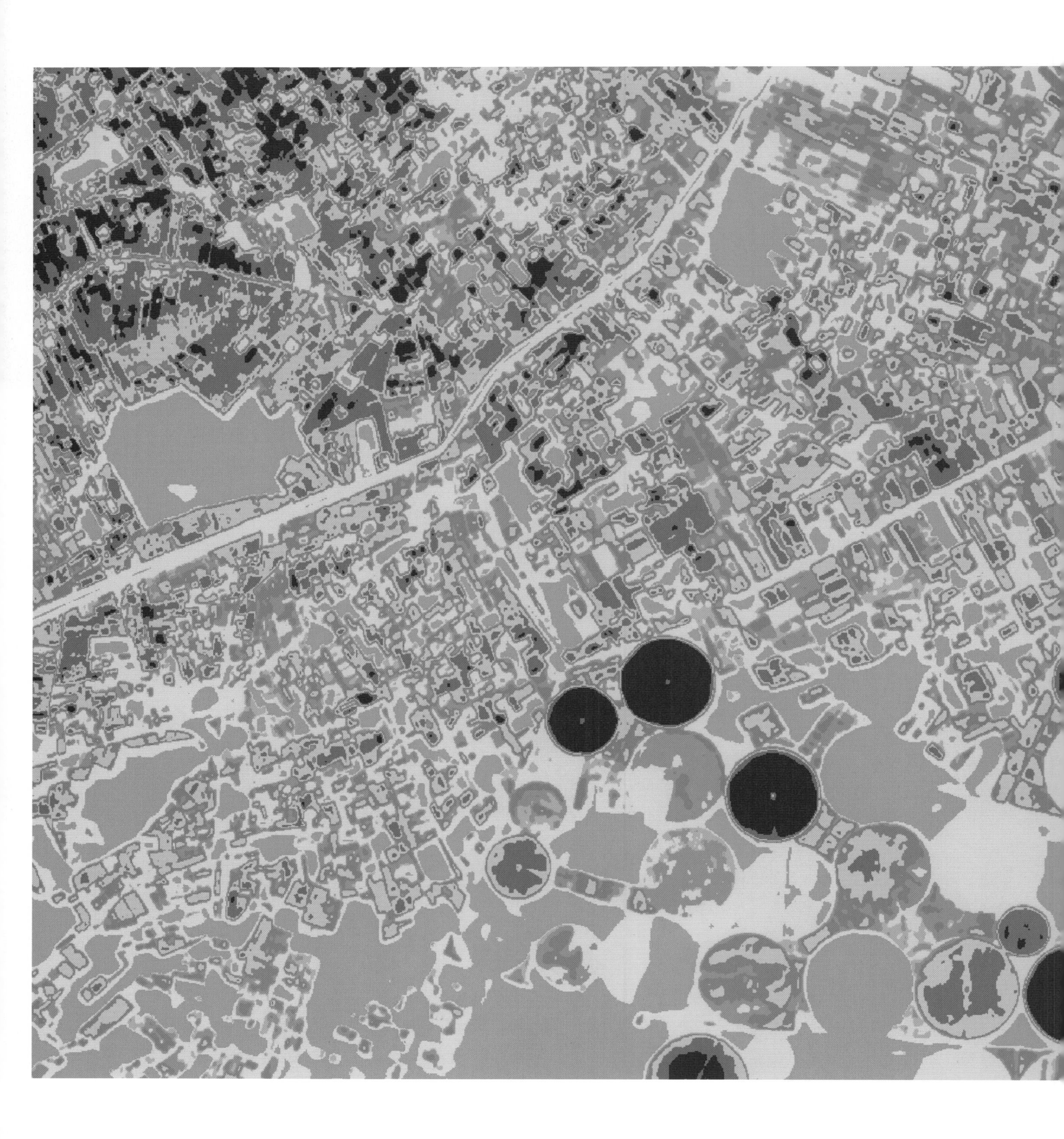

尼罗河三角洲灌溉农田
Irrigated Farmland in the Nile Delta

埃及尼罗河三角洲是由尼罗河干流汇入地中海形成的，是世界上最大的三角洲之一。区域地势低平，土壤肥沃，人口密集，水源充足，灌溉农业发达，基本不依赖降水。从农田蒸散耗水量的空间分布特征可以看出，尼罗河三角洲内部（图中左侧）人均耕地少，以小农场经营为主，田块较小。尼罗河三角洲外围（图中右侧）新开垦的农田以大型喷灌农田为主，田块边界清晰。

The Nile Delta in Egypt is formed by the confluence of the main Nile River into the Mediterranean Sea and is one of the largest deltas in the world. The regional terrain is low and flat, the soil is fertile, the population is dense, and the water resource is sufficient. The highly developed irrigation agriculture basically does not rely on precipitation. From the spatial distribution characteristics of evapotranspiration (ET) in farmland, it can be seen that within the Nile Delta (on the left side of the figure), there is less per capita arable land, mainly operated by small farms, with smaller fields. The newly cultivated farmland on the outskirts of the Nile Delta (on the right side of the figure) is mainly large-scale sprinkler irrigation farmland, with clear boundaries of the fields.

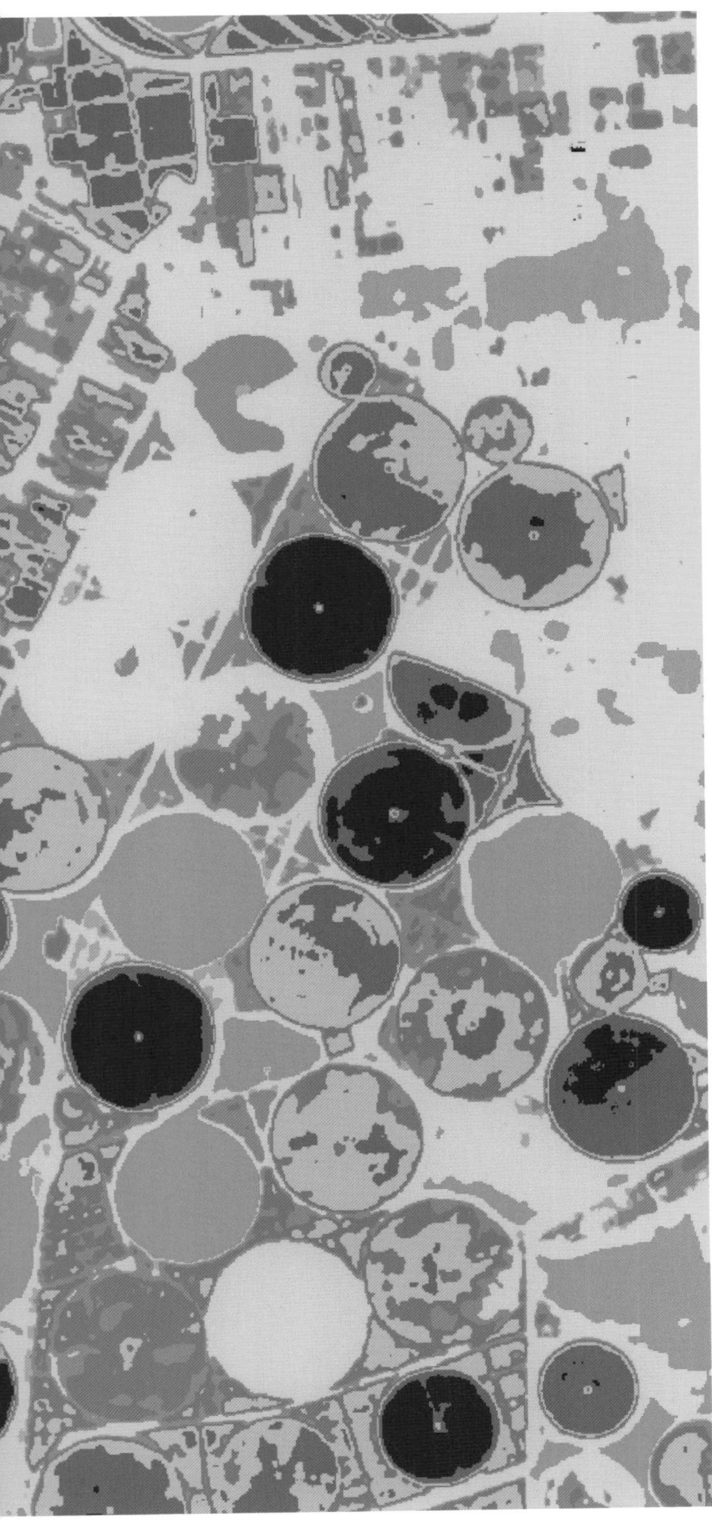

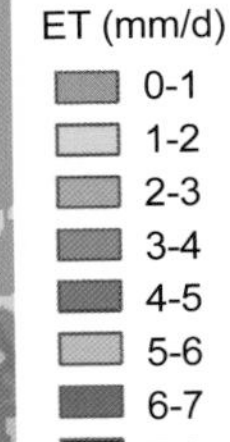

数据获取时间：2023-08-04

Data capture time: 2023-08-04

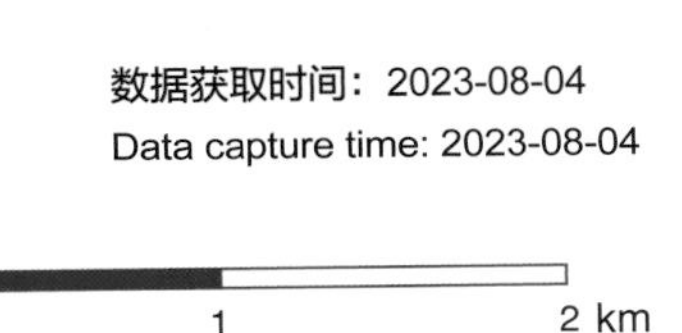

河西走廊
Hexi Corridor

河西走廊地处黄河以西、祁连山和巴丹吉林沙漠中间的甘肃省西北部，是国家生态安全屏障、全国特色高效农业示范区和丝绸之路重要开放廊道。从蒸散发（ET）空间分布特征可以看出，图中上部为河西走廊平原，地势平坦开阔，依赖水利开发和农业灌溉使其成为甘肃省主要粮食产区，也是区域水资源消耗区。下部为祁连山北麓，气候寒冷湿润，蒸散发明显低于河西走廊灌溉绿洲区，为区域水源涵养区和水资源形成区。

Hexi Corridor is located in the northwest of Gansu Province, west of the Yellow River, between the Qilian Mountains and the Badain Jaran Desert. It is a national ecological security barrier, a national special and efficient agricultural demonstration zone, and an important open corridor of the Silk Road. From the spatial distribution characteristics of evapotranspiration (ET), it can be seen that the upper part of the figure is the plain of the Hexi Corridor, with a flat and open terrain. Relying on water conservancy development and agricultural irrigation, it has become the main grain producing area in Gansu Province and also a regional water resource consumption area. The lower part of the figure is located at the northern foot of the Qilian Mountains, with a cold and humid climate and significantly lower evapotranspiration than the irrigated oasis in the Hexi Corridor, and it is a regional water resource conservation area and a water resource formation area.

数据获取时间：2022-07-22

Data capture time: 2022-07-22

0 5 10 km

ET (mm/d)
0-1
1-2
2-3
3-4
4-5
5-6
6-7
7-8

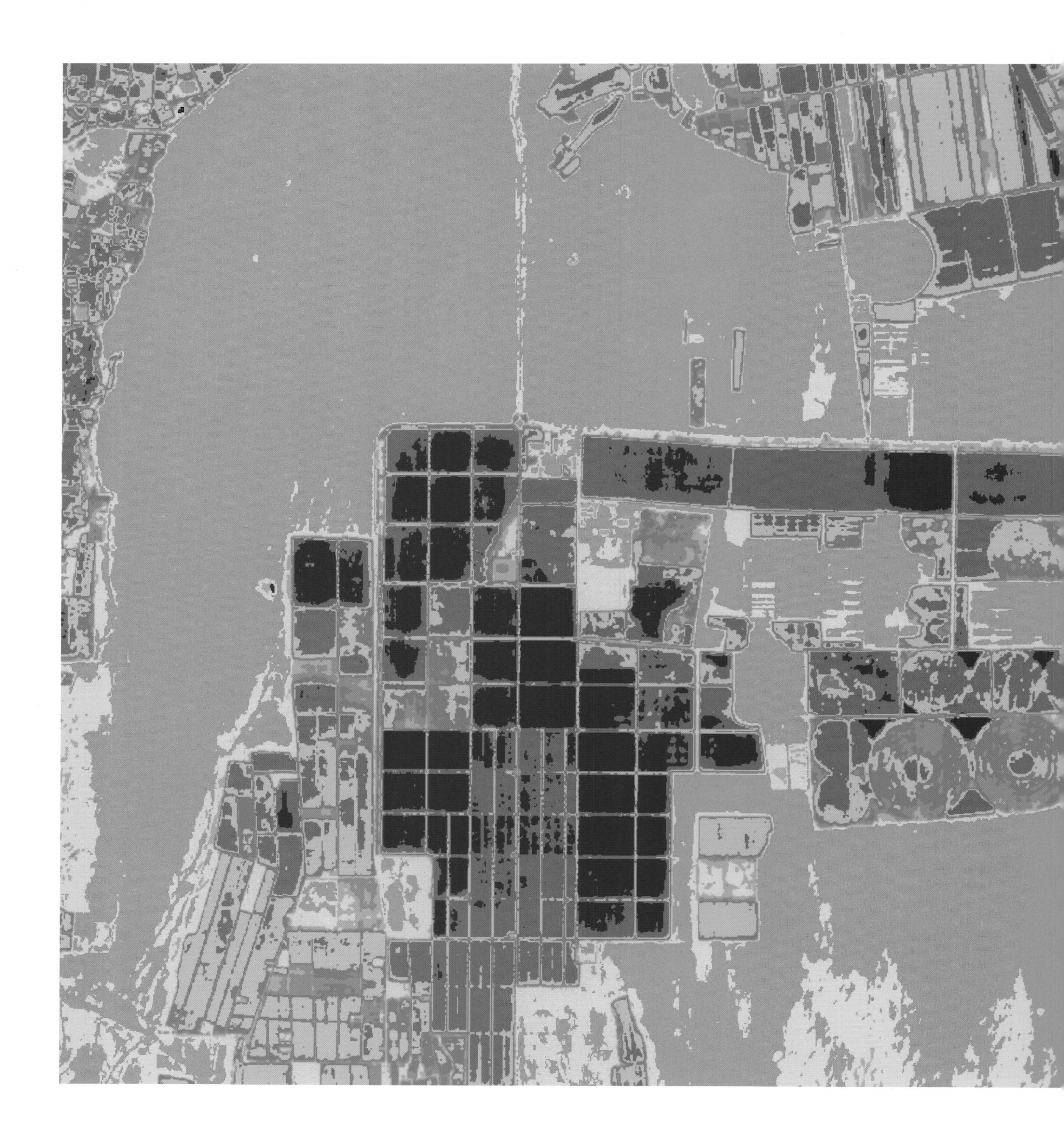

张掖绿洲
Zhangye Oasis

黑河是中国第二大内陆河，流经张掖全境。依靠河流和地下水的灌溉，形成荒漠背景下的灌溉绿洲景观，成为戈壁荒漠地带上丝绸之路的中间驿站。张掖绿洲是河西走廊最大的绿洲，灌溉农业发达，灌溉水源主要消耗于农田蒸散发。从蒸散发空间分布特征可以看出，绿洲植被覆盖度高，水热资源丰富，有利于植物光合作用和蒸腾作用的进行，该地区的蒸散耗水量显著高于戈壁荒漠地区。

The Heihe River is the second largest inland river in China, flowing through the entire territory of Zhangye. Relying on the irrigation of rivers and groundwater, an irrigated oasis landscape is formed in the desert, becoming an intermediate station on the Silk Road in the Gobi Desert area. Zhangye Oasis is the largest oasis in the Hexi Corridor, with highly developed irrigation agriculture. The irrigation water resource is mainly consumed by farmland evapotranspiration. From the spatial distribution characteristics of evapotranspiration (ET), it can be seen that oases have high vegetation coverage and abundant water and heat resources, which are conducive to plant photosynthesis and transpiration. The water consumption of evapotranspiration in oases is significantly higher than that of Gobi Desert area.

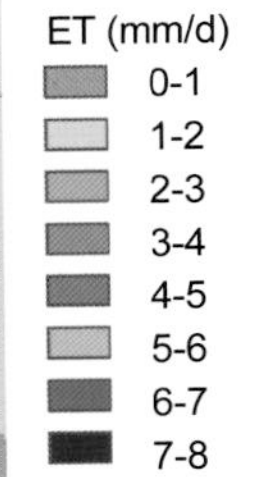

数据获取时间：2022-07-22

Data capture time: 2022-07-22

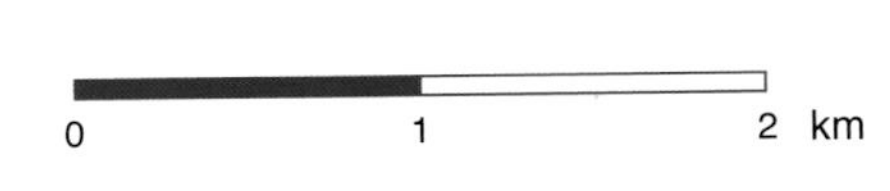

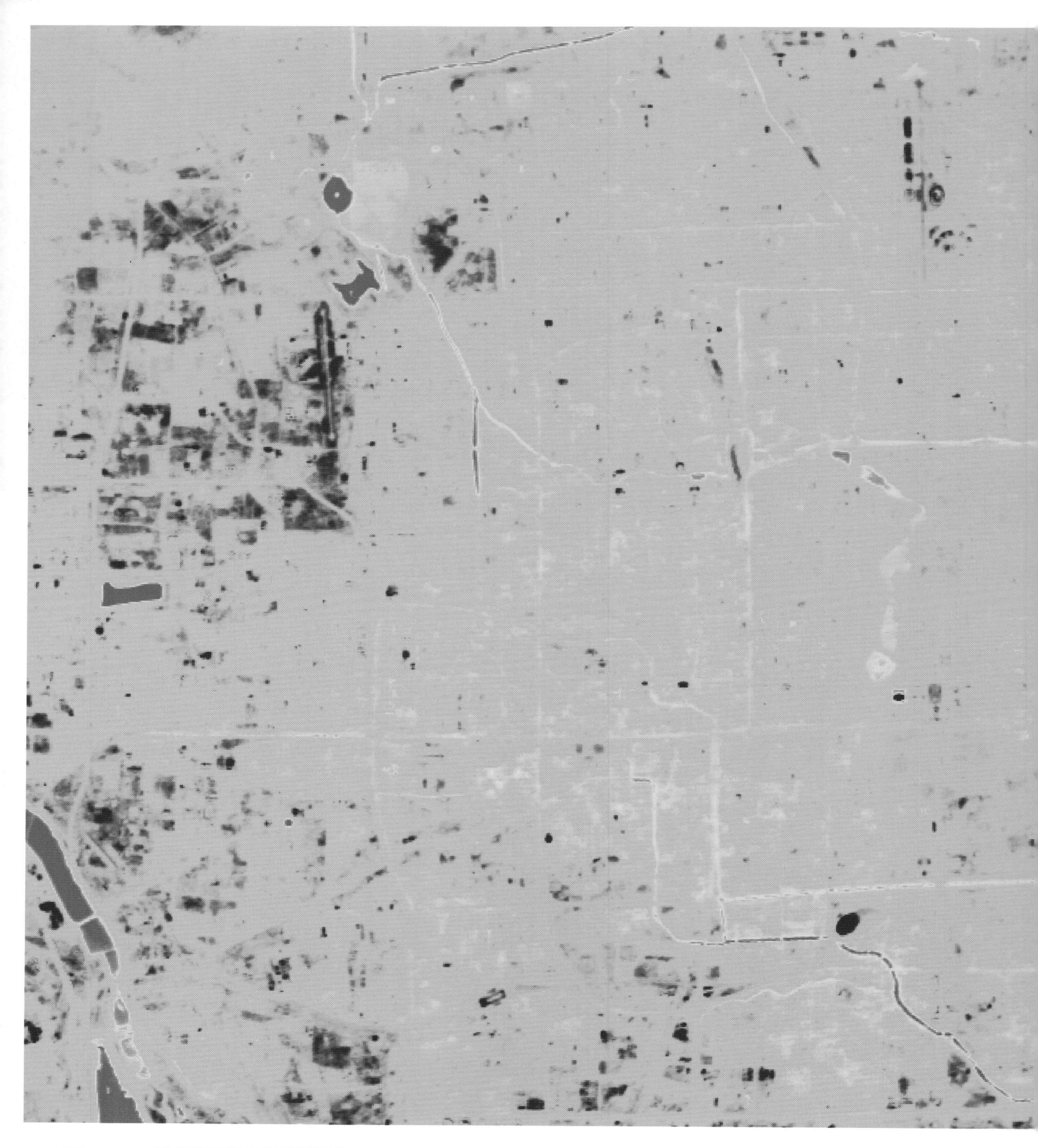

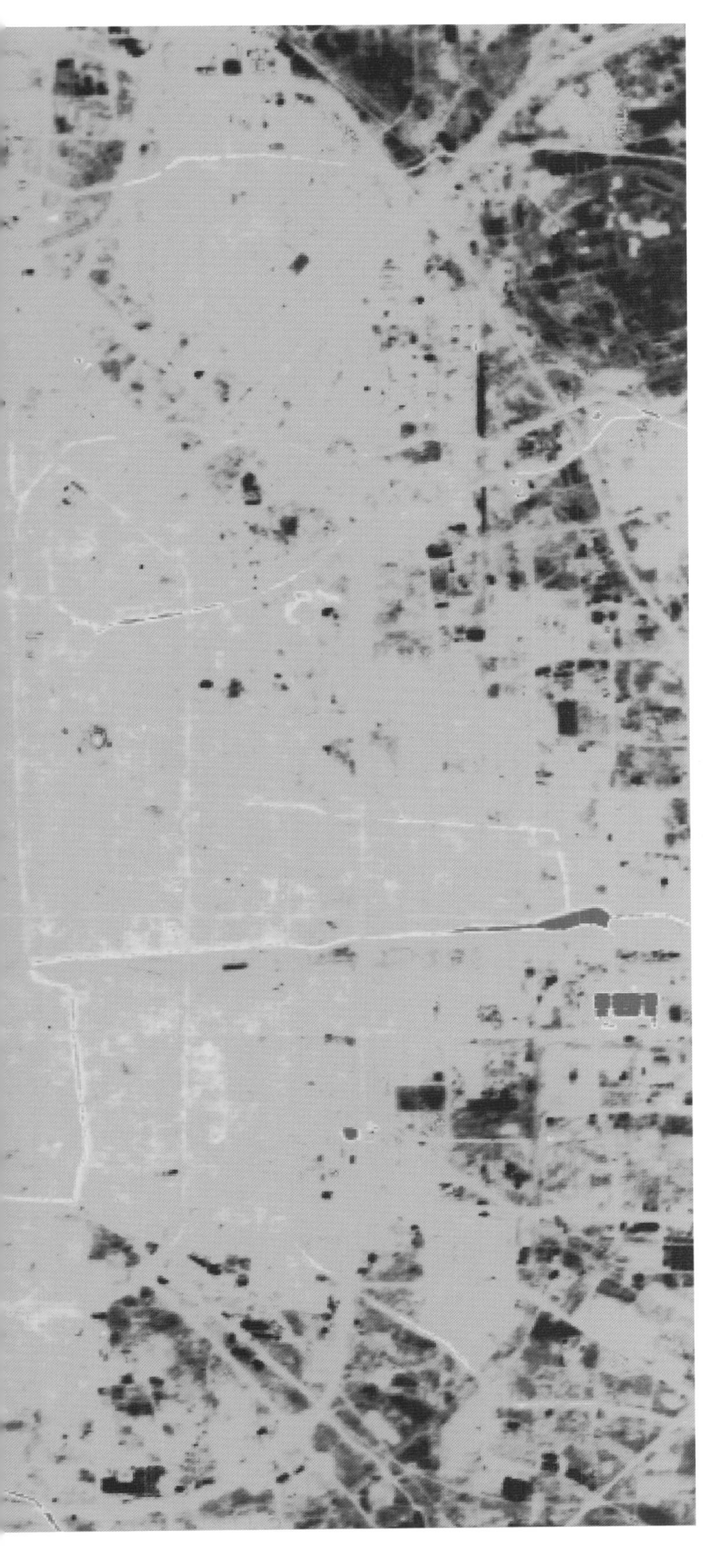

北京
Beijing

北京城区覆盖范围广，组成复杂，有用砖石材料建成的老城区，也有大范围使用水泥以及玻璃幕墙的新建筑。人类活动以及各类建筑的集中分布导致了可观的城市热岛效应。基于夜间热红外图像的地表温度数据反演表明城市中心以及水体在夜间温度较高，但城区间也存在差异，而图像边缘的郊区地带温度较低。

The urban area of Beijing covers a wide range of areas and has a complex composition, ranging from old urban areas built with brick and stone to new buildings with extensive use of concrete and glass curtain walls. The concentration of human activities and buildings of all kinds leads to a considerable urban heat island effect. The inversion of surface temperature data based on nighttime thermal infrared images shows that the city center and water bodies are warmer at night, but there are also differences between urban areas, while the suburban areas at the edge of the image are cooler.

温度（K）
Temperature (K)

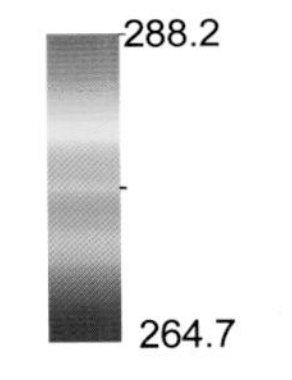

数据获取时间：2023-11-26 夜间
Data capture time: 2023-11-26 Nighttime

0 2 4 km

洱海
Erhai Lake

洱海，位于中国云南省大理白族自治州大理市，是一个风光秀媚的高原淡水湖泊，湖水面积约 250km^2，为云南第二大湖，呈狭长形，南北长约 40km，海拔 1972m。洱海西侧为苍山断块山地，走势呈南北向，洱海南、西、北三面受河流冲积影响，形成了广袤的洪积平原。该热红外地表温度图中包括了洱海及周边区域，相较于山地，湖水温度均匀，产生了明显对比，有效展示了自然湖泊区域的热环境情况。

Erhai Lake, located in Dali City, Dali Bai Autonomous Prefecture, Yunnan Province, China, is a beautiful plateau freshwater lake with a water area of about 250km^2. It is the second largest lake in Yunnan, with a narrow and elongated shape, about 40km long from north to south, and an altitude of 1972m. To the west of Erhai Lake is the Cangshan fault block mountainous area, with a north-south trend. The southern, western and northern sides of Erhai Lake are affected by river erosion, forming a vast alluvial plain. The land surface temperature map includes the Erhai Lake and its surrounding areas. Compared with the mountains area, the lake water temperature is uniform, creating a clear contrast and effectively displaying the thermal environment of the natural lake area.

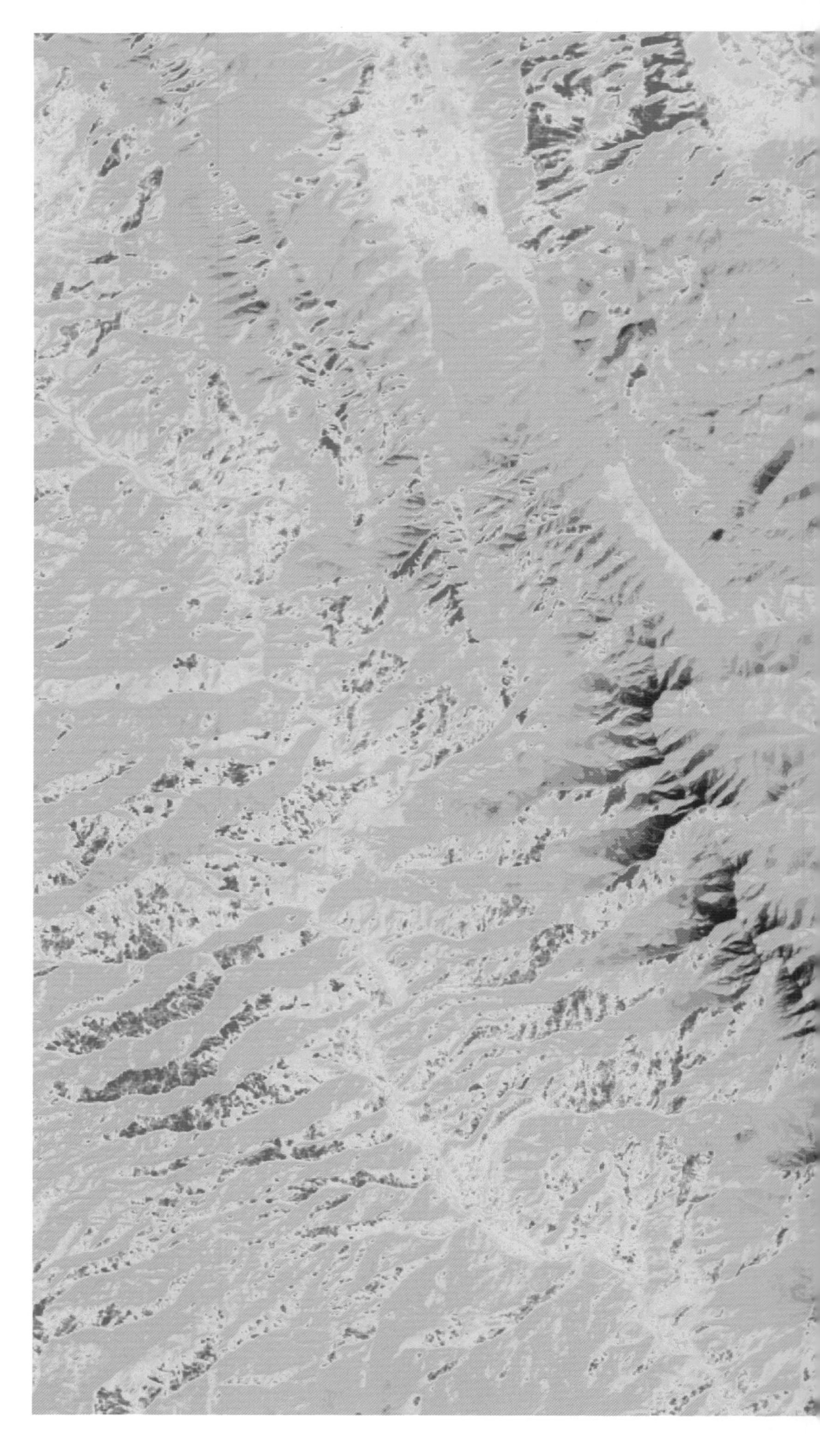

温度（K）
Temperature (K)

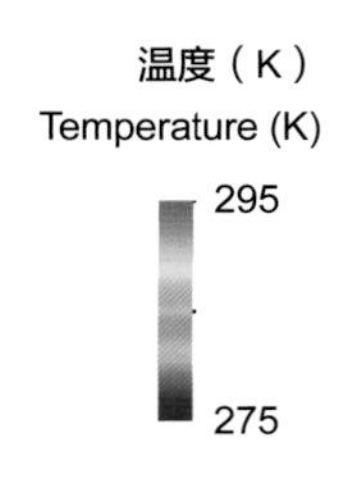

数据获取时间：2024-01-03 日间
Data capture time: 2024-01-03 Daytime

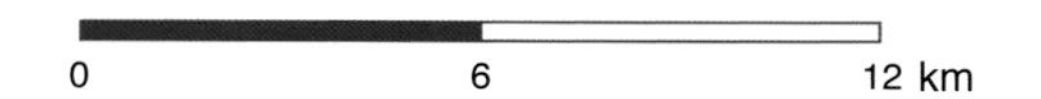

印度河灌溉农田
Irrigated Farmland by the Indus River

印度河是巴基斯坦主要河流，也是巴基斯坦重要的农业灌溉水源。流域内降水较少，分布有大片沙漠。印度河沿岸灌溉农业发达，是巴基斯坦主要人口聚集区之一。从蒸散发的空间分布特征可以看出，图中右侧的河流水面蒸发量最高，其次是其沿岸的灌溉农田，图中左侧距离河流约为 6km 的农田外围即为干旱的荒漠区，地表蒸散量最低。

The Indus River is a major river in Pakistan and an important agricultural irrigation water resource. There is less precipitation in the watershed, and there are large deserts distributed. The developed irrigation agriculture along the Indus River is one of the main population gathering areas in Pakistan. From the spatial distribution characteristics of evapotranspiration (ET), it can be seen that the river on the right side of the figure has the highest water evaporation, followed by irrigated farmland along its banks. The outer edge of farmland on the left side of the figure, which is about 6km away from the river, is an arid desert area with the lowest evapotranspiration.

数据获取时间：2023-08-29
Data capture time: 2023-08-29

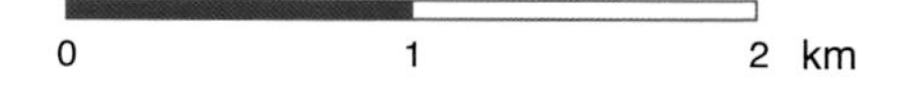

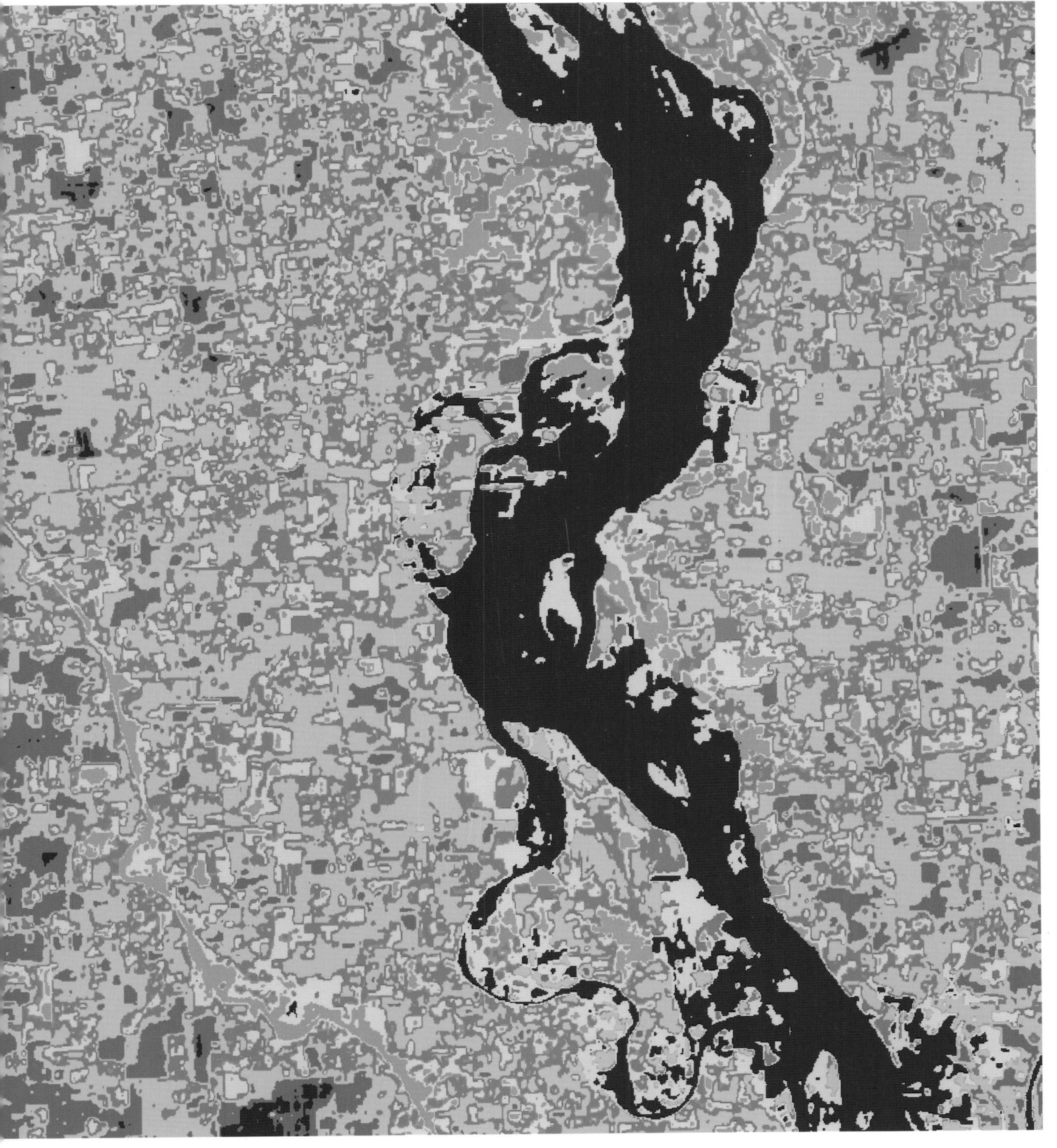

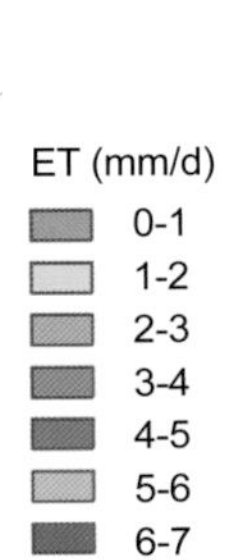
ET (mm/d)
0-1
1-2
2-3
3-4
4-5
5-6
6-7
7-8

编后说明

本图集的内容囊括了全球各地自然景观和人类活动的 SDGSAT-1 卫星热红外图像。因图像的选择受到篇幅和气象条件制约，图集无法罗列涉及到所有国家及地区的昼夜热红外图像。

考虑到不同地物的特征需要在相适应的比例尺上展示，图集每章节的选材和图像截取的规划各不相同。具体如下：“水体”部分中的水系与入海口地点通常没有严格的边界，热红外图像截取了各地点温度差异明显的部分；受幅宽限制，每景热红外图像并不能囊括山脉整体，因此，“山脉”部分中的图像展示了各山脉较有代表性的部分；因农业产地的分布常常横跨整个行政区，“农业用地”部分选取了数个著名产粮地区中农田密集的部分区域；为了更好展示沙漠中的细节，“沙漠”的图像常为各干旱地区的局部图像；“火灾”的图像比例尺较小，专注于火灾现场，以配合展示热红外图像对于火点的敏感性；为了展示图像对于冰间裂隙的探测能力，“海冰”部分聚焦于海冰集中地区的部分地点；“工业热源”包括了具体工厂以及工业区的较小比例尺图像；为了展示宏观的城市图像以及部分基础设施的细节，“城市”的图像运用了一种大比例尺图像和局部放大的组合；“船只”的图像展示了在较宽幅场景下的船只分布；“数据产品”部分在较小比例尺上展示了基于 SDGSAT-1 热红外数据的蒸散发与地温反演。

Editor's Note

This atlas encompasses SDGSAT-1 satellite thermal infrared images of natural landscapes and human activities from around the world. Due to limitations in space and meteorological conditions, the atlas cannot include day and night thermal infrared images of every country and region.

Considering the need to display different land features at appropriate scales, each part of the atlas has different selection criteria and image planning, as detailed below:

- The "Water Bodies" part includes water systems and estuary locations that typically lack strict boundaries. Thermal infrared images capture sections of these locations where temperature differences are significant.
- Due to the width limitation, each thermal infrared image cannot encompass entire mountain ranges. Therefore, the "Mountain Ranges" part showcases the most representative parts of each mountain range.
- Given that agricultural production areas often span entire administrative regions, the "Agricultural Land" part selects portions of well-known grain-producing regions where farmland is densely concentrated.
- To better display desert details, the images in the "Deserts" part often focus on sections of arid regions.
- The image scale for "Fire Incidents" is smaller, concentrating on the fire sites to highlight the thermal infrared images' sensitivity to fire points.
- To demonstrate the ability to detect fissures in ice, the "Sea Ice" part focuses on parts of regions where sea ice is concentrated.
- The "Industrial Heat Sources" part includes specific factories and smaller scale images of industrial areas.

- The “Cities” part uses a combination of large-scale images and localized enlargements to display comprehensive urban images and details of certain infrastructure.
- The “Boats and Ships” part displays the distribution of ships in wider scene images.
- The “Data Products” part showcases evapotranspiration and ground temperature inversion based on SDGSAT-1 thermal infrared data at a smaller scale.

The content selection and image extraction planning for each chapter are tailored to effectively present the characteristics of different land features, providing a detailed and comprehensive view of the Earth’s thermal environment as captured by SDGSAT-1.

地物列表

地物	地点
水体	大盐湖、密西西比河、亚马孙河、卡累利阿地峡、第聂伯河、日内瓦湖、咸海、尼罗河三角洲 - 苏伊士运河、乍得湖、青海湖、长江、鄱阳湖、恒河三角洲、伊洛瓦底江三角洲、湄公河三角洲、艾尔湖
山脉	阿巴拉契亚山脉、安第斯山脉、阿尔卑斯山脉、阿特拉斯山脉、高加索山脉、青藏高原
农业用地	得克萨斯州、奇瓦瓦、马托格罗索州、罗斯托夫州、布海拉省、北方邦、黑龙江省、维多利亚州
沙漠	死亡谷、索诺兰沙漠、纳斯卡荒漠、阿塔卡马沙漠、阿尔及利亚沙漠、利比亚沙漠、撒哈拉之眼、戈巴贝布、纳米布沙漠、拉蒙凹地、内盖夫沙漠、西奈半岛、甘肃冲积扇、巴丹吉林沙漠、腾格里沙漠、塔克拉玛干沙漠、艾尔湖盆地
火灾	西北地区（加拿大）野火、东非热带稀树草原火灾、西伯利亚森林火灾、甘孜藏族自治州森林火灾、昆士兰州森林火灾
海冰	喀拉海、拉普捷夫海、东西伯利亚海、波弗特海、巴芬湾、伊丽莎白女王群岛海岸
工业热源	亚利桑那州核电站、密西西比河沿岸化工厂、夸察夸尔科斯河沿岸石化厂、布良斯克州发电站、扎波罗热州核电站、波斯湾炼油厂、拉斯拉凡炼油厂、波洛夸内炼油厂、约翰内斯堡矿井、京津冀地区工业热源、惠州市工业热源、防城港市工业热源、泰米尔纳德邦发电站、澳大利亚西部氧化铝精炼厂
城市	丹佛、旧金山、洛杉矶、墨西哥城、芝加哥、多伦多、纽约、华盛顿、布宜诺斯艾利斯、柏林、阿姆斯特丹、伦敦、巴黎、丹吉尔、斯德哥尔摩、莫斯科、维也纳、布鲁塞尔、达曼、拉各斯、北京、首尔、东京、上海、广州、深圳、香港、澳门、黑德兰港、悉尼、墨尔本
船只	纽约港、休斯顿港、墨西哥湾、阿姆斯特丹港、直布罗陀海峡、苏伊士运河、东京港、南海
数据产品	摩洛哥杜卡拉灌区、尼罗河三角洲灌溉农田、河西走廊、张掖绿洲、北京、洱海、印度河灌溉农田

Imaging Locations

Chapter	Locations
Water Bodies	Great Salt Lake, Mississippi River, Amazon River, Karelian Isthmus, Dnieper River, Lake Geneva, Aral Sea, Nile Delta-Suez Canal, Lake Chad, Qinghai Lake, Yangtze River, Poyang Lake, Ganges Delta, Irrawaddy Delta, Mekong Delta, Lake Eyre
Mountain Ranges	Appalachian Mountains, Andes Mountains, Alps, Atlas Mountains, Caucasus Mountains, Qinghai-Tibet Plateau
Agricultural Land	Texas, Chihuahua, Mato Grosso, Rostov Oblast, Beheira Governorate, Uttar Pradesh, Heilongjiang Province, Victoria
Deserts	Death Valley, Sonoran Desert, Nazca Desert, Atacama Desert, Algerian Desert, Libyan Desert, Eye of the Sahara, Gobabeb, Namib Desert, Makhtesh Ramon, Negev Desert, Sinai Peninsula, Gansu alluvial fan, Badain Jaran Desert, Tengger Desert, Taklamakan Desert, Lake Eyre Basin
Fire Incidents	Wildfires in Northwest Territories (Canada), Fires in tropical Savanna of East Africa, Forest fires in Siberian, Forest fires in Garze Tibetan Autonomous Prefecture, Bushfires in Queensland
Sea Ice	Kara Sea, Laptev Sea, East Siberian Sea, Beaufort Sea, Baffin Bay, Coast of the Queen Elizabeth Islands
Industrial Heat Sources	Nuclear power plant in Arizona, Chemical plant by Mississippi River, Petrochemical plant by Coatzacoalcos River, Power station in Bryansk Oblast, Nuclear power plant in Zaporizhzhia Oblast, Oil refinery in Persian Gulf, Oil refinery in Ras Laffan, Oil refinery in Polokwane, Mine in Johannesburg, Industrial heat sources in Beijing-Tianjin-Hebei region, Industrial heat sources in Huizhou city, Industrial heat sources in Fangchenggang region, Power station in Tamil Nadu region, Alumina refinery in western Australia
Cities	Denver, San Francisco, Los Angeles, Mexico City, Chicago, Toronto, New York, Washington D.C., Buenos Aires, Berlin, Amsterdam, London, Paris, Tangier, Stockholm, Moscow, Vienna, Brussels, Dammam, Lagos, Beijing, Seoul, Tokyo, Shanghai, Guangzhou, Shenzhen, Hong Kong, Macao, Port Hedland, Sydney, Merlbourne
Boats and Ships	The Port of New York, The Port of Houston, The Gulf of Mexico, The Port of Amsterdam, The Strait of Gibraltar, The Suez Cana, The Port of Tokyo, The South China Sea
Data Products	Doukkala Irrigation District in Morocco, Irrigated Farmland in the Nile Delta, Hexi Corridor, Zhangye Oasis, Beijing, Erhai Lake, Irrigated Farmland by the Indus River